T0206167

BIOPROCESSING OF VIRAL VACCINES

The concept of this book originates from the delivery of the series of Cell Culture-based Viral Vaccines Course sponsored by the European Society of Animal Cell Technology (ESACT) https://esact.org/cell-culture-based-viral-vaccines-course/ and its content includes some of the material that has been developed and adapted over many years.

The book focuses on cell-culture–produced viral vaccines to meet the needs of the rapidly expanding research and development in academia and industry in the field. This book introduces the basic principles of vaccination and the manufacturing of viral vaccines. *Bioprocessing of Viral Vaccines* will provide an overview of the advanced strategies needed to respond to the challenges of new and established viral infection diseases. The first few chapters cover the basics of virology and immunology as essential concepts to understand the function and design of viral vaccines. The core of the content is dedicated to process development, including upstream processing and cell culture of viral vaccines, downstream processing, and analytical technologies as applied to viral vaccines. Advanced process analytical technologies (PAT) and quality by design (QbD) concepts are also introduced in the context of vaccine manufacturing. The case studies included cover inactivated, attenuated vaccines exemplified by influenza vaccines; sub-unit vaccines exemplified by virus-like particles (VLPs: HPV vaccines) and recombinant protein vaccines (Flublock); vectored vaccines: adenoviruses (AdV) and vesicular stomatitis virus (VSV) vectored vaccines; genomic vaccines (DNA and mRNA), vaccines as developed for COVID-19 response in particular; and a review of all COVID-19 vaccines approved or in advanced clinical trials. Primarily this book is aimed at graduate engineers and professionals in the fields of vaccinology, bioprocessing, and biomanufacturing of viral vaccines.

BIOPROCESSING OF VIRAL VACCINES

Edited by

Amine Kamen and Laura Cervera

CRC Press is an imprint of the
Taylor & Francis Group, an **informa** business

First edition published 2023
by CRC Press
6000 Broken Sound Parkway NW, Suite 300, Boca Raton, FL 33487-2742

and by CRC Press
4 Park Square, Milton Park, Abingdon, Oxon, OX14 4RN

CRC Press is an imprint of Taylor & Francis Group, LLC

ISBN: 978-1-032-13211-2 (hbk)
ISBN: 978-1-032-13551-9 (pbk)
ISBN: 978-1-003-22979-7 (ebk)

DOI: 10.1201/9781003229797

Typeset in Times
by MPS Limited, Dehradun

Contents

Editors

Amine Kamen is a professor of bioengineering at McGill University, and Canada Research Chair in bioprocessing of viral vaccines. He is a researcher emeritus of the National Research Council of Canada (NRC) where he was employed until early 2014, as head of the Process Development section of the Human Health Therapeutics Portfolio. At NRC, he established one of North America's largest and most advanced governmental centers for animal cell culture addressing process development and scale-up of biologics. Also, he developed with his team and licensed to industry multiple technology platforms for efficient manufacturing of recombinant proteins and viral vectors and vaccines and led technology transfer to manufacturing sites for clinical evaluation and commercialization. His current research activities focus on uncovering mechanisms associated with cell production of viral vectors and viral vaccines, cell and metabolic engineering, process control and monitoring, and process analytical technologies of high-yield productions of viral vectors for gene delivery and vaccination. He has published over 170 papers in refereed international journals and acts as a consultant for several national and international private and public organizations.

Laura Cervera is a Chemical Engineer from Universitat Autònoma de Barcelona (Barcelona, Spain). After graduating she did her PhD in Biotechnology on the topic "Strategies for improving production levels of HIV-1 VLPs by transient transfection of HEK 293 suspension cultures". Then she moved to McGillUniversity (Montreal, Canada) to pursue her research on VLP production, this time using Insect cells as a platform. She came back to Barcelona to join a project on AAV production for gene therapy applications using HEK 293 cells.

List of abbreviations

RNA	Ribonucleic acid
DNA	Deoxyribonucleic acid
ss	Single stranded
ds	Double stranded
HIV	Human Immunodeficiency Virus
AIDS	Acquired Immunodeficiency Syndrome
TMV	Tobacco Mosaic Virus
mRNA	messenger RNA
vRNA	viral RNA
DdDp	DNA dependent DNA polymerase
RdRp	RNA dependent RNA polymerase
CME	Clathrin mediated endocytosis
PIP2	Phosphatidylinositol 4,5-bisphosphate
GTP	Guanosine triphosphate
ESCRT	Endosomal sorting complex required for transport
RT	Reverse Transcriptase (enzyme)/ Transcription (process)

List of abbreviations

1 Bioprocessing of viral vaccines—Introduction

Amine Kamen
Viral Vectors and Vaccines Bioprocessing Group, Department of Bioengineering, McGill University, Montréal, QC, Canada

Laura Cervera
Grup d'Enginyeria Cel·lular i Bioprocés, Universitat Autònoma de Barcelona, Bellaterra, Barcelona, Spain

CONTENTS

1.1 AN ABBREVIATED HISTORICAL BACKGROUND OF VACCINES

From a broad universal perspective there are clear references in the Chinese and Indian history of ideas that suggest knowledge of vaccination principles [1,2]. However, the *Occidental Modern History of Vaccines* dates back the concept of vaccination to 1796 referring to Edward Jenner. Building on the observation that milk maids who were exposed to infected cows were resistant to smallpox infection, Edward Jenner was the first to demonstrate protection against smallpox infection by exposing the individual's immune system to material from cowpox pustules to provide protection [3].

In the nineteenth century, vaccination became a cause of national prestige, and the first vaccination laws were passed. The leading figure of Louis Pasteur [1822–1895] [4] developed the vaccine against rabies infection and contributed to the global promotion of vaccination through the initiative of the institutes Pasteur network [5]. Some of the most fascinating events over this period were captured in the entertaining book titled: *Plague & Cholera* [6], describing the pioneering humanitarian action of Alexandre Yersin, underlining the dominant role he played in

DOI: 10.1201/9781003229797-1

the discovery of the plague pathogen *Yersinia pestis* which exemplifies the need of a global perspective for vaccine distribution.

The development of vaccines reached its golden age during the twentieth century with the implementation and widespread use of many successful vaccines. As a result, smallpox has been eradicated (WHO declaration of global eradication by October 1979) and many other infectious diseases that have threatened humanity for centuries have virtually disappeared [7].

Eradication of the polio virus infection was targeted by year 2000 through the WHO Global Eradication Initiative [8]; however, cases persist in war areas not accessible to vaccination.

Accumulated data since 1800 shows a decrease of the global child mortality under the age of 5 years from 43% (1800) to 4.3% (2015) [9]. The decrease of global child mortality has been largely attributed to vaccination. The World Bank estimated that a combination of vaccines, malaria prevention, and improved newborn health care has helped reduce under-5 child mortality globally from 20 million in 1960 to 6.6 million in 2012. Consequently, the Bill and Melinda Gates Foundation is building its strategy to deliver vaccines to low-income countries to achieve in 2035 the goal of reducing the deaths by 1,000 births to 15, a ratio achieved in the United States in 1980.

Overall, these contributions to humanity are major and vaccines are making a great difference in human health, yet they are taken for granted by the public until challenged by a pandemic situation, as illustrated by the current COVID-19 pandemic situation and global exceptional measures implemented.

Table 1.1 shows the vaccines for preventable diseases since their first introduction in the United States in 1798, starting with smallpox. Within the list of vaccines shown here, in white background are microbial types of infections for which vaccine production use microbial fermentations.

Among the vaccines listed, some are for travelers in specific countries where the infectious disease is endemic. For example, yellow fever vaccination is mandatory when traveling to countries in Western Africa where there is a risk of infection. The number of marketed vaccines is increasing, but compared to pharmaceutical drugs, this number remains low.

1.2 ROLE OF PUBLIC HEALTH ORGANIZATIONS AND INDUSTRY

Vaccines are considered commodities and they fall under public health priorities. Governments are engaged globally aiming to implement solutions to address public health emergencies in their countries and globally in cases of pandemic situations. The COVID-19 unfolding pandemic situation since January 2020 is a live demonstration of the needs of these precious commodities to control the global public health situation and reduce the emergence of SARS-CoV-2 variants.

Because of its public health importance, historically, vaccine manufacturing and delivery was managed by public health organizations such as the National Institute of Allergy and Infectious Diseases in the United States; Connaught

TABLE 1.1
List of infectious diseases for which vaccines have been licensed

Disease	Year	Disease	Year
Smallpox	1798*	Bacterial Meningitis	1975&
Rabies	1885*	Pneumonia: polysaccharide	1977&
Typhoid fever	1896*	Adenovirus type 4 and 7	1980&
Cholera	1896*	Hepatitis B	1981&
Plague	1897*	Invasive Hib	1985&
Diphtheria	1923*	Japanese Encephalitis	1992&
Whooping cough: Pertussis	1926*	Hepatitis A	1995&
Lockjaw: Tetanus	1927*	Chickenpox: Varicella	1995&
Tuberculosis	1927*	Lyme disease	1998&
Tick-borne Encephalitis	1937*	Pneumonia: conjugate	2002&
Influenza	1945&	Rotaviral diarrhea	2006&
Yellow Fever	1953&	Shingles: Zoster	2006&
Poliomyelitis	1955&	Papillomavirus	2006&
Measles	1963&	Dengue	2019&
Mumps	1967&	Ebola	2019&
Rubella	1969&	COVID-19	2021&
Anthrax	1970&		

Notes
* Year of first reported use.
& Year of U.S. licensure.
Light gray background: egg-based production of vaccine.
White background: bacterial production of vaccine.
Dark gray background: cell-culture production of vaccine.
The list is Adapted from [10] and updated with data from [11].

Laboratories in Canada; the network of Institutes Pasteur and affiliated organizations; Butantan institute in Brazil; Serum Institute in India; Robert Koch Institute in Germany; Academy of Sciences in China; and many other organizations worldwide involved in the manufacturing, procurement, and distribution of vaccines. The Institutes Pasteur International Network with headquarters in Paris is still operating as a network involving many institutes in the Americas, Asia, and Africa, with a mandate that shifted from the original mission of providing vaccines wherever needed to centers and research organizations responding to broad national priorities.

In recent history and driven by high regulatory manufacturing standards, key industries, such as GlaxoSmithKline (GSK), Sanofi-Pasteur, Merck & Co, Pfizer, and a number of other small and medium size companies have become active in the field of manufacturing and commercialization of vaccines. The current situation of India and China with important governmental and private vaccine manufacturing capacities is of particular interest in observing the evolution of the vaccine global market. It is very likely that new strategic investments following the COVID-19 pandemic situation will be made by many countries to create sustainable vaccine biomanufacturing capacities, which will probably change the overall global vaccine market shape.

The WHO role has been critical in providing guidance to country national health authorities and regulatory bodies especially in time of pandemic or situation of "global concern," as it has been the case with the Ebola epidemics in 2013–14. Each country relies on a national regulatory authority such as the US-FDA in the United States, Health Canada, EMA in Europe, China-CDC etc. to approve vaccines in each country. Additionally, the WHO has the responsibility to provide "pre-qualification" of vaccine manufacturing capacity to commercialize vaccines in other countries. Details on the mandate of the WHO, organizational structure, and associated centers can be found on the WHO website (WHO, https://www.who.int).

1.3 THE VACCINE MARKET AND ECONOMIC DRIVERS

Vaccine economics is also very specific to the field. Vaccines as commodities are made available at the global scale with the support of public and governmental organizations, charity organizations such as the Bill and Melinda Gates Foundation: https://www.gatesfoundation.org, the Welcome trust: https://wellcome.org, the Rockefeller Foundation: https://www.rockefellerfoundation.org, and many others. Distribution of vaccines at global scale is managed through vaccine alliances such as GAVI: https://www.gavi.org and international bodies like UNICEF: https://www.unicef.org, the World Bank: https://www.worldbank.org/en/home and the World Health Organizations (WHO): https://www.who.int).

The vaccine market, in terms of total value used to be very small corresponding to 2/3 of a percent of the global pharmaceutical market. It is important to note that these figures are pre-COVID-19 pandemic and will be probably significantly revised post-pandemic. However, the importance of vaccines in preventing diseases largely outbalances the cost of the vaccines.

The vaccine market is generally segmented essentially in pediatric vaccines as opposed to adult vaccines, where growth is becoming very sustained with vaccination of aging populations in many countries.

The distribution in the revenue market shares and volume of vaccine sales is monitored by the WHO [12] through the Market Information for Access to Vaccines (MI4A) initiative to enhance vaccine market transparency and understand global vaccine market dynamics. A report of 2019 indicates that a small number of manufacturers dominates the global market with many products in the following order: GSK, Sanofi-Pasteur, Serum Institute of India (SII), Microgen, Merck, Bharat Biotech-Vaccines International Ltd (BBIL), Bio Farma, and Pfizer with five or more licensed vaccines in their portfolio. In perspective, the top five selling vaccines by 2020 are Pfizer's Prevnar 13 for prevention of pneumococcal infection with a sustained sale's growth generating nearly US$6 billion; Pentacel against diphtheria/pertussis/whooping cough/tetanus/polio/Haemophilus influenza type B from Sanofi-

Pasteur; Gardasil against Human papillomavirus (HPV) infection from Merck; Fluzone/Vaxigrip against influenza infection from Sanofi-Pasteur; and Pediatrix for prevention of diphtheria/tetanus/pertussis/whooping cough/hepatitis B/polio from GSK. The sales of the last four vaccines range from US$1.7 to 2.3 billions. Other well-positioned vaccines in the top ten list such as Varivax against varicella, Zostavax against Zoster virus infection, Rotateq against rotavirus infections and Pneumovax 23 for pneumonia infection are all produced by Merck, whereas Twinrix for prevention of Hepatitis A&B is produced by GSK. It is expected that the global vaccine market in 2021 and beyond will be restructured with the approval and sales of new COVID-19 vaccines.

Often referred to as the vaccine market north and south gap, figures collected by the WHO compare the burden of disease (how much diseases are present in population), which is present at 93% in the developing countries and 7% in industrialized country to the vaccine sales as being 82% in industrialized countries and only 18% in developing countries. These figures need to be interpreted taking into consideration the many different vaccine procurement processes put in place by the different countries. For example, high-income countries' (HICs) self-procurement of vaccines account for most of the market value, at over US$12 billion annually representing less than 5% of the global market volume [13].

As a consequence of the 2009 H1N1 influenza pandemic declaration, which raised concerns especially about aging populations, a sense of urgency dominated. The influenza vaccine market was estimated at $2.9 billion in 2011 and increased to $3.8 billion by 2018. Only in the U.S. market, the value increased from $1.6 billion in 2011 to $2.2 billion in 2018. The global market is projected to rise to US $ 58.4 billion by 2024 [14] with more than 120 new products in the development pipeline, among which 60 are of importance for the developing countries.

Clearly, vaccines are becoming an engine for the pharmaceutical industry and new business models are emerging. For example, many companies that did not have vaccines in their portfolio have started making alliances or acquiring smaller companies that could produce vaccines. It is also expected that the global vaccine market will be dramatically reshaped following the COVID-19 pandemic that revealed several challenges additionally to existing ones such as increased competition and narrowed traditional vaccine markets, new vaccine opportunities, and markets and threats driving rapid development of novel technologies within a new regulatory framework. Consequently, the vaccine field has a predicted spectacular growth rate of 10.7% a year for the forecast period 2019–2027 [15].

1.4 SAFETY AND REGULATION OF VACCINES

Safety of vaccines is paramount as they are prophylactic interventions that are delivered to healthy people including infants. Confidence of the population on the safety of vaccines and their efficacy is a priority for the public health regulatory bodies and vaccine manufacturers. It is well understood by the vaccine community that one single failure will impact public faith in the entire product class. Post-marketing surveillance assesses the effectiveness and safety of vaccines and begins after vaccines are approved for use and includes the monitoring of adverse events following immunization.

Vaccines are regulated by different, specialized branches within the public regulatory bodies. As an example, the process for approval of vaccine in Canada is presented in the following reference [16].

Safety and efficacy are first evaluated in animal studies or preclinical studies. Afterwards, the four phases of the clinical studies are initiated:

> Phase 1 studies a vaccine on a small group of people (usually fewer than 100) for the first time, examining its safety including dosage range and side effects.
>
> Phase 2 studies a vaccine on a larger group of people (usually several hundred or more), to see how effective the vaccine is in preventing a disease, confirming its safety and its optimum dosage.
>
> Phase 3 studies a vaccine on a larger group of people (usually many thousands) to confirm that it is both effective and safe by monitoring its side effects and any adverse reactions.
>
> Phase 4 occurs after the vaccine has been approved for use and is incorporated into immunization surveillance programs. This is also known as post-marketing surveillance. This includes ongoing safety monitoring, assessing vaccine effectiveness in specific population groups, and determining the duration of immunity to inform future decisions on the need for booster doses.

As it is the case for any biologics, any significant change in the vaccine manufacturing process might require clinical demonstrations of safety and eventually efficacy. Manufacturing of vaccines on different sites or countries might require bridging clinical trials. The role of the WHO through "pre-qualification" of vaccine manufacturing sites worldwide is an improved process for facilitating commercialization of vaccines in the different regions of the world.

1.5 BASIC PRINCIPLES OF VIRAL VACCINE DESIGN AND TRADITIONAL PRODUCTION

This textbook focuses only on viral vaccines; therefore, although designed using the same principles to activate immune response, other microbial infections and vaccination strategies will not be discussed. Vaccines might be classified taking into consideration their design and mode of exposure of the dominant antigen to the human immune system.

Figure 1.1 captures the principles of vaccination and provides a simplified view of the possible interaction pathways with key mediators of the immune response. This section will be detailed in Chapter 3 of this textbook by reviewing the basic principles in immunology that are directly applicable to vaccine design and development.

Traditional viral vaccines involve the whole virus as **inactivated** or **live attenuated,** such as influenza vaccines (Chapter 9), or a **sub-unit,** representing **the**

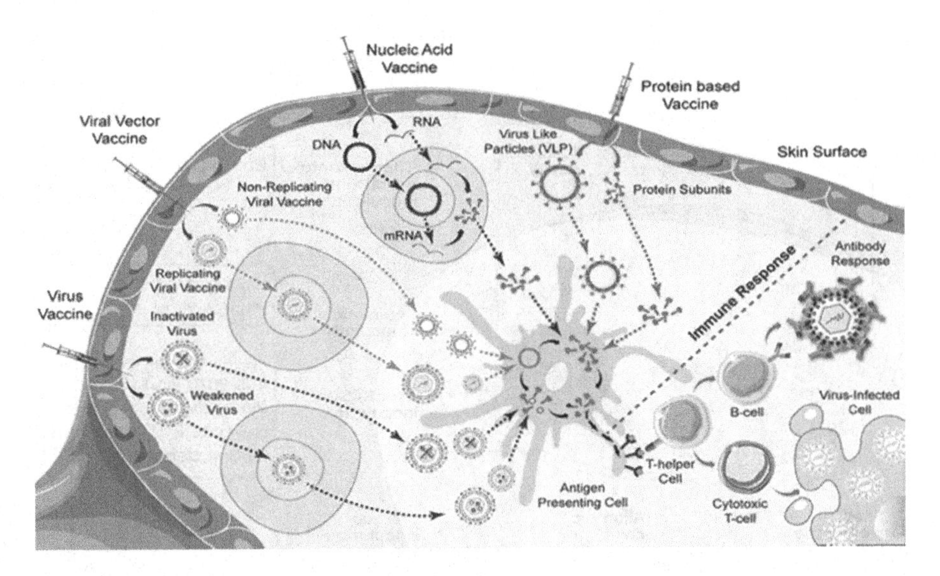

FIGURE 1.1 Description of the different types of vaccines and their possible interaction with the immune system. Credit for the design: Kumar Subramaniam.

dominant antigen extracted from the viral structure, such as Hepatitis B vaccines. Recombinant DNA technology enabled the design of **vectored vaccines** (Chapter 11) such as adeno-vectored vaccines against Ebola, SARS-CoV-2 infections, **virus like-particle vaccines** (Chapter 10) such as human papilloma virus vaccine, and **recombinant protein vaccines** such as recombinant hemagglutinin as the first approved influenza vaccine of its kind. An emerging class of vaccines based on delivery of genomic components of the virus include DNA vaccines and **mRNA vaccines** (Chapter 12). Over the COVID-19 pandemic, a remarkable demonstration has been made on safety, efficacy, and effectiveness of mRNA vaccines against SARS-CoV-2 infection, establishing this vaccine technology platform as a major technological jump in vaccinology. As data collected for phase 4 following the vaccine approval and commercialization of COVID-19 mRNA vaccines are compiled, more insights will be provided on the long-term protection of this class of vaccines. This textbook will detail the design, development, manufacturing of these different classes of vaccines within the core chapters and will illustrate with several case studies each class of these cell-culture produced vaccines.

New approaches based on vaccination using cells as antigen presenter cells (dendritic cells) are evaluated in pre-clinical and clinical trials showing some effectiveness in treatment of cancer, which falls in this case under the umbrella of therapeutic vaccines that is extensively documented in a number of reviews [17], but will not be discussed in this first textbook edition.

Traditionally, virus productions used live animals such as chickens' embryonated eggs to grow the virus in specific egg cavities and collect the virus thereafter in a specific cavity, as illustrated in Figure 1.2.

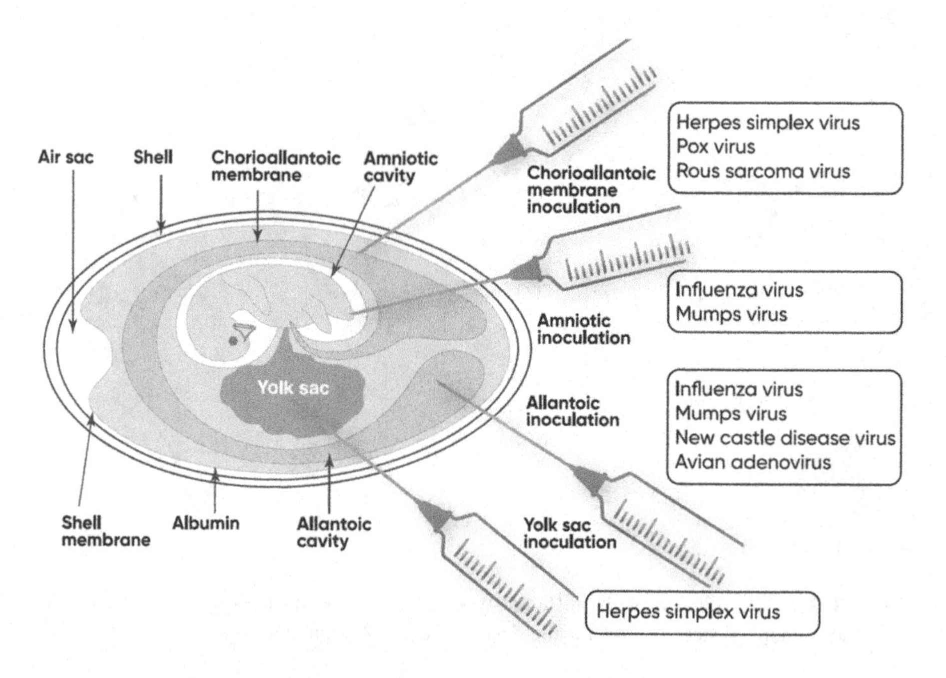

FIGURE 1.2 An embryonated chicken egg showing the different compartments in which viruses may grow. The different routes by which viruses are inoculated into eggs are indicated.

Although 80% of the influenza vaccines are still produced using embryonated egg technology [18], there is a trend to use cell culture technologies in all other cases of traditional viral vaccines including rabies, yellow fever, poliomyelitis, measles, mumps, rubella, adenovirus, Japanese encephalitis, Hepatitis A, varicella, rotavirus, human papilloma virus, dengue, zoster, and Ebola vaccines.

With the exception of mRNA vaccines, all COVID-19 vaccines approved or in advanced clinical trials are cell-culture produced vaccines indicating a solid trend for establishing cell culture advanced technologies for manufacturing viral vaccines.

1.6 CELL-CULTURE PRODUCTION PROCESSES

The cell-culture production process is designed integrating the basic knowledge on virology including the viral structure, its biology, and the viral replication cycle will be extensively described in Chapter 2 as basic virology for process design. Viruses are biological structures with sizes ranging from 20–300 nm, containing DNA or RNA coding for proteins necessary for replication, including non-structural and structural proteins contributing to a "shell" capsid that might be an icosahedron or helical asymmetrical capsid, and eventually an envelope as it is the case for HIV, SARS-CoV, and influenza viruses, for example, or non-enveloped (adenovirus, adeno-associated virus, or human papilloma virus) (Figure 1.3).

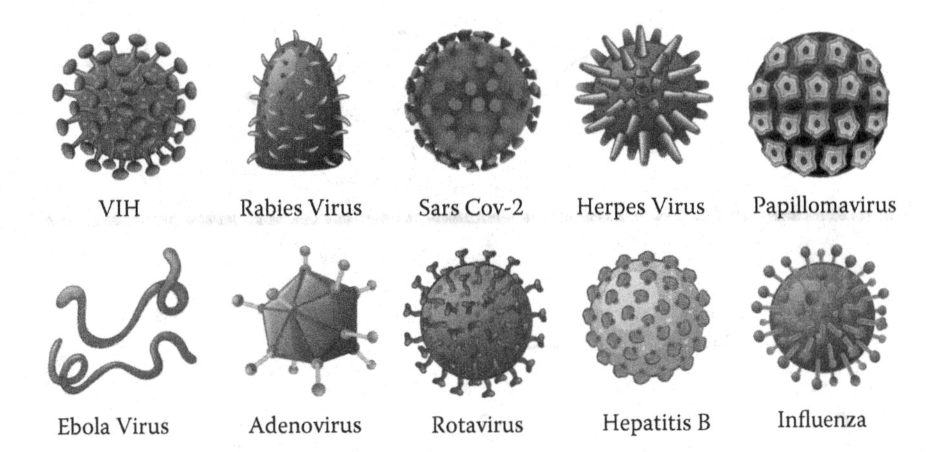

FIGURE 1.3 Example of different types of viruses.

David Baltimore (Nobel Prize in 1975) proposed a virus classification in 1971, focusing on the mode of viral replication. The classification was adopted by the virologists in parallel to standard virus taxonomy, which is based on evolutionary history (Chapter 2).

The general steps involved in virus replication include transport of the virus to the cellular membrane, attachment that might be mediated by specific receptors on the cell membrane, direct penetration, or endosome-mediated entry in the cell cytoplasm and uncoating of the virus structure as early events. Within the cells, the viral genomic coding sequences will determine the early gene transcription-the genome replication-late gene transcription/translation as middle events. Finally, late events include assembly and egress of the virion. Any of these steps and events are viral specifics but remain essential in determining the kinetics of replication in the cell-culture production process.

From a processing standpoint, there are two types of viruses, the ones that end with the lytic cycle (non-enveloped) and the ones that will bud from the cell surface (enveloped). For example, adenovirus has a lytic infection cycle and would infect the cells, replicate, accumulate, and then the cell will lyse and the virus will be released. An example of an enveloped virus is the coronavirus that will bud off the cells at maturation of the virion. The enveloped or non-enveloped nature of the virus has critical effects on the selection of upstream and downstream processing steps of the associated viral vaccines. Therefore, the nature of the infectious viral unit and its interaction with the host cell needs to be well integrated in the design of viral-structure-based vaccines (inactivated, attenuated, and vectored-vaccines) and the overall manufacturing stream. These mechanisms will be described in detail in the virology chapter as well as in the specific case study chapters (Figure 1.4).

Cell culture technology is the most effective mode of production of viral vaccines. Primary cell lines such as chick embryo fibroblasts are used in the production of measles and mumps vaccines, whereas human diploid cell lines

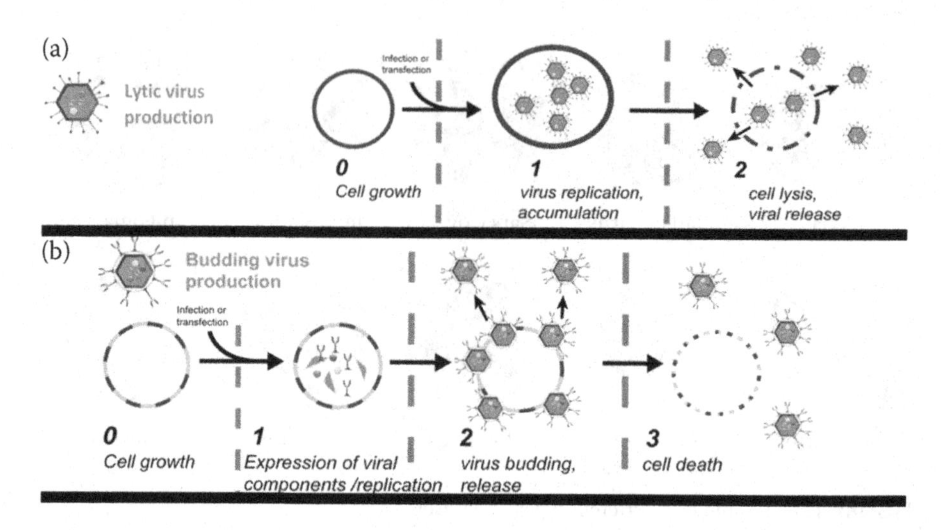

FIGURE 1.4 Non-enveloped lytic virus cycle (A) and enveloped virus cycle (B) replication. Credit to Sven Ansorge for the design of the image.

with limited generations such as Wi-38 and MRC-5 derived human lung fibroblasts are used to produce rubella and varicella vaccines. Vero cell line, a continuous cell line derived from the kidney of the African Green Monkey is used to produce Salk polio, rabies, rotavirus, and influenza vaccines exploiting microcarrier technology to enable industrial scale-up of adherent cell cultures. Other continuous cell lines that have been adapted to suspension cultures include the human embryo kidney cells, HEK293 and the primary embryonic retina cells, PER.C6, two transformed human cell lines, that are used to produce adenovectored vaccines currently licensed for immunization against SARS-CoV-2 infections. Madin-Darby Canine Kidney, MDCK, continuous cell line has been used as a preferred host for replication of influenza strains and has been adapted to suspension culture for industrial manufacturing of influenza vaccines. The development of these different cell culture platforms and the regulatory framework associated with their use for manufacturing viral vaccines is discussed in Chapter 4, dedicated to cell lines for viral vaccines production.

1.7 MANUFACTURING CHALLENGES OF VIRAL VACCINES

The complex structure of viral vaccines is highly associated with the structure and biology of the virus causing the infectious diseases. A primary challenge relates to the size distribution of the viral structures. For example, a measle virus has dimensions ranging from 250 to 400 nm, which is largely beyond the 200 nm sterilizing filter pore size. Consequently, cell-culture production of the measles virus cannot benefit from a final sterilization step, which imposes a complete process under totally sterile conditions. With recombinant vaccines in the class of the most advanced and characterized vaccines, human papilloma virus (HPV) vaccine, a virus like particle (VLP) vaccine, has a mean size diameter of 40 nm, as compared to an immunoglobin (IgG) with a size of about 10 nm. HPV vaccine manufacturing requires an extensive post-production disassembly/reassembly reprocessing of the capsid structural proteins to achieve a high level of purity of VLPs within a narrow distribution size range.

As detailed in the virology section (Chapter 2), viruses have different structures and properties and analytical methods to monitor their production as vaccines are often under-developed. The potency of the vaccine product that might involve an adjuvant in the final formulation is determined by its interaction with the immune system which is incompletely understood (Chapter 3), contributing to the complexity of viral vaccines design and manufacturing. Viral vaccine production processes are characterized by 1) a lack of platform technologies requiring multiple cell types, mostly operated in adherent cell culture mode; 2) different virus/cell interactions; 3) different scales for production contributing to further complexity in the bioprocessing and biomanufacturing of viral vaccines. This is particularly emphasized when compared to the well-established process for manufacturing monoclonal antibodies using a CHO platform in the broad context of biomanufacturing of biologics.

Manufacturing and releasing a cell-culture produced viral vaccines is lengthy and complex (Figure 1.5). The current good manufacturing process (cGMP) requires establishing and extensively documenting a master cell bank and a master virus seed bank derived from the virus isolates according to regulatory guidelines [19].

In short, the manufacturing process is initiated by cell amplification from a working cell bank, derived from the master cell bank. The cell culture process stream will depend on the cell substrate type that would grow in adherent or suspension cell cultures and would determine the mode of operation and type and scale of the bioreactor for production. In the case of adherent cell lines such as Vero cells, supports such as T-Flasks, roller-bottles, cell factories, or packed-bed bioreactors are required to sustain cell growth. To mitigate surface limitation to produce large quantities of vaccines as is the case for polio vaccines, microcarrier technology as a support might be used to facilitate the scalability up to 3,000 L operational volume. Cell cultures in suspension are generally more amenable to streamlined scale-up to larger volumes up to 10,000 L. As it has been the trend for production of biologics such as recombinant proteins and monoclonal antibodies, single-use equipment is deployed more and more frequently as a rapid response to surge manufacturing of viral vaccines [20].

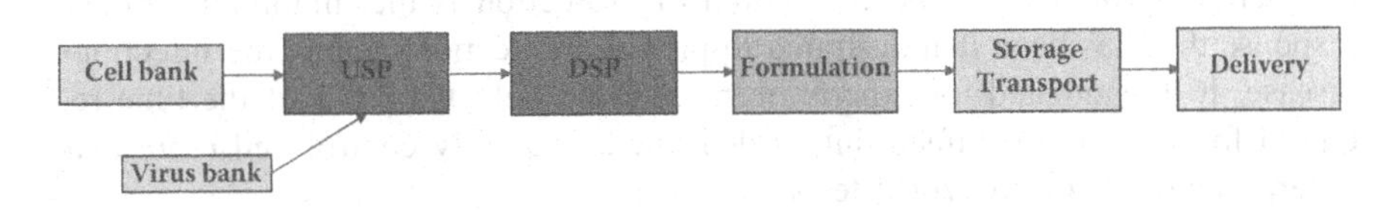

FIGURE 1.5 Typical scheme on cell-based vaccine manufacturing.

Following the cell amplification phase, a viral production phase is initiated. In a first cell culture stream, cells are infected for viral production with a working virus seed stock, derived from the master virus bank according to pre-established process parameters. A second uninfected cell culture stream is processed under similar operating conditions and monitored in parallel to the cell culture viral production stream as a control to demonstrate the absence of non-adventitious agents in the whole cell-culture production process. Depending on the infection/replication kinetics, the *virus bulk* is harvested after days to weeks.

The virus bulk harvest is purified following pre-established downstream process steps optimized and validated during the early process development phase and involve multiple filtration/ultra-filtration or chromatographic steps highly depending on the virus type, its structure, and stability. It is not uncommon to use large-scale ultracentrifugation in the vaccine manufacturing process. For example, hundreds of millions of influenza vaccine doses are produced using large-scale ultracentrifugation.

The upstream, downstream, and analytical protocols are detailed in dedicated core chapters in this textbook (Chapters 5 to 8) and examples are extensively presented and discussed in the case study reports of cell-culture produced vaccines and associated processes (Chapters 9 to 12). A special emphasis is placed on the development and execution of different sets of assays for vaccine lot release. As an example, single-radial immunodiffusion assay (SRID) (which is the only validated assay approved by regulatory agencies for the release of influenza vaccine lots) would require the preparation of a standardized antigen and specific polyclonal antibody. The preparation of the antibody might take weeks to months, and it is produced through immunization of animals, such as mice, rabbits, or sheep, collecting the serum from animals after immunization with the specific antigen. The overall process for preparation of these reagents is very time-consuming, taking weeks to months and might delay the approval of a vaccine for eventual evaluation in humans. A report from the WHO [21] identified the timelines required for the deployment of the SRID assay as one of the many reasons that made the response to the 2009 H1N1 pandemic and the timely availability of vaccine inadequate. Although the vaccine was manufactured, delays due to unavailability of standardized SRID reagents in a timely fashion were observed, raising significant concerns on readiness to respond to pandemic situations. The search for alternative assays to SRID assays fueled significant research in this area underlining the critical role of assays in vaccine development and manufacturing. Ideally, when the mechanism of action of the **vaccine** is known, a **potency assay** that is predictive of the clinical response should be available, serving as a "clinical correlate of protection." The correlation of protection is the minimum immune response that has been demonstrated to provide protection against the infectious disease. It is estimated by experts in the vaccine field that 70% of the time required for vaccine manufacturing is dedicated to quality control and represents several hundreds of analytical tests.

Specific to the vaccine field regulation, virus bulk batches require formal release by agents of the territory competent regulatory agency such as Health Canada, FDA, or

EMA within the premises of the manufacturing facility. The virus bulk is then formulated, filled in product containers, and then packaged in product containers. A second formal release by the regulatory agency agents is conducted prior to shipping. Additional regulatory assessments are conducted by regulatory agencies following shipping and distribution through extensive testing and documentation of release criteria [10].

The formulation of vaccines is an important step in the manufacturing process. This would combine in the bulk vaccine, the antigen product with eventually adjuvants in the case of subunit vaccines and excipients that would contribute to vaccine thermostability enabling delivery to remote areas. This step will also involve a final fill, a sterile filtration, and eventually a freeze-drying in the case of lyophilized vaccines prior to release and storage according to a predetermined cold chain. Adjuvants, as immune potentiators, play a role in the delivery of the antigen as particulates or molecular structures acting as depot/carriers of the antigen to immunostimulants, inducing a broader immune response and providing a better protection while maintaining safety of the vaccine product.

In the case of freeze-dried vaccine formulations, reconstitution of the vaccine prior to injection is required. The principles of freeze-drying to remove the solvent, usually water, from dissolved or dispersed vaccine will be described and exemplified in case studies.

1.8 PANDEMIC PREPAREDNESS AND OUTLOOK

The current COVID-19 pandemic has put the focus on the development process of vaccines and their safety and efficacy. However, this is not the first pandemic that has occurred. By the end of World War I, in 1918, the Spanish Flu (which was an H1N1 influenza-type virus) caused between 30 and 100 million deaths. SARS (Severe Acute Respiratory Syndrome), with a mortality of 10%, appeared in China in 2002. The H1N1 pandemic (2009), called the swine flu, showed that viruses can transmit via animals. All these viruses are RNA viruses, meaning that evolution is possible. MERS, Middle East Respiratory Syndrome, with a mortality of 30%, appeared in Saudi Arabia in 2012. MERS has been thereafter associated with close contact of humans with dromedary camels as the vehicle of disease transmission and has spread to other countries [22].

The field of vaccines, particularly in the context of pandemic preparedness, requires the development of new tools and methods to enable accelerated process and product development and manufacturing. Eventually, all the vaccines that are under development build on prior knowledge on vaccine design and manufacturing technologies with emergence of paradigm shifts such as the messenger RNA technology that led to the rapid development of COVID-19 mRNA vaccines within unprecedented timelines. Importantly, the scale-up potential and the robustness of the manufacturing technology for global delivery remain key drivers in the field of vaccine development, requiring alliances of experts in vaccinology, virology, cell biology, engineering, and medicine.

REFERENCES

[1] "History of Vaccines - A Vaccine History Project of The College of Physicians of Philadelphia | History of Vaccines," https://www.historyofvaccines.org/

[2] A. Boylston, "The origins of inoculation," *J. R. Soc. Med.*, vol. 105, no. 7, p. 309, Jul. 2012, doi: 10.1258/JRSM.2012.12K044

[3] M. R. Hilleman, "Vaccines in historic evolution and perspective: A narrative of vaccine discoveries," *Vaccine*, vol. 18, no. 15, pp. 1436–1447, Feb. 2000, doi: 10.1016/S0264-410X(99)00434-X

[4] G. L. Geison, "The private science of Louis Pasteur," *Priv. Sci. Louis Pasteur*, vol. 306, pp. 22–50, Dec. 1996, doi: 10.1515/9781400864089/HTML

[5] "Pasteur International Network association | Institut Pasteur," https://www.pasteur.fr/en/international/pasteur-international-network-association

[6] P. Deville, *Plague and Cholera,* 2014.

[7] J. B. Ulmer, U. Valley, and R. Rappuoli, "Vaccine manufacturing: Challenges and solutions," *Nat. Biotechnol.*, vol. 24, pp. 1377–1383, 2006, doi: 10.1038/nbt1261

[8] "GPEI – Global Polio Eradication Initiative," https://polioeradication.org/

[9] M. Roser, H. Ritchie, and B. Dadonaite, "Child and Infant Mortality," "https://ourworldindata.org/child-mortality," 2013.

[10] M. C. Flickinger, "Encyclopedia of biotechnology: Bioprocess, bioseparation, and cell technology. Chapter: Viral vaccine production in cell culture (John Aunins)," pp. 7 zv. (XXVII, 5051 str.) TS-WorldCat T4-Biop, 2010.

[11] "Vaccines Licensed for Use in the United States | FDA," https://www.fda.gov/vaccines-blood-biologics/vaccines/vaccines-licensed-use-united-states

[12] "Global Vaccine Market Report Overview of MI4A," 2018, Accessed: Nov. 08, 2021. [Online]. Available: http://who.int/immunization/MI4A

[13] M. Kaddar, "Global Vaccine Market Features and Trends," https://www.who.int/influenza_vaccines_plan/resources/session_10_kaddar.pdf, 2013.

[14] "Vaccines Market | Including & Excluding COVID-19 vaccines | Global Forecast to 2026," https://www.marketsandmarkets.com/Market-Reports/vaccine-technologies-market-1155.html?gclid=CjwKCAiA78aNBhAlEiwA7B76p3fHFVjr3dO5EaXlxFj-m0mxqxkiGyuX5k3QX6pOk14HpuSc4P8KZbxoCBjEQAvD_BwE

[15] "Vaccines Market Size, Share, Growth | Global Industry Report [2027]," https://www.fortunebusinessinsights.com/industry-reports/vaccines-market-101769

[16] P. Health Ontario, "At a Glance: Vaccine regulatory process in Canada 1 AT A GLANCE Vaccine Regulatory Process in Canada."

[17] Y. Gu, X. Zhao, and X. Song, "Ex vivo pulsed dendritic cell vaccination against cancer," *Acta Pharmacol. Sin. 2020 417*, vol. 41, no. 7, pp. 959–969, May 2020, doi: 10.1038/s41401-020-0415-5

[18] E. Sparrow *et al.*, "Global production capacity of seasonal and pandemic influenza vaccines in 2019," *Vaccine*, vol. 39, no. 3, pp. 512–520, Jan. 2021, doi: 10.1016/J.VACCINE.2020.12.018

[19] "General Principles for the Development of Vaccines to Protect Against Global Infectious Diseases | FDA," https://www.fda.gov/regulatory-information/search-fda-guidance-documents/general-principles-development-vaccines-protect-against-global-infectious-diseases

[20] E. S. Langer and R. A. Rader, "Biopharmaceutical Manufacturing is Shifting to Single-Use Systems. Are the Dinosaurs, the Large Stainless Steel Facilities, Becoming Extinct? | American Pharmaceutical Review – The Review of American Pharmaceutical Business & Technology," *BioPlan Associates Inc*, 2018. https://www.americanpharmaceuticalreview.com/Featured-Articles/

354820-Biopharmaceutical-Manufacturing-is-Shifting-to-Single-Use-Systems-Are-the-Dinosaurs-the-Large-Stainless-Steel-Facilities-Becoming-Extinct/
[21] World Health Organization (WHO), "PANDEMIC INFLUENZA A (H1N1) Donor Report," https://www.who.int/csr/resources/publications/swineflu/h1n1_donor_032011.pdf, 2011.
[22] "Middle East Respiratory Syndrome Coronavirus (MERS-CoV)." https://www.who.int/news-room/fact-sheets/detail/middle-east-respiratory-syndrome-coronavirus-(mers-cov)

2 Introduction to virology

Shantoshini Dash
Viral Vectors and Vaccines Bioprocessing Group,
Department of Bioengineering, McGill University, Montréal,
QC, Canada

CONTENTS

2.1 INTRODUCTION

2.1.1 WHAT ARE VIRUSES?

Viral infections are known to cause diseases that may or may not require hospitalization; but also, they are the leading cause of a heavy mortality rate, including other health complications. Viral infection proves a greater threat to human health that can be least controlled [1]. Viruses, the causative agents of viral infections, are small, subcellular microorganisms, extremely dependent on host cells and are known as obligate intracellular and parasites. They can infect any form of life (plants, animals, bacteria, fungi). While the first virus discovered was in 1892, known as the TMV

DOI: 10.1201/9781003229797-2

(Tobacco Mosaic Virus) followed by foot and mouth disease virus in 1898, the first human virus was discovered only in 1901 as the yellow fever virus. But scientists did not see an actual virus until 1930. In 1915, Frederick Twort, a bacteriologist, discovered a bacteriophage, the virus that can infect bacteria, notifying it as a microorganism that would kill bacteria. Hence, it established a unique feature of the virus as their size could vary within a range between 20 nm to 1 micron. They are much smaller than the cells they could infect.

2.1.2 Virus Characteristics

Viruses are among the most symmetrical biological objects. They can be either helical, spherical, icosahedral, or have more complex structure. They could be filamentous with elongated structures. Viruses can be visualized by x-ray crystallography or electron microscopy. Looking at the structure of TMV, it has given the concept that viruses are structurally composed of repeating subunits. The viral structure consists of some key features; namely, the capsid encapsulating the viral genome. The virus may or may not have an envelope layer.

2.1.3 Virus Evolution and Classification

Evolution of viruses has remained very speculative as they do not fossilize. They do not have a common ancestor. There have been different theories of virus evolution. First is the devolution or regressive theory, meaning they could have originated from free-living cells. Second is the escapist or progressive theory, which explained they might have originated from RNA and DNA molecules that escaped from the host cell. The third is the self-replicating theory, explaining a system of self-replication involving evolution alongside the host cell.

Viruses are classified based on their morphology, chemical composition, host organism, or mode of replication. But, since the discovery of viruses, the classification system has been modified to the system that is currently being followed. The first was Holmes' classification, who suggested a first complete taxonomic system. He proposed the order "virales" composing three suborders, namely, Phaginae (virus that infects bacteria), Phytophaginae (virus that infects plants), and Zoophaginae (virus that infects animals). He further created 13 families, 32 genera, and 248 species. Then came the classification system that gained the community support known as the LHT (Lwoff, Horne, and Tournier) system. This system grouped the viruses into one phylum called "vira" with two subphyla defining the genetic material, which is either DNA or RNA. This was further classified into classes based on the symmetry of the viral capsids.

Later, an urgent need to have an official system for taxonomy led to the establishment of the International Committee on Taxonomy of Viruses (ICTV). Thereafter, in 1971, David Baltimore published a classification system that is still in use in parallel. He grouped all viruses into seven groups based on the type of genome. From then until the present day, virus taxonomy has been considered by this committee following the Baltimore classification system (Figure 2.1).

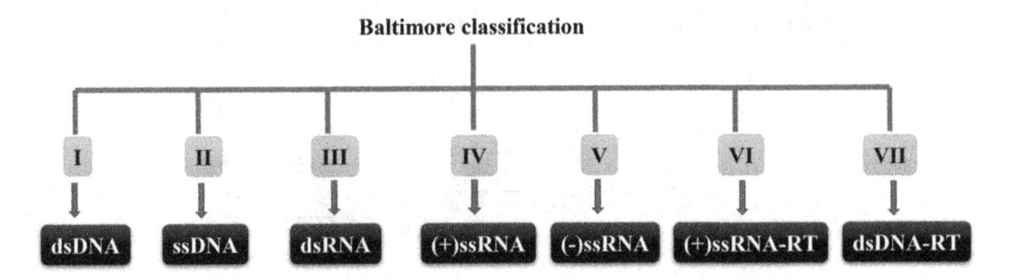

FIGURE 2.1 Baltimore Classification: This classification is based on the genome of the virus. **+ve RNA**: similar to mRNA and can be readily translated by host cell; **–ve RNA**: complementary to mRNA and requires conversion to +ve RNA by RNA polymerase prior to translation; **+ve DNA**: also called sense/coding strand. Sequence corresponds to the mRNA transcript ready to be translated; **–ve DNA**: also called anti-sense/template strand. Reverse complementary to the sense strand as well as to the mRNA transcript.

2.2 STRUCTURE AND GENOME

2.2.1 Virus Structure and Function

2.2.1.1 Capsid

A capsid is a protein shell required to protect the viral genome from host nucleases. For some viruses, during infection, the capsid is responsible for attachment to the specific receptors exposed on the host cellular surface. A capsid can be either single- or double-protein shells containing few structural proteins. Hence, multiple copies of the capsid must self-assemble to form the 3D capsid structure, allowing different viruses to have wide ranges of shape and structure.

2.2.1.2 Envelope

Some viral families have an additional protective coat, called the envelope. The envelope is a lipid bilayer which is partly obtained from the host cell membrane. The lipid composition of this envelop closely reflects that of the specific host cell plasma membrane. The exterior of this bilayer exhibits protruding structures known as "spikes," containing virus-coded glycosylated trans-membrane proteins. In enveloped viruses, spikes also assist in the attachment of the virus to the host cell surface.

The viral envelope serves the function of protecting the viral genetic material. It also helps in facilitating the entry of the virus while infecting a host cell along with evading the host immune response. However, due to the envelope's fragile nature, non-enveloped or also known as naked viruses, could be more resistant to parameters such as temperature, pH, and few common chemical disinfectants (Figure 2.2).

2.2.2 Viral Genome

A virus could have either an RNA or DNA genome. This RNA or DNA genome could be again categorized into being either positive or negative sense and single or

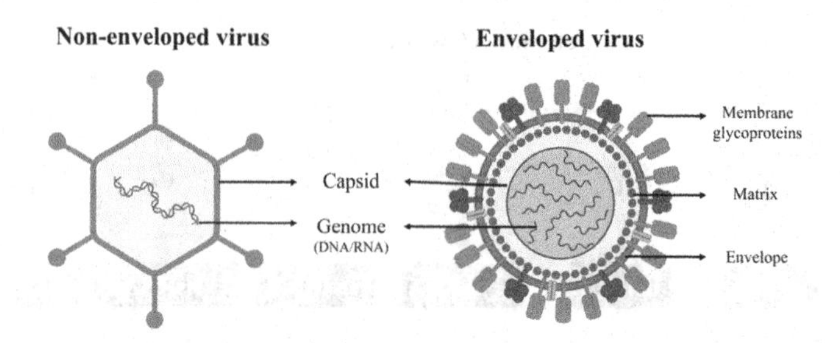

FIGURE 2.2 A non-enveloped virus (example: adenovirus) versus an enveloped virus (example: influenza virus).

double stranded. Because of its limited size, this genome codes for a minimum number of proteins necessary to allow its multiplication by the host cell. A fully assembled infectious virus is called a virion. For example, the influenza A virus consists of eight single-stranded (negative sense) RNA segments encoding for a total of 11 viral proteins.

2.2.3 Viral Genome Transcription Via Intermediates

The seven groups divided by the Baltimore classification explain the virus mechanism to facilitate its genome replication, transcription, and translation using the host cellular machinery. The seven groups have been described in detail in the following (Figure 2.3):

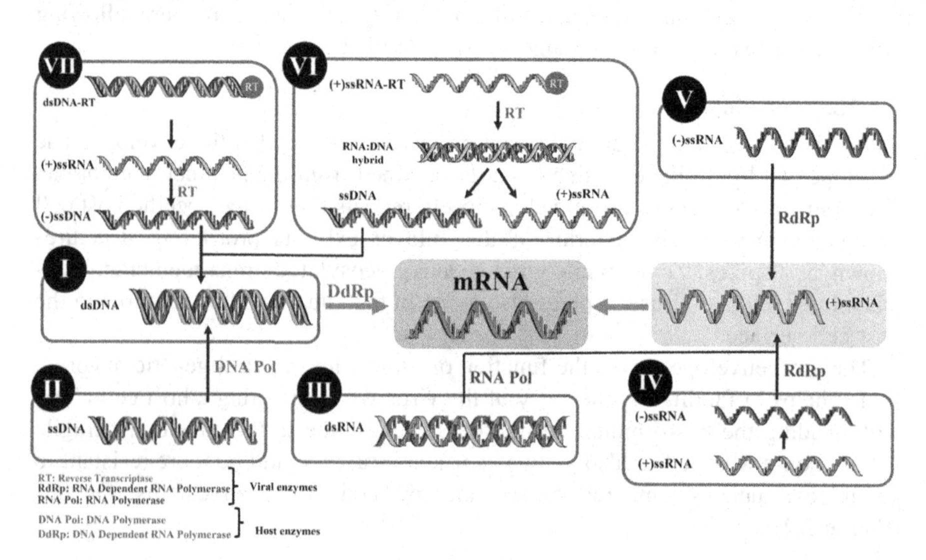

FIGURE 2.3 Virus transcription via intermediates: Transcription mechanisms followed by different groups of viruses defined by the Baltimore classification system in a eukaryotic host. The final goal is to produce a functional mRNA to be translated by the host ribosome to synthesize viral proteins.

1. **Group I – dsDNA virus:** The viruses belonging to this group follow the simplest mechanism of transcription. They use DNA-dependent DNA polymerase (DdDp) from the host cell to generate mRNA, which could late translate into functional viral proteins.
   Examples: Adenovirus, Herpesvirus
2. **Group II – ssDNA virus:** This group first needs to create a complementary strand to form a dsDNA. This is achieved by using DNA polymerase from the host to synthesize the strand which is complementary to their ssDNA genome. After the formation of dsDNA, it makes mRNA using the host enzyme DdDp.
   Example: Parvovirus
3. **Group III – dsRNA virus:** This group of viruses contains an RNA polymerase that would transcribe the dsRNA to (+) ssRNA, which would ultimately be the mRNA. This mRNA would serve two purposes. First, it would get translated to the required viral proteins, and second, it would also serve as a template to synthesize (–) ssRNA to form the dsRNA genome for packaging.
   Examples: Reovirus, Rotavirus
4. **Group IV – (+) ssRNA virus:** The genome of this group of viruses is similar to the host mRNA sequence. But the only hurdle is that the host ribosomes do not recognize the viral RNA. Hence, the viral polymerase moves along the (+) strand of the RNA template and elongates the (–) stranded RNA molecule. Now this (–) ssRNA serves as a template for polymerizing new (+) ssRNA, serving as the genome of the virus.
   Examples: Picornavirus, Togavirus, Coronavirus
5. **Group V – ssRNA virus:** This group makes for the largest family of viruses. They carry the RNA-dependent RNA polymerase (RdRp). This enzyme makes two types of (+) RNA strand: 1) a short viral mRNA that would be translated into necessary viral proteins and 2) a full-length RNA that would be replicated to make (–) ssRNA genome for packaging into progeny visions. Here, viral RNA is replicated separately and has a highly organized and regulated packaging process to make sure that one of each distinct RNA has been received by each virion.
   Examples: Orthomyxovirus, Paramyxovirus
6. **Group VI – (+) ssRNA-RT virus:** This group of viruses contains an enzyme called reverse transcriptase (RT). They go opposite or reverse to the normal transcription process. The (+) ssRNA undergoes reverse transcription to make a complementary DNA strand. This DNA strand acts as a template from a dsDNA in the nucleus of the host. This dsDNA is covalently linked to host chromosomal DNA and, therefore, replicates as a host genome. Viruses of this group benefit from the error-prone RT enzyme. It provides them with their ability to evade the immune system by minor changes to their protein capsules.
   Example: Retrovirus
7. **Group VII – dsDNA-RT virus:** This group of viruses replicates through RNA intermediates. The members of this group have a very different genome because of two facts. First, one strand has a protein at the 5' end

and is considered a complete strand. Second, the other strand has a short RNA at the 5' end, making it an incomplete strand. Hence, one strand being complete and the other being incomplete, they have a gapped DNA genome. The mechanism followed by the members of this group must consider the facts that the 1) transcription occurs in the host nucleus, 2) only the RNA part would be exported out of the nucleus, and 3) viral assembly occurs in the host cytoplasm.

Therefore, the gapped genome is first repaired/filled by viral polymerase forming a circular DNA. This DNA is transcribed into RNA by host polymerase. The RNA goes out of the nucleus. Finally, the viral RT enzyme makes DNA from RNA, making it possible to be packaged into virion capsid.

Example: Hepatitis B (Figure 2.3).

2.3 VIRAL INFECTION CYCLE

The viral genome, despite its limited size, will have to follow the general rule of the "Central Dogma" to transfer the information from its genes to the proteins. The viral genome needs to first replicate itself. Second, it must transcribe to form a functional mRNA, and third, this mRNA must undergo translation to generate a functional protein, including all post-translational modifications, producing an infectious virion, and for all these it depends on the host system. An infection cycle needs to follow a series of events ultimately leading to the production of new virions using the host cellular machinery. There are three types of commonly known infection cycles.

2.3.1 Lytic Cycle

A lytic cycle is also known as virulent infection. This type of infection cycle is mostly followed by viruses infecting bacteria (bacteriophages, Figure 2.4). This cycle results

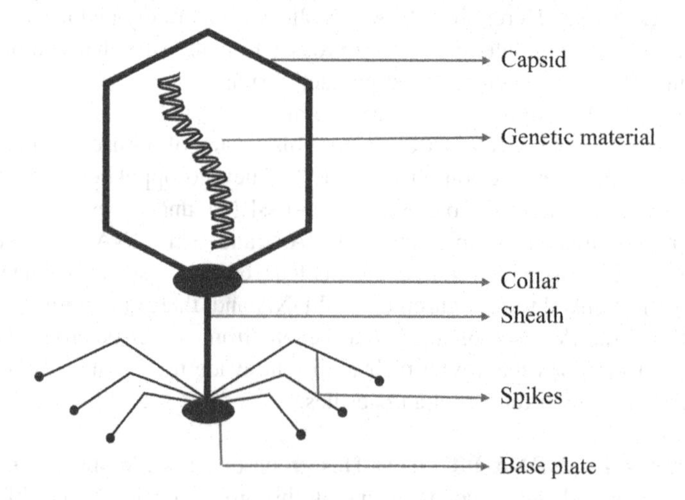

FIGURE 2.4 **Bacteriophage:** A virus that infects bacteria.

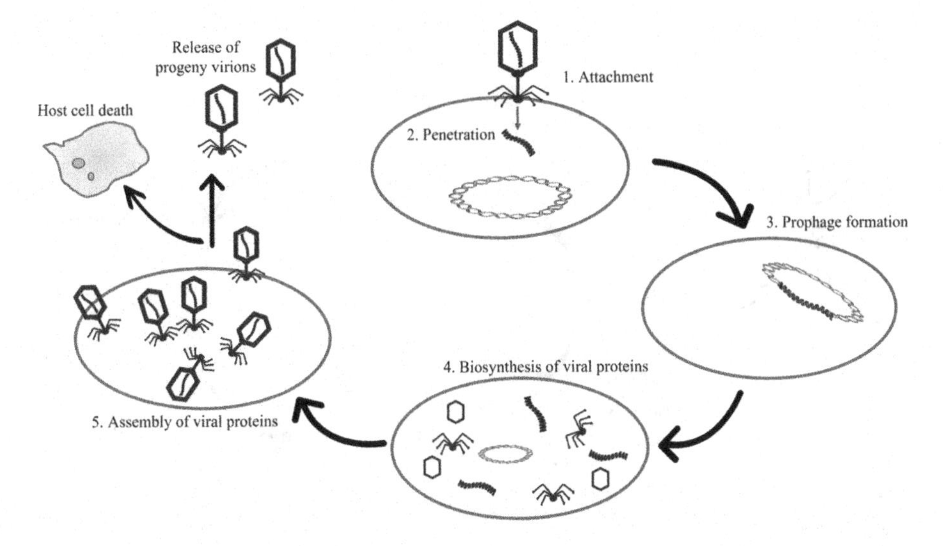

FIGURE 2.5 Lytic cycle: Schematic representation of lytic infection cycle followed by bacteriophage resulting in subsequent death of the host cell releasing newly formed virions.

in the death or lysis of the infected cell. The cycle begins with the process of **attachment** to the host cells, following which the viral genome is injected into the host cell, termed **penetration**. Then replication of the viral genome and **biosynthesis** of viral proteins occurs. This newly formed viral genome is **assembled,** followed by the maturation and release of the progeny virions.

During this process of the lytic cycle, the host cell gradually weakens due to viral enzymes and consequently burst/lyse, releasing the newly formed virions into the surrounding environment (Figure 2.5).

2.3.2 Lysogenic Cycle

The lysogenic cycle is also known as a non-virulent infection. This does not kill the host cell; rather, it remains in a dormant state. The lysogenic cycle also starts with **attachment** and **penetration** phases as described in the lytic cycle. But then the genome integrates itself into the host genome with the help of phage-encoded integrase enzymes. This integrated genome is termed a "prophage." The prophage gets replicated with the host genome and remains there if the host cell is dividing.

But, when the virus following the lysogenic cycle encounters any environmental stress, the prophage extracts itself from the host genome and enters the lytic cycle, resulting in the lysis of the host cell. This conversion from the lysogenic to the lytic cycle is termed "**induction.**"

Sometimes, during this induction process, the prophage might be left behind, or take a portion of the host genome when they re-circularize. As a result, during the next infection they might transport bacterial genes from one strain to another: known as **transduction**. It's one of the reasons resistance virulence traits spread through the bacterial population (Figure 2.6).

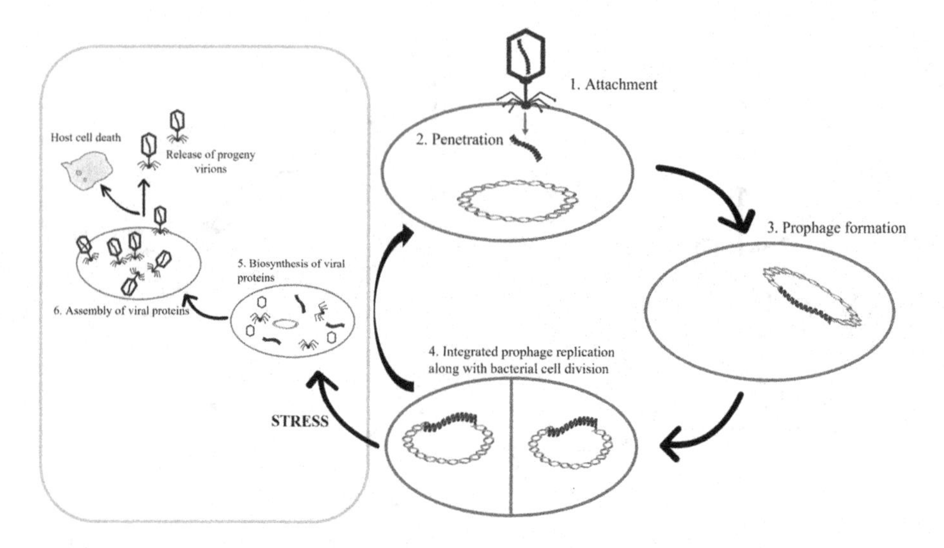

FIGURE 2.6 Lysogenic cycle: Schematic representation of lysogenic infection cycle followed by bacteriophage, which could either persist in the host cell for a longer duration or lyse the host cell under environmental stress.

A similar infection cycle is followed by viruses belonging to Herpesviridae family. Herpesviruses are DNA viruses that could cause infections in animals and humans. This group of viruses largely depends on their ability of latency and reactivation for an efficient infection. They maintain their latency period in non-replicating cells such as neurons. The viral genome undergoes circularization, forming an episome. During the latency period, the virus must limit its protein expression to a minimum in order to keep evading the host immune system. This is achieved by either generating long non-coding RNA transcripts [2] or through modulation of chromatin insulators [3]. However, it is necessary to express those viral proteins that would help the viral genome remain attached to the host chromosome during cell division of that particular cell.

Reactivation from latency occurs on encountering extreme stress such as suppression of the host immune system, neural trauma, or exposure to UV radiation [4]. Some allogenic transplantation could also induce the reactivation from latency such as in human cytomegaloviruses [5] (Figure 2.7).

2.3.3 Infection in Eukaryotic Host

Viral infection in eukaryotic hosts is rather specific depending on the type of viral genome and the host species. This is due to the fact that the host itself is a multicellular complex organism. There are several steps for the virus to undergo for a successful infection (Figure 2.8).

1. **Attachment:** This is the first event where the virus recognizes the host cell. This depends on the attachment factors present on the virus (e.g., glycoproteins) and host cellular receptors that facilitate the host-virus

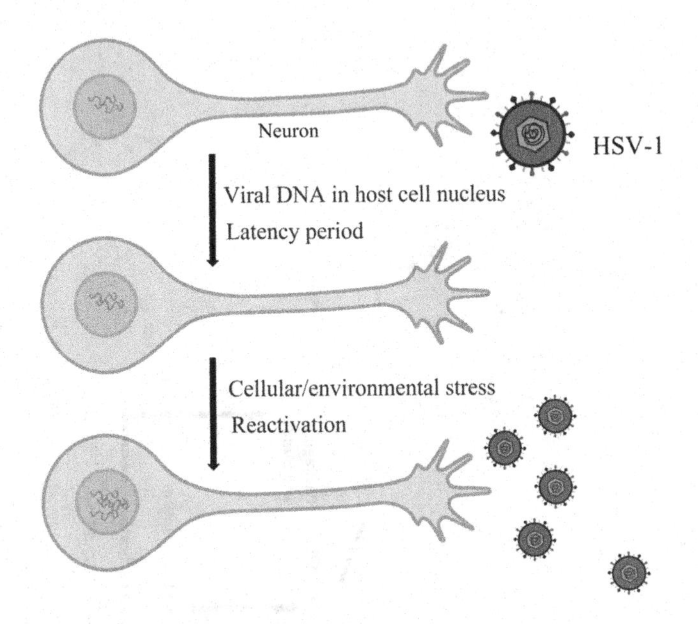

FIGURE 2.7 HSV-1 infection cycle: HSV-1 latency period and reactivation from latency in neurons.

interaction, giving access to the virus to enter the host cell. The virus binds first to the host cell surface. This binding is mediated by the recognition of specific protein sequences present on the host cell surface by the viral glycoproteins.

2. **Internalization:** After the attachment is established through receptor binding, the virus is taken into the host cell cytoplasm. This could be through direct fusion (fusion of viral envelope and host cellular membrane) or through receptor-mediated endocytosis (followed by most viruses). The entry of the virus into the host cell is commonly dependent on receptor-mediated endocytosis. In this process, the virus particle is engulfed by the host cell membrane, which then fuses with the intracellular compartments, known as endosomes. Consequently, the virus is released inside the endosomal lumen. Generally, "substances" taken up by the endosomes, ultimately end up in lysosomes, degraded by some hydrolytic enzymes. But most viruses can escape this degradation process.

The most common way of viral internalization is through clathrin-mediated endocytosis (CME). This process requires a protein coat to be assembled on the membrane to induce membrane curvature and form a spherical invagination. Along with clathrin, which is a scaffolding protein, other factors like epsin, ampiphisin, adaptor protein 2, dynamin, actin, and PIP2 are also important in the process of CME [6]. These proteins assemble and form a complex on the cytoplasmic side of the plasma membrane, which leads to the formation of a coat at the site of endocytosis. The phospholipid PIP2 is found at a higher concentration at the

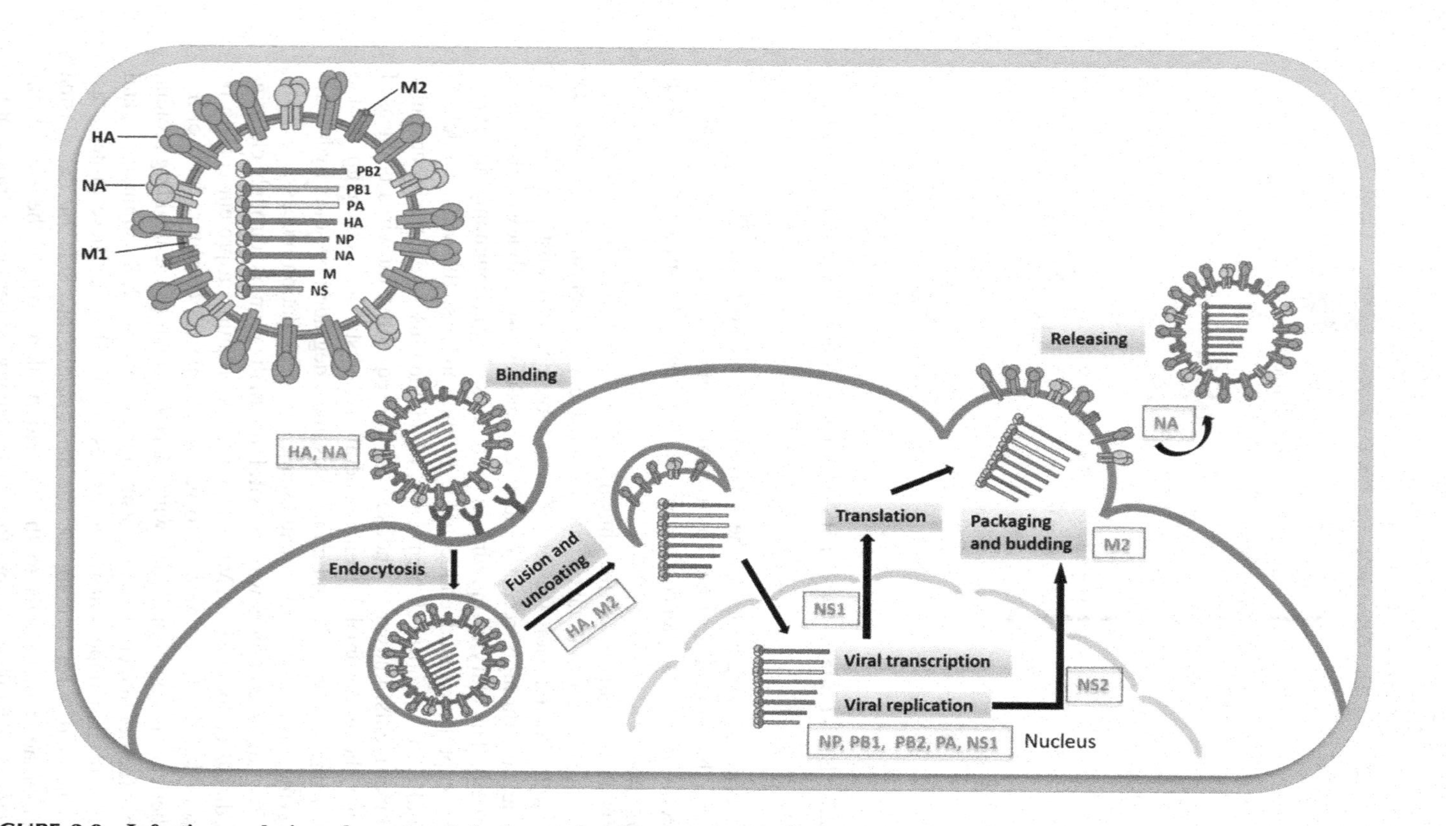

FIGURE 2.8 Infection cycle in eukaryotes: Infection cycle of an enveloped virus (influenza A virus). Different steps of viral infection and the participating viral proteins are indicated. [*Source: Zhu W, Wang C, Wang BZ. From variation of influenza viral proteins to vaccine development. Int J Mol Sci. 2017;18(7):E1554.*]

site of endocytosis, where it can interact with multiple membrane proteins. Electron micrograph studies have shown CME pits with well-defined clathrin lattice forming the backbone of the budding vesicle [7].

3. **Uncoating:** To gain access to the cytoplasm, the virus needs to be released from the endosome. Indeed, the low pH of the endosomal vesicle induces conformational changes of the viral glycoprotein, inducing the fusion of the virus envelope and the endosomal membrane. Following this fusion event, in some cases, there is an activation of the viral ion channel. Proton pumps of the endosomal membrane regulate the low pH of the endosomes. This pH initiates the fusion between the viral envelope and the endosomal membrane, which is a determinant step of the infection cycle. At this acidic pH, the viral glycoprotein undergoes a conformational change, followed by the process which induces the merging of the viral membrane with that of the endosome of the host. This process also results in formation of some fusion pores, which facilitate the transfer of the viral genome into the host cell cytosol.
4. **Viral gene expression and genome replication:** The viral genome replication depends on the type of genome as discussed earlier. For the genome replication and transcription, some of the virus families may perform it independently without the need of the host cell machinery. However, all the viruses absolutely depend on the host translation machinery, i.e., the ribosomes for their protein synthesis without an exception.
5. **Assembly and budding:** In the case of enveloped viruses, assembly and budding are the final steps of the virus replication, before the release of infectious progeny virions. Most of the viruses are not in favor of forming a long-term stable host-virus relationship inside the infected host. The survival of these viruses depends on host-to-host transmission, which, in turn, depends on the release of the progeny virus from the infected host cell. They might take the advantage of the lipid rafts present in the cell membrane of infected cells as site of assembly and budding of the neosynthesized virion [8–10] Lipid rafts are membrane domains of variable size, enriched in cholesterol and sphingolipids [11]. Lipid rafts have been described in the budding process of viruses such as human immunodeficiency virus-1 (HIV-1), influenza virus, and Ebola virus [11–14]. To form an infectious virion, at least one copy of each vRNA should be packaged. For example, in case of the influenza A virus, one copy of each of the eight segments must be present to form an infectious particle. This infectious particle should have the ability to infect and enter a host cell but would replicate depending on the access to host cellular factors. This process starts with capsid assembly followed by genome packaging. A packaging signal present on the viral genome is specifically recognized by the viral capsid which does selective packaging of the viral genome.
6. **Release:** There are different strategies followed by viruses to be released from the host cell. First, they could get released by cell lysis. In this case, there is no specific mechanism being followed except the host cell membrane being disrupted by the accumulation of viral proteins and

enzymes. Second, for enveloped viruses, the capsid gets coated by the host lipid bilayer before the virus release. This could go through two different mechanisms: i) the lipid bilayer coating can occur after the capsid assembly has taken place. This needs the interaction between the viral capsid and its glycoproteins present on the envelop. ii) The lipid bilayer coating can occur simultaneously with the capsid assembly, as seen in retroviruses.

Following the lipid bilayer coating around the viral capsid, the host cell plasma membrane must undergo conformational changes, resulting in the formation of a bud-like structure that facilitates the extracellular release of the virion. This process is commonly called "budding" of the newly formed viral particles. Most of the viruses have one peptide motif or a domain that triggers this budding process. These viral proteins contain the specific peptide motif or domain interact with various host cellular factors such as members of small GTPases or actin family to regulate the formation of the bud, ultimately releasing the virion. For example, viruses like HIV-1 and Ebola virus use the Endosomal sorting complex required for transport (ESCRT) complex to achieve membrane scission and subsequent release of the newly formed virus depends on VPS4 protein [15,16]. This process is mediated by the late domain sequences present in viral proteins (p6 for Gag of HIV-1) that recruit the ESCRT-I and ESCRT-II complex at the site of budding (Figures 2.8 and 2.9).

2.4 VIRUS-HOST INTERACTIONS

As discussed earlier, the virus requires access to the host cellular machinery for its survival, specifically for the translation of viral proteins. Hence, there is an absolute need for the virus to interact with the host cell and its various components. Moreover, the parameters of an infection depend on the virus involved as well as

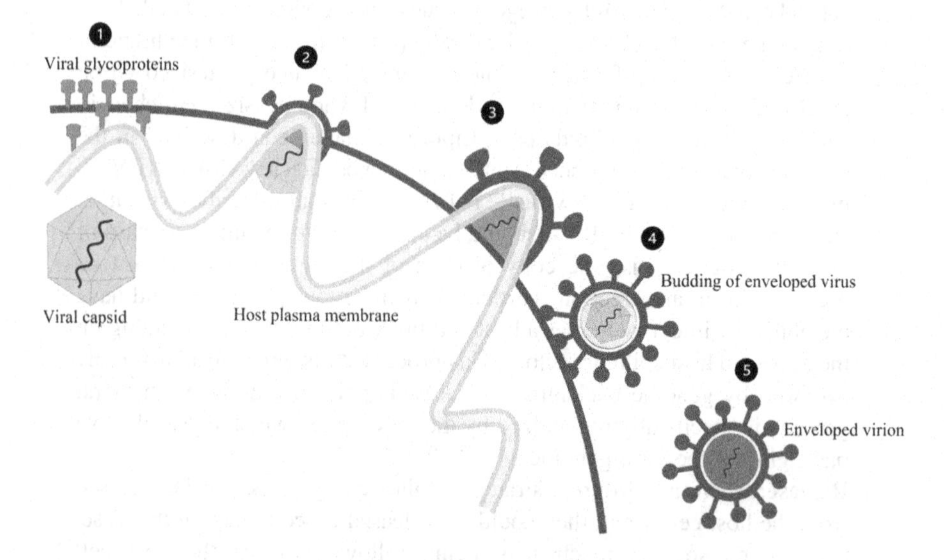

FIGURE 2.9 Budding and release mechanism of an enveloped virus.

the type of cell it infects. This virus-host interaction will also determine the severity of an infection. The same virus could have different consequences depending on the cell type it infects. Despite belonging to different families, viruses have a few mandatory requirements when inside a host cell:

- Ensure their genome replication: The virus has limited genes and lacks enzymes needed for genome replication and transcription.
- Viral protein translation and packaging: As the virus is devoid of ribosomes for mRNA formation, they need to alter the host cellular machinery for their purpose. Some viruses also need to derive part of their envelope from the host cell plasma membrane.
- Evasion of host immune response: This is a very crucial step not only for the survival of the virus, but also for release of infectious viral particles.

2.4.1 Types of Virus-Host Interactions

1. Persistent and productive: Majorly seen in the case of RNA viruses. This type of infection does not show severe cytopathic effects but alters some of the host cellular mechanisms such as, proliferation of cell membranes or endoplasmic reticulum membranes, formation of inclusion bodies. These viruses do not kill the cell but would replicate inside the infected cell, resulting in production of infectious particles.
2. Persistent and non-productive: This type of interaction does not kill the host cell but is also unable to produce infectious virions. They could only be able to induce the virus.
3. Cytocidal: Severe cytopathic effects observed in the host cells, such as morphological changes, cell membrane ruffling, and cell death. The cytocidal mechanism could be via alteration in host cell transcription or translation, inhibition of host mRNA processing or serious change in host cell membrane. This type of interaction results in the production of infectious virions that are ultimately released in the environment after host cell death.
4. Transformation: This type of interaction also leads to changes in host cell morphology. These viruses could pass indefinitely within the host cell creating a niche for tumor development and could be termed *oncogenic viruses*. The oncogenic viruses with DNA genomes would not be able to form infectious virions, but oncogenic retroviruses would be able to produce progeny virions.

2.4.2 Components of a Host Cell Interacting with Viral Elements

A virus needs to use the host cellular machinery for more than one purpose to survive and be persistent. To accomplish this, it would need to interact with the host cell components at various levels, such as:

1. Cytoplasm: Most of the viruses need to gain access to host cytoplasm for either transcribing their genome in the host nucleus or while viral assembly

process. In the cytoplasm, the viral structural proteins interact with cellular factors such as enolase (ENO1) or heat shock proteins (HSP90/70).

2. Nucleus: In the nucleus, the virus interacts with many of the elongation factors (eEF1G, eEF1B) and mRNA decay factors (SMG9, DDX5) to control the host DNA and protein metabolism, therefore promoting its replication.
3. Plasma membrane: Interaction with the host cell plasma membrane is the first (entry of the virus) and the last step (exit of virions) in an infection cycle. To achieve this, the virus interacts with the components of the host cell lipid bilayer such as phospholipids, sphingolipids, and annexin (ANXA2).
4. Cytoskeleton: Structural changes in the host cell plasma membrane are required during cell entry, viral assembly, and extracellular release of the virus. The host cell cytoskeleton, namely, actin filaments (globular and filamentous actin), microtubules (tubulin chains), and small GTPases (Rac, Rho, Cdc42) are the major players in these events.

2.4.3 Types of Infection

There are different degrees of a viral infection, depending on the outcome and transmission of an infection:

1. **Abortive infection:** The virus would enter the host cell but cannot complete the replication successfully. This could be a result of non-permissive host cell or formation of defective virions. Example: MP strain of herpes simplex virus with canine kidney cells
2. **Asymptomatic infection:** In this type, the host is considered the carrier of the virus without experiencing any symptoms of the infection. This type of infection can also be called an inapparent or subclinical infection. Asymptomatic infections do not prompt the infected host to take proper precautions and care, leading to higher chances of transmitting the infection. Example: Most cases of Human papillomavirus infections
3. **Symptomatic (Active) infection:** In this, the virus successfully completes its genome expression in the host cell, producing infectious virions. Symptomatic infections are clinical and apparent, showing more than one symptom related to the specific infection. Examples: Fever in most cases, cough in case of respiratory infections such as SARS, influenza A virus
4. **Self-limiting infection:** These kinds of viral infections run their course (generally seven days) and get cleared out from the host system without severe medications. Examples: Common cold, chickenpox
5. **Latent infection:** This kind of infection remains hidden or inactive or dormant. They remain static, persistent, and can exist for a very longer duration inside the host cell before becoming active. This is achieved through either escaping the cell-mediated immune response or by infecting the cells of immune-privileged sites such as the brain and eyes. The viral genome can remain as an episome or integrated into the host genome. The virus maintains its latency by expressing very few viral genes that keep the

viral genome silent and hidden from the host immune system. The latency can be retracted if the host cell is introduced to stress signals. Example: Herpesvirus, retrovirus (HIV)

2.4.4 Key Elements of a Viral Infection

Also, there are a few key elements to consider along with the type of infection to establish the outcome of an infection:

1. **Virus strain:** The virus strain determines the host cellular receptors for attachment and release of the newly formed particles from the host cell surface.
2. **Inoculum size:** The virus load, that is the inoculum size, is a key factor in determining the virus spread kinetics and outcome of the infection. A lower dose of the inoculum might get cleared out from the host system whereas, a higher viral load could lead to severe immunopathology and eventually could cause mortality.
3. **Route of exposure:** The first occurrence of virus-host interaction is the exposure of the susceptible host to the virus. The infection could be via respiratory droplets or aerosols, contaminated food, and water, or through body fluids and tissue. Sometimes the exposure could be directly through contaminated needles or even through the bite of an animal vector. In some cases, the virus could be transmitted from an infected mother to the infant.
4. **Susceptibility of the host:** Susceptibility to a virus is greatly affected by the age and sex of the host. For some viruses, younger hosts could show a higher mortality rate and for some older animals could have a higher susceptibility. The sex of the animal could also contribute to the outcome of an infection, depending on their genetics, respectively.
5. **Immune status of the host:** The host uses its wide range of immune defense mechanisms to either avoid infection or to reduce the degree of tissue damage done by the virus. But in the case of an immunocompromised host, the host has a lower ability to respond to the virus due to its weak immune system, ultimately leading to a higher mortality rate compared to a healthy host. This could be caused by several conditions such as cardiovascular or lung disease, HIV infection, diabetes, or even obesity.

2.4.5 Disease Occurrence

Different terms have been used to describe the occurrence of a specific disease at a given time. This also depends on the disease-causing agent, its background, and risk factors. In the case of viruses, they are mostly termed infectious diseases that could be emerging (newly identified) or re-emerging (reoccurrence). Some commonly used terms are:

1. **Sporadic:** Diseases that are irregular and non-frequent within a given population and time.

2. **Endemic:** This is termed the baseline level of a disease in a community, at any given time. It refers to the constant presence of the disease in that population.
3. **Hyperendemic:** It is like endemic, but the only difference is the constant presence of a higher level of the disease. Hence, a higher baseline of disease occurrence.
4. **Clusters:** Aggregated or grouped cases, which are generally higher than cases expected from the specific population at a given time.
5. **Outbreak:** Mostly a sudden increase in the number of cases of a disease; more than what could normally be expected from the population of a particular area. Could be further divided into point-source and common-source outbreaks. Example: A single case of an infectious disease could be termed an outbreak.
6. **Epidemic:** Same as an outbreak, but not limited to a geographical area. This could result from several factors such as recent increase in viral load, change in mode of transmission, factors leading to increased host susceptibility, or host exposure. Examples: Smallpox (1633-34); HIV/AIDS (Acquired Immunodeficiency Syndrome)(1980- present)
7. **Pandemic:** An epidemic that has spread over several countries or continents; usually affecting many people. Examples: Spanish flu (1918), Swine flu (2009), SARS-CoV-2 (COVID-19)

REFERENCES

[1] S. Baron, *Medical Microbiology*. 4th edition. University of Texas Medical Branch at Galveston, Galveston, Texas, 1996.

[2] W. Ahmed and Z. F. Liu, "Long non-coding RNAs: Novel players in regulation of immune response upon herpesvirus infection," *Front. Immunol.*, vol. 9, pp. 1–8, 2018, doi: 10.3389/fimmu.2018.00761

[3] P. M. Lieberman, "Epigenetics and Genetics of Viral Latency," *Cell Host Microbe,* vol. 19, no. 5, pp. 619–628, 2016, doi: 10.1016/j.chom.2016.04.008

[4] D. S. Y. Ong *et al.*, "Epidemiology of multiple herpes viremia in previously immunocompetent patients with septic shock," *Clin. Infect. Dis.*, vol. 64, no. 9, pp. 1204–1210, 2017, doi: 10.1093/cid/cix120

[5] C. Söderberg-Naucleŕ, K. N. Fish, and J. A. Nelson, "Reactivation of latent human cytomegalovirus by allogeneic stimulation of blood cells from healthy donors," *Cell*, vol. 91, no. 1, pp. 119–126, 1997, doi: 10.1016/S0092-8674(01)80014-3

[6] T. Kirchhausen, "Three ways to make a vesicle," *Nat. Rev. Mol. Cell Biol,* vol. 1, no. 3, pp. 187–198, 2000, doi: 10.1038/35043117

[7] S. Saffarian, E. Cocucci, and T. Kirchhausen, "Distinct dynamics of endocytic clathrin-coated pits and coated plaques," *PLoS Biol.*, vol. 7, no. 9, 2009, doi: 10.13 71/journal.pbio.1000191

[8] M. Takeda, G. P. Leser, C. J. Russell, and R. A. Lamb, "Influenza virus hemagglutinin concentrates in lipid raft microdomains for efficient viral fusion," *Proc. Natl. Acad. Sci. U. S. A.*, vol. 100, no. 25, pp. 14610–14617, 2003, doi: 10.1073/pnas.2235620100

[9] B. J. Chen, M. Takeda, and R. A. Lamb, "Influenza Virus Hemagglutinin (H3 Subtype) Requires Palmitoylation of Its Cytoplasmic Tail for Assembly: M1 Proteins of Two Subtypes Differ in Their Ability To Support Assembly," *J. Virol.*, vol. 79, no. 21, 13673–13684, 2005, doi: 10.1128/jvi.79.21.13673-13684.2005

[10] G. P. Leser and R. A. Lamb, "Influenza virus assembly and budding in raft-derived microdomains: A quantitative analysis of the surface distribution of HA, NA and M2 proteins," *Virology*, vol. 342, no. 2, pp. 215–227, 2005, doi: 10.1016/j.virol.2005.09.049

[11] D. A. Brown and J. K. Rose, "Sorting of GPI-anchored proteins to glycolipid-enriched membrane subdomains during transport to the apical cell surface," *Cell*, vol. 68, no. 3, pp. 533–544, 1992, doi: 10.1016/0092-8674(92)90189-J

[12] K. Simons and E. Ikonen, "Functional rafts in cell membranes," *Nature*, vol. 387, no. 6633, pp. 569–572, 1997, doi: 10.1038/42408

[13] K. Simons and D. Toomre, "Lipid rafts and signal transduction," *Nat. Rev. Mol. Cell Biol*, vol. 1, no. 1, pp. 31–39, 2000, doi: 10.1038/35036052

[14] M. Suomalainen, "Lipid rafts and assembly of enveloped viruses," *Traffic*, vol. 3, no. 10, pp. 705–709, 2002, doi: 10.1034/j.1600-0854.2002.31002.x

[15] S. Welsch, B. Müller, and H. G. Kräusslich, "More than one door – Budding of enveloped viruses through cellular membranes," *FEBS Lett.*, vol. 581, no. 11, pp. 2089–2097, 2007, doi: 10.1016/j.febslet.2007.03.060

[16] B. J. Chen, G. P. Leser, D. Jackson, and R. A. Lamb, "The influenza virus M2 protein cytoplasmic tail interacts with the M1 protein and influences virus assembly at the site of virus budding," *J. Virol.*, vol. 82, no. 20, pp. 10059–10070, 2008, doi: 10.1128/jvi.01184-08

3 Introduction to basic immunology and vaccine design

Alaka Mullick
National Research Council of Canada, Montreal, Quebec, Canada

Shantoshini Dash
Viral Vectors and Vaccines Bioprocessing Group, Department of Bioengineering, McGill University, Montréal, Canada

CONTENTS

DOI: 10.1201/9781003229797-3

3.1 INTRODUCTION

We are continuously exposed to pathogens, many of which could cause serious disease. If it were not for our immune system, that is constantly surveying the environment, recognizing pathogens, and defending us from attack, we would succumb to infection more frequently. This is a formidable task given the large number and diversity of pathogens. The immune system needs not only to recognize them but also to distinguish one from the other. Moreover, some pathogens share molecular features with humans and so the immune system has the other major challenge of distinguishing "foreign" from "self." This chapter will describe the toolkit and strategies that the immune system uses to accomplish this daunting task.

The response to a pathogen takes place in two stages, depicted schematically in Figure 3.1. The first, a "rapid-acting" innate immune system, recognizes the threat and swiftly triggers a response, although the recognition and therefore the response is not specific to the pathogen. However, the importance of this phase cannot be overstated, since, not only the intensity, but also the quality of the second or adaptive phase

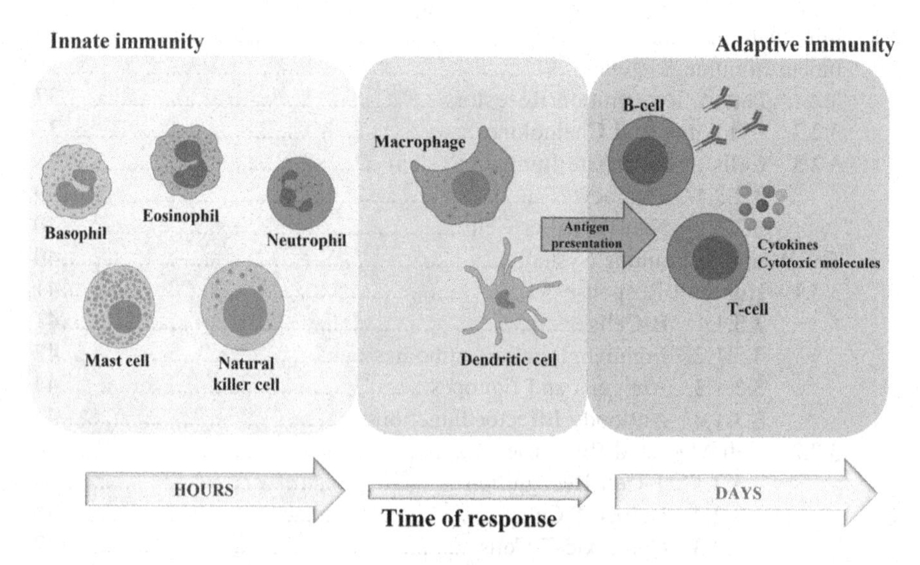

FIGURE 3.1 The immune system. This figure is a schematic representation of the response of the immune system to a foreign organism. Starting from the rapid response of the cells of the innate immune system, information about the risk is conveyed by the bridging elements (macrophage and dendritic cells) to the adaptive immune system that mounts both a humoral (production of antibodies by the B cells) and a cellular (T cells that can either support the humoral response or produce cytotoxic molecules) response that attacks the pathogen with exquisite specificity.

depends upon the information it receives from the innate phase. The second phase is triggered when the information gathered by the innate system is conveyed to the adaptive immune system, which in turn, uses this information to mount a response specific to the pathogen. It takes longer than the innate immune system, but the adaptive system makes up in specificity what it lacks in the rapidity of response.

3.2 INNATE IMMUNE SYSTEM

3.2.1 Pattern Recognition Receptors

The innate immune system recognizes molecular patterns associated with pathogens. Such pathogen-associated molecular patterns or PAMPs include various lipopolysaccharide molecules on bacterial surfaces, yeast cell wall components, and structures such as flagella and microbial nucleic acids. The cells of the innate immune system possess special receptors called pattern recognition receptors (PRRs) that can distinguish these general molecular patterns, thereby detecting the presence of foreign organisms in the environment [1]. One such family of receptors is named toll-like receptors (TLR), referring to "toll," a drosophila receptor, that was the first member of this family to be identified. As shown in Figure 3.2, TLRs 2, 4, 6, and 10

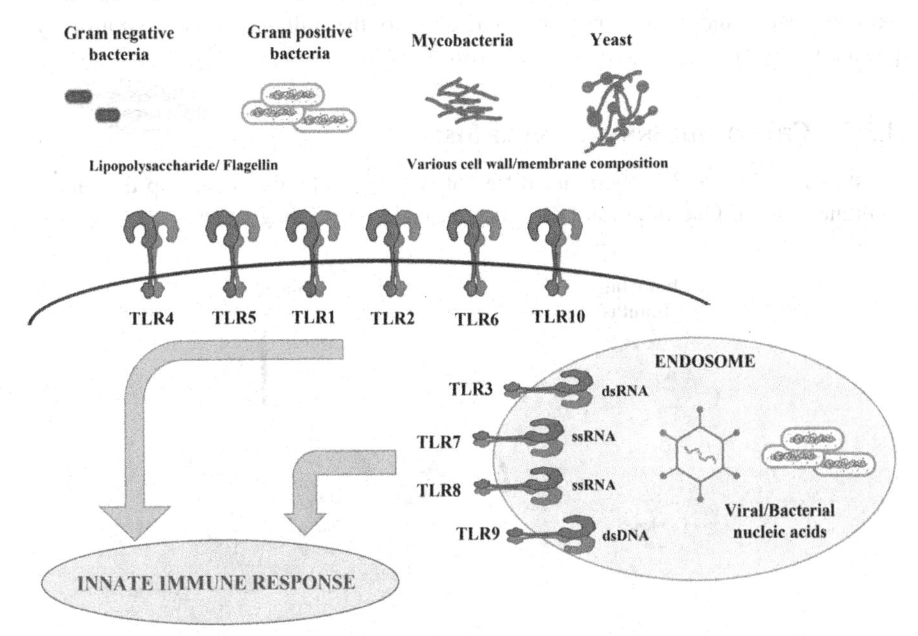

FIGURE 3.2 The human toll-like receptor family. This figure provides examples of the microbes and their components that are recognized by the members of the human toll-like receptor (TLR) family. Receptors localized to the plasma membrane, TLRs 1, 2, and 6 recognize various components of microbial cell membrane and cell walls such as lipopolysaccharides, peptidoglycans and zymosan. TLR5 recognizes bacterial flagellin. Intracellular TLRs 3, 7, 8, and 9 are present on the endolysosomal membrane and recognize bacterial and viral nucleic acids.

are present on the cell surface and recognize extracellular pathogens. Others, such as TLR 7, 8, and 9 are present on the surface of endosomes and can recognize nucleic acids from intracellular pathogens [2]. Other families of pattern recognition receptors include C-type lectin receptors (CLRs), NOD-like receptors (NLRs), and RIG-like receptors (RLRs) that detect the presence of molecules derived from pathogens in the cytoplasm [1],

3.2.2 Cytokines and Chemokines

As shown in Figure 3.3, upon recognition of a foreign organism, cells of the innate immune system secrete small molecules called cytokines to send a danger signal to the other elements of the immune system. These molecules are responsible for the communication between different cell types, stimulating certain functions, repressing others, thereby orchestrating a coordinated response to the external threat. A subset of cytokines called chemokines can induce chemotactic activity, recruiting various cell types to the site of infection. The recruited cell type also secretes cytokines and chemokines, thereby amplifying the response [3]. It is important to note that, in addition to the cells of the innate immune system, other cell types such as epithelial cells, endothelial cells, and fibroblasts also express PRRs and thus can also contribute to the inflammatory response by producing cytokines.

3.2.3 Cells of the Innate Immune System

As shown in Figure 3.1, there are different types of cells that make up the innate immune system. One important category is made up of phagocytes.

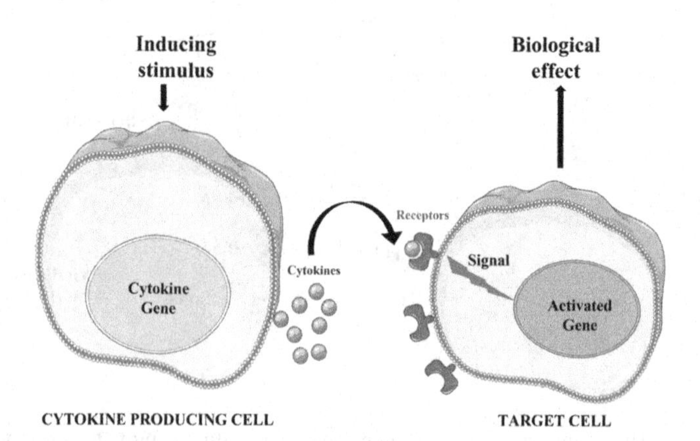

FIGURE 3.3 Cytokines. Microbes, damaged or infected cells, are recognized by, and serve, as a stimulus for healthy cells to induce small signalling molecules known as cytokines. Cytokine-binding to receptors on immune cells convey a signal that triggers a response in the form of a biological activity.

3.2.3.1 Phagocytes

Phagocytes, are cells that ingest and destroy pathogens, a process called phagocytosis from the Greek "phagos" to eat. As shown in Figure 3.4, a foreign body is internalized in a phagosome, which then fuses with the lysosome, resulting in degradation of its contents. Phagocytes are distributed all over the body and can be quickly recruited to the site of infection or injury. The examples of phagocytes are neutrophils, macrophages, and dendritic cells.

a. ***Neutrophils***

Neutrophils are specialized phagocytes that contain intracellular granules containing several microbicidal molecules including reactive oxygen species, lysozymes, and defensins. These molecules are released upon degranulation that is triggered by a foreign organism. Neutrophils also attack microbes by extruding a network of chromatin fibers containing antimicrobial peptides, thereby trapping and neutralizing them [4].

b. ***Antigen-Presenting Cells***

Antigen-presenting cells (APCs) such as macrophages and dendritic cells are phagocytes that ingest foreign matter, and then chop the pathogen into small bits which get displayed on their cell surface. By so doing, APCs display these fragments to cells of the adaptive immune system, making a bridge between innate and adaptive immunity.

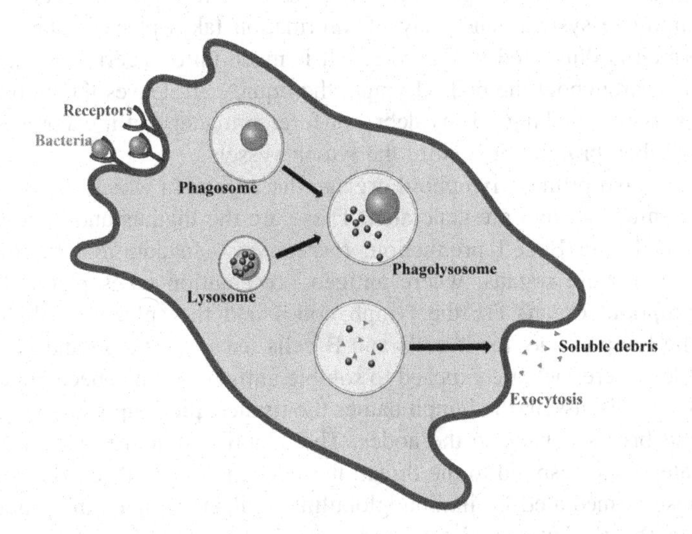

FIGURE 3.4 Phagocytosis. A bacterium is engulfed by extensions formed by the phagocyte. It is internalized in a vacuole called the phagosome. The fusion of the phagosome with the lysosome gives rise to the phagolysosome wherein the lysosomal enzymes degrade the contents.

3.2.3.2 Natural Killer Cells

APCs, described previously, deal with pathogens that can be phagocytosed, but there are intracellular pathogens, such as viruses that are typically not phagocytosed. In a typical viral infection, the virus recognizes a cellular cell surface protein, and uses this contact to, either enter or release genetic material into the cell. The virus then uses the cellular machinery to make progeny viruses, which are released. So the majority of the life-cycle of the virus is spent within the cell, protected from phagocytosis. TLRs such as TLR8 and 9 and NLRs such as RIG-1 and MDA5 can recognize the distinct characteristics of viral nucleic acids and this recognition leads to activation of the inflammatory pathway. In addition to virally infected cells, abnormal cells, such as cancerous, apoptotic, or stressed cells also need to be removed from the body and can be recognized by virtue of certain damage-induced molecular patterns (DAMPs). These too are recognized and eliminated by natural killer or NK cells.

Virus-infected cells produce interferons which activate NK cells. Upon activation, NK cells can kill target cells by releasing cytotoxic molecules such as perforin, which perforates the membrane of the target cells and granzyme that induces apoptosis. In addition, activated NK cells produce interferon γ (type II IFN-γ), which stimulates other cells to orchestrate defenses against viruses and stressed or cancerous cells.

3.3 THE ADAPTIVE IMMUNE SYSTEM

The innate immune cell types recognize the threat and convey the information to the adaptive immune system by presenting fragments of the foreign organism to the adaptive immune system. The relay of information takes place in the lymphatic system, which is illustrated in Figure 3.5. It is made up of a series of vessels that carry lymph throughout the body. Lymph, the liquid part, leaves the blood vessels and bathes tissues, picking up any debris or foreign matter that it encounters in the tissues, and then bringing it back to the lymph vessels.

There are also primary lymphoid organs, the organs in which the cells of the adaptive immune system are generated. These are the thymus and bone marrow, the sites of T and B cell production, respectively. In addition, there are the secondary lymphoid organs, where antigen presentation takes place. The secondary lymphoid organs are the lymph nodes and the spleen. As blood goes through the lymph node, the T-cells and B-cells leave the blood and go into the lymph node, where they are exposed to soluble antigen that has been collected by the lymph. As discussed, the lymph bathes the tissues, picks up whatever is in the tissues, and brings it back to the nodes. The adaptive immune system uses two major strategies to respond to the threat: the first, the humoral response, wherein the response is mediated by immunoglobulins, molecules that can recognize the pathogen with exquisite specificity and second, the cellular response, in which T-cells play a pivotal role not only in attacking the infected cell themselves, but also in supporting the humoral response.

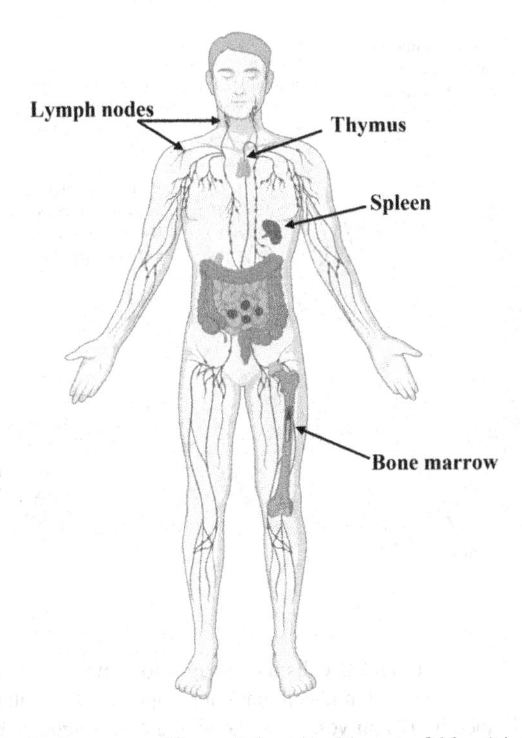

FIGURE 3.5 Lymphatic system. The cell-free liquid part of blood is transported out of blood vessels and bathes tissues, providing nourishment and removing waste matter. This liquid returns to the circulation via the open ends of lymph vessels, which then deliver lymph to lymph nodes. The primary lymphoid organs where lymphocytes are generated are the bone marrow (B cells) and thymus (T cells).

3.3.1 Humoral Response

3.3.1.1 B-Cells

B-cells recognize soluble antigens. However, not all B cells can recognize every antigen. At birth, everyone possesses a large number of naïve B-cells. They are called naïve because they have not been exposed to any antigen yet. Each one of these naïve B-cells produces a specific antibody and therefore recognizes only a specific antigen. If a naïve B cell is exposed to its target antigen, antigen-antibody binding triggers a series of events that results in the proliferation and expansion of this specific B-cell. This is called the clonal selection theory, and this is how the body generates its repertoire of antibodies, each one recognizing a specific antigen. As shown in Figure 3.6, when lymph-carrying soluble antigen enters the lymph node, B-cells are exposed to these antigens, and the B-cell that recognizes that particular antigen, migrates to the follicle where clonal expansion takes place.

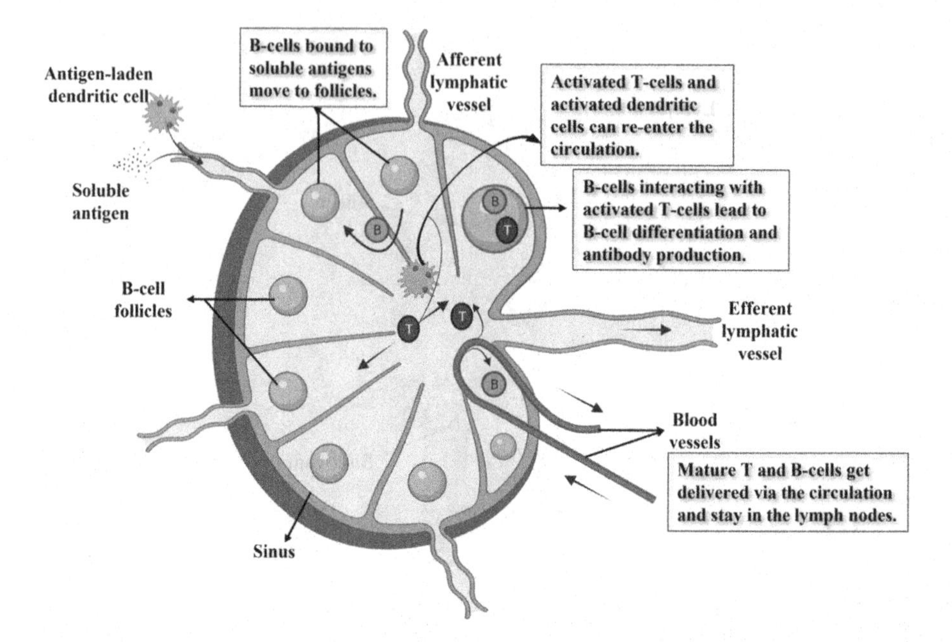

FIGURE 3.6 Lymph node. T- and B-cells are brought to lymph nodes by blood vessels. This figure shows the steps involved in B-cell maturation upon binding soluble antigen that is brought to the lymph node by lymph vessels. Following antigen-binding, B-cells migrate to the follicles, the site of clonal expansion. Antigen-laden dendritic cells also enter the lymph node and present their antigens to T-cells, thereby activating them. Activated T-cells can enter the follicle where they stimulate B-cells towards maturation.

3.3.1.2 Immunoglobulin/Antibodies

What are antibodies? As shown in Figure 3.7, antibodies are molecules made up of two heavy chains (50 kDa each) and two light chains (25 kDa each) held together by disulfide bonds. These molecules recognize an antigen with very high specificity. Both heavy and light chains have variable and constant regions. The variable region makes up the antigen binding site. The constant region of the antibody, also known as the Fc domain, interacts with the rest of the immune system to bring about effector function [5].

3.3.1.2.1 Heavy and Light Chain Isotypes

There are different kinds of both heavy (μ, δ, γ, α ε) and light chains (κ and λ), and this gives rise to a diverse set of antibodies since each heavy chain isotype can associate with either light chain isotype. This diversity in structure is also reflected in distinct characteristics and functions. Relevant for this discussion are the IgG and IgM antibody isotypes. The structures of these antibody types are shown in Figure 3.8. The first antibody to appear in the circulation as a response to an infection, is a pentameric molecule wherein all 10 heavy chains are of the type μ, hence the name IgM. At this point the affinity of antibody recognition is not very

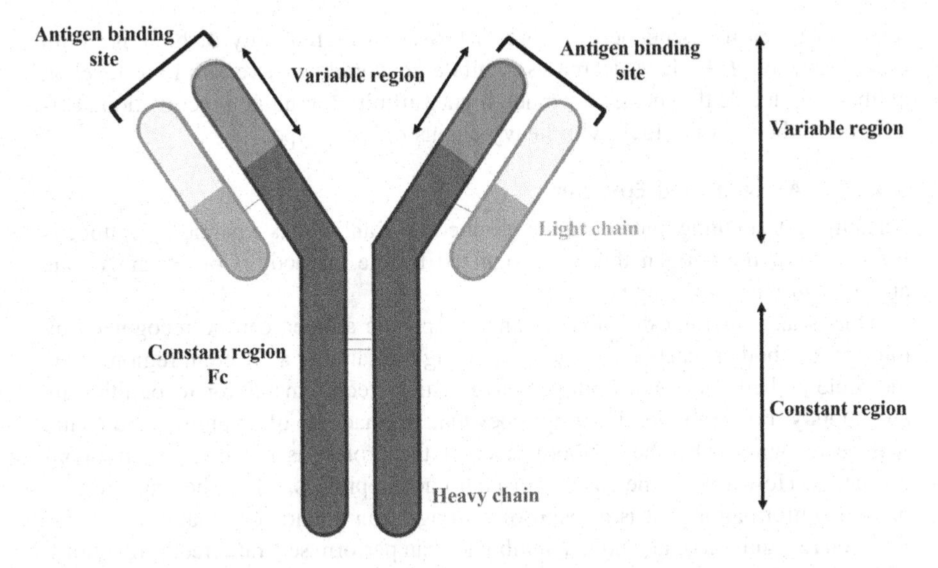

FIGURE 3.7 Antibody molecule. Antibodies are composed of two identical heavy chains and two identical light chains, held together by disulfide bonds. The variable region of the heavy and light chains create the antigen-binding capacity of the molecule. The constant region, also referred to as the Fc domain of the antibody, interacts with other elements of the immune system to trigger various effector functions.

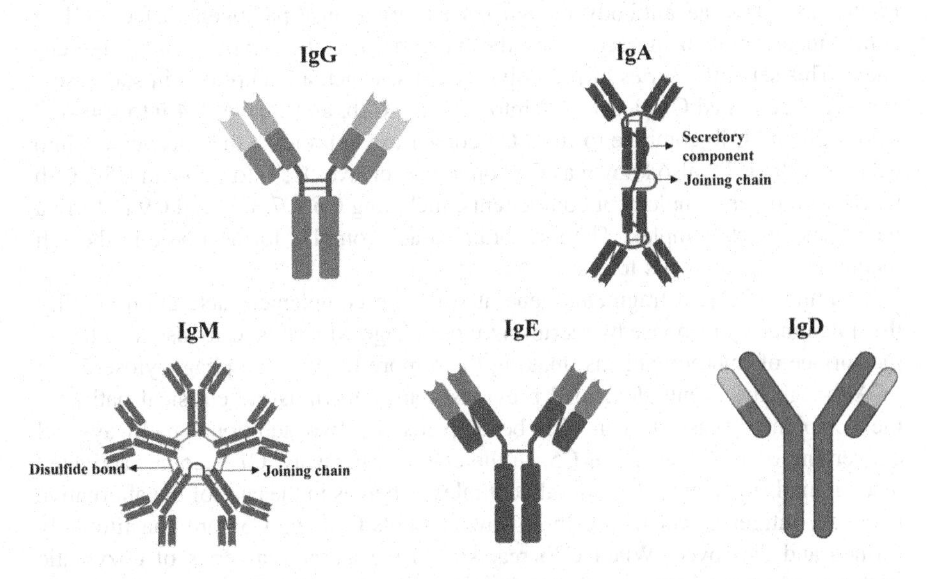

FIGURE 3.8 Antibody diversity. Antibodies are divided into different classes and their nomenclature is based on the heavy chain isotype. Therefore, antibodies made up of γ, υ, δ, α and e chains are named IgG, IgM, IgD, IgA, and IgE, respectively. An additional chain, the joining or J chain stabilizes the dimeric structure of IgA and pentameric structure of IgM.

high, but given the pentameric nature, it possesses high avidity of binding. With time, the level of the IgM decreases, while a concomitant increase in the level an antibody molecule that possesses much higher affinity for antigen recognition. This type of antibody is the IgG, with heavy chains of the γ-type.

3.3.1.3 Antigens and Epitopes

An antigen is anything that elicits the production of antibodies whereas an epitope is the region of the antigen that is recognized by the antibody. Consequently, one antigen can have many epitopes.

This is an important distinction, since a specific antigen can be recognized by multiple antibodies; each antibody recognizing the same or a different region of the molecule (called epitopes). Epitopes have distinct requirements for recognition by an antibody. For example, linear epitopes that are made up of an amino acid chain, will be recognized by the antibody even if the protein is not in its native conformation. However, in the case of discontinuous epitopes, antibodies recognize a protein conformation that is a result of folding of an amino acid chain.

Upon recognition of an antigen, antibodies can perform several effector functions.

3.3.1.4 Antibody Effector Functions

a. ***Complement cascade***

One such effector function is the activation of a proteolytic cascade that is shown in Figure 3.9. Once an antibody recognizes its target on a pathogen surface, C1, a complement component recognizes the antigen-antibody complex and gets activated. This sets off a series of proteolytic events, which are amplified in successive steps. The activated C1 cleaves C2 into C2a and C2b, and cleaves C4 into C4a and C4b. C2b and C4b combine to form C3 convertase, a protease that cleaves C3 into C3a and C3b. C3b, also known as C5 convertase cleaves C5 into C5a and C5b. C5b reacts with other complement components, including C6, C7, C8, and C9 to form a membrane attack complex. The membrane attack complex forms a hole in the cell membrane, causing cells to lyse.

C3a and C5a, two fragments generated during complement activation amplify the inflammatory response by recruiting more phagocytic cells. C3b also attaches to the surface of microorganisms, making them more likely to be phagocytosed.

Although the events described previously are known as the classical pathway, the complement cascade can also be activated by two additional pathways, all converging on the cleavage of C5. In the classical pathway, C1 becomes activated when it binds to an antigen-antibody complex, whereas in the case of the alternative pathway, antigens react with C3b, as low amounts C3a and C3b are constitutively formed and destroyed. When C3b reacts with the antigen, a series of enzymatic steps are triggered involving factor B, factor D, and properdin and results in the formation of an activated complex that cleaves C5. The third pathway is triggered by lectins that recognize pathogens containing mannose-rich cell walls. This complex activates two proteases mannose-associated serine proteases 1 and 2

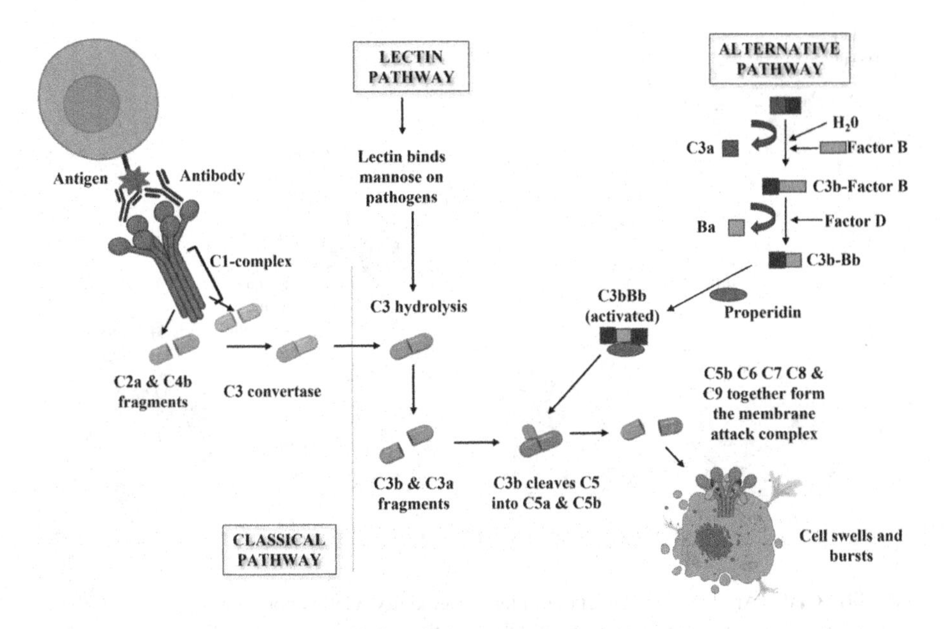

FIGURE 3.9 Complement cascade. This figure is a schematic representation of the steps involved in the three pathways that result in activation of the complement pathway. The classical pathway is activated by the antigen-antibody complex. The binding of lectins to mannose-rich cell surface molecules triggers the lectin pathway. The alternative pathway involves the formation of an activated form of a C3 fragment, by the sequential action of proteins B, D and properidin. All three pathways converge on the proteolytic cleavage of C5, which is followed by the formation of the membrane attack complex.

(MASP1 and MASP2), which also activate C3 cleavage. After C5 is cleaved, all three pathways follow the same steps [6].

b. ***Antibody dependent cellular cytotoxicity***

In addition to recruiting the complement cascade, the Fc portion of the antibody also mediates antibody-dependent cellular cytotoxicity (ADCC). As shown in Figure 3.10, if a virally infected or cancer cell is bound by an antibody, the NK cells recognize the Fc region of the antibody and this triggers NK cells to release cytotoxic molecules such as the pore-forming perforin and the apoptosis-inducing granzyme. The target cell is thus destroyed. It is important to note the distinction between NK cell-mediated cytotoxicity that was discussed in the section on innate immunity and what is being discussed here. In the previous section, NK cells did not rely on specific antibodies for recognizing their targets, whereas for ADCC, as the name suggests, antibodies are central to the recognition process.

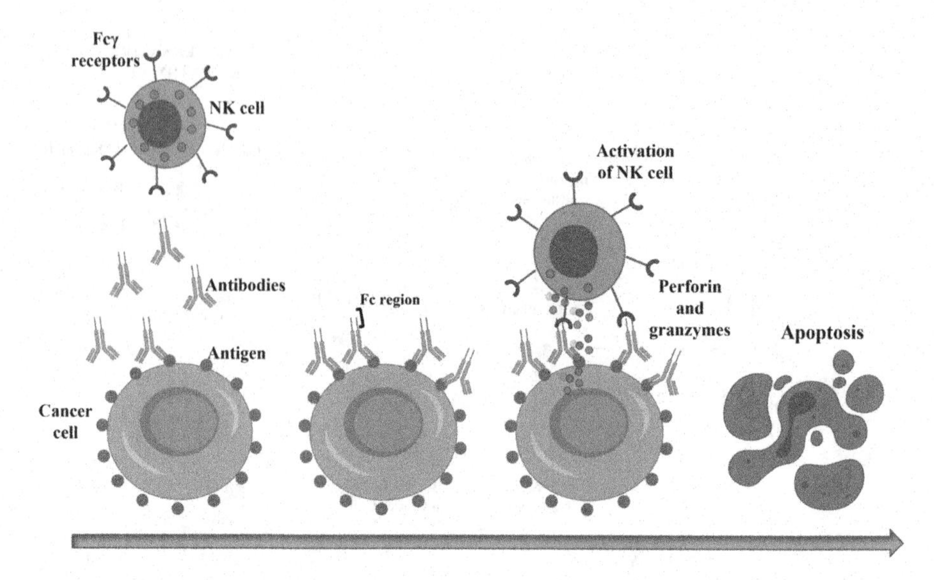

FIGURE 3.10 Antibody-dependent cellular cytotoxicity. Cells expressing antigens on their cell surface that are recognized by antibodies are the targets of antibody-dependent cellular cytotoxicity, when the Fc portion of the antibody is bound by specific receptors on NK cells. This triggers the release of perforin that generates pores in the membrane of the target cell, allowing the entry of granzyme, an apoptosis-causing molecule.

c. ***Antibody dependent cellular phagocytosis***

Similarly, antibody-dependent phagocytosis a virally infected cell is recognized by an antibody and the phagocyte recognizes the antibody (not the cell) and phagocytoses the cell.

3.3.2 Cell-Mediated Response

3.3.2.1 T-Cell Recognition

Each T-cell has a T-cell receptor, and each T-cell receptor recognizes a different epitope of an antigen. T-cell receptors, similar to antibodies, recognize an antigen with very high specificity. As shown in Figure 3.11, the T-cell receptor is made up of two 43 kDa chains, α and β, held together by a disulfide bond. Each chain spans the plasma membrane and has a short cytoplasmic tail. Much like the heavy and light chains making up antibodies, α and β chains also possess variable and constant regions and the antigen-binding region is made up by the variable region.

3.3.2.2 Helper T-Cells

The next phase of immune activation takes place in the lymph node, which is represented diagrammatically in Figure 3.6. As discussed earlier, B-cells become activated upon antigen-binding. This triggers a number of events starting with their

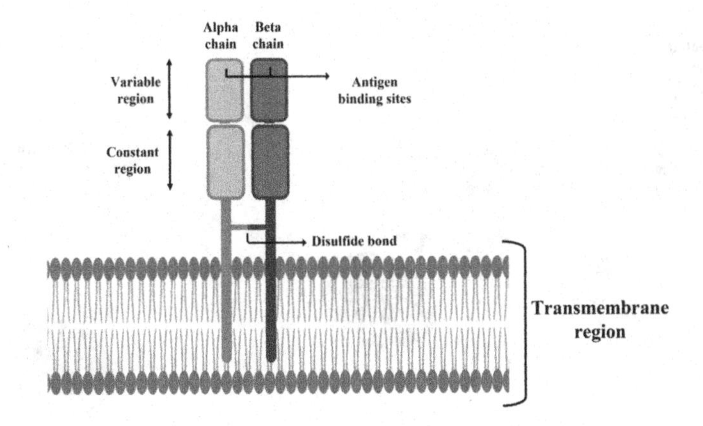

FIGURE 3.11 T-cell receptor. The T-cell receptor is made up of two chains, α and β held together by a disulfide bond. Both chains consist of variable and constant regions, with the variable region harboring an antigen-binding site. Both chains span the plasma membrane and have a short cytoplasmic domain.

migration to the follicles in the lymph node. Also coming to the lymph node are antigen-laden dendritic cells, which present the antigen to T-cells. This in turn causes T-cell activation followed by migration into the follicle. In the follicle, activated T-cells stimulate B-cells to initiate a process of antibody maturation involving genetic rearrangements, that results in the production of higher-affinity binders. They also trigger the formation of memory cells that can be re-activated upon re-infection. These T-cells are thus called T-helper cells.

3.3.2.3 Cytotoxic T-Cells

The other important type of T-cells are the cytotoxic T-cells. This type of T-cell attacks abnormal cells, such as those infected by a virus. Intracellular proteolytic fragments of a pathogen or abnormal protein are displayed on the surface of the cell in question, and are recognized by T-cells expressing receptors that specifically recognize those fragments. As shown in Figure 3.12, upon activation by antigen-binding, cytotoxic T-cells produce perforin, a molecule that will kill the cell by making holes in it. They also produce granzymes, molecules that trigger apoptosis in the target cell. Cytotoxic T-cells induce cell death very much like NK cells, but their action is much more specific because T-cells recognize their targets with the help of very specific receptors. Figure 3.13 shows the difference in function of the two types of T-cells, one that supports antibody formation in B cells (T-helper) and the other that triggers cell death (killer T) in abnormal cells such as those that are virally infected or cancerous [7].

3.4 VACCINES

The aim of vaccination is to induce protection from the pathogen, without making the individual experience the sickness. Therefore, the strategy is to initiate the innate

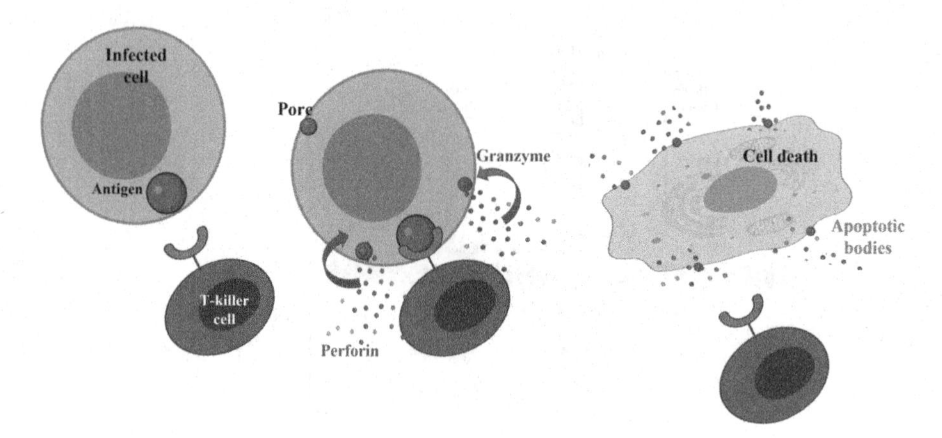

FIGURE 3.12 Cytotoxic T-cell. Cells expressing antigens on their cell surface that are recognized by specific receptors on T-cells. This triggers the release of perforin that generates pores in the membrane of the target cell, allowing the entry of granzyme, an apoptosis-causing molecule.

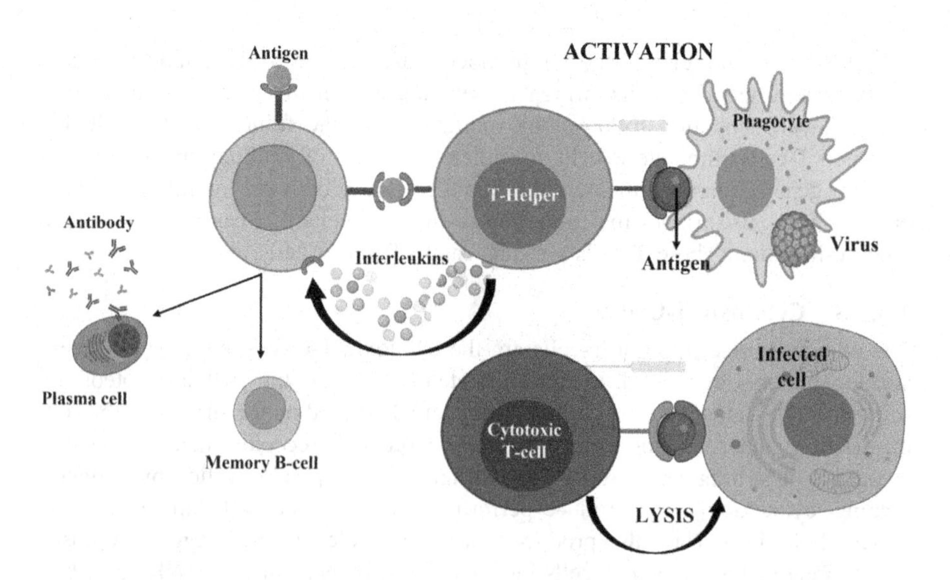

FIGURE 3.13 Helper vs cytotoxic T-cells. This figure illustrates the differences in the action of the two types of T-cells. Upon activation, helper T-cells promote B-cell maturation, whereas cytotoxic T cells release perforin and granzyme that results in apoptosis of the target cell.

immune response (and therefore activate an antigen-specific adaptive immune response), but in a very measured fashion, making sure that the vaccine will not cause severe disease. The idea is to create immunological memory, such that, faced with the same infection, the body will be to reactivate the response that is specific to that infectious agent. In an ideal situation, once vaccinated, the individual should stay immune for life. Indeed, certain antigens can induce a robust immune response that is maintained over a long period. However, for most vaccines, immunity starts waning after a certain period. At this point it is important to administer a booster which can stimulate both the B- and T-cell populations specific to the pathogen.

3.4.1 Vaccine Design

Key to the success of a vaccination strategy is the design of the antigen and several common strategies are outlined in Figure 3.14. For instance, to create a response that is as close as possible to the infection, it would be best to inject the same pathogen, but in an attenuated state so that it cannot induce severe illness. Alternatively, the pathogen can be inactivated or killed. In this case it will not replicate, reducing the amount of antigen in the system and therefore the response.

Going a step forward, it is possible to identify a part of the pathogen that can elicit a response that can protect the individual from the infection. This is called a subunit vaccine. The subunit could be a protein, purified by fractionation of the pathogen, or produced as a recombinant [8].

When a vaccine is developed, it must be tested, and part of that test measures efficacy, the ability of the response to neutralize the effect of the pathogen. An important element of the response is the production of antibodies and such antibodies are referred to as neutralizing antibodies. Although most antibodies developed during vaccination are protective, it is important to be aware of and test for

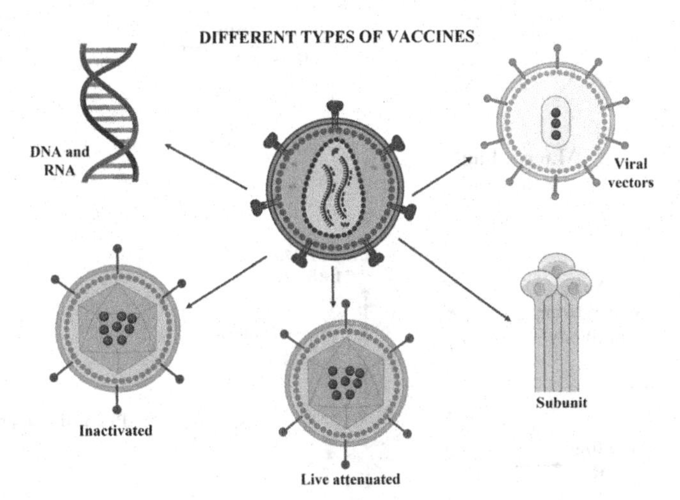

FIGURE 3.14 Types of vaccines. This figure shows the different strategies that are used to generate the antigen for a viral vaccine. It can be the virus, itself, either in an attenuated or inactivated form. It can also be a component of the virus and this is referred to as a subunit vaccine. The subunit vaccine can be delivered as a protein, or in the form of a nucleic acid (DNA or RNA) fragment that codes for the subunit. The subunit can also be delivered by a viral vector.

antibodies that can actually enhance infection, referred to as antibody-dependent enhancement (ADE). This has been observed for certain viral infections and occurs when the specificity, affinity or titer is too low to be neutralizing and as shown in Figure 3.15, the Fc portion of the antibody binds to cell surface receptors on a target

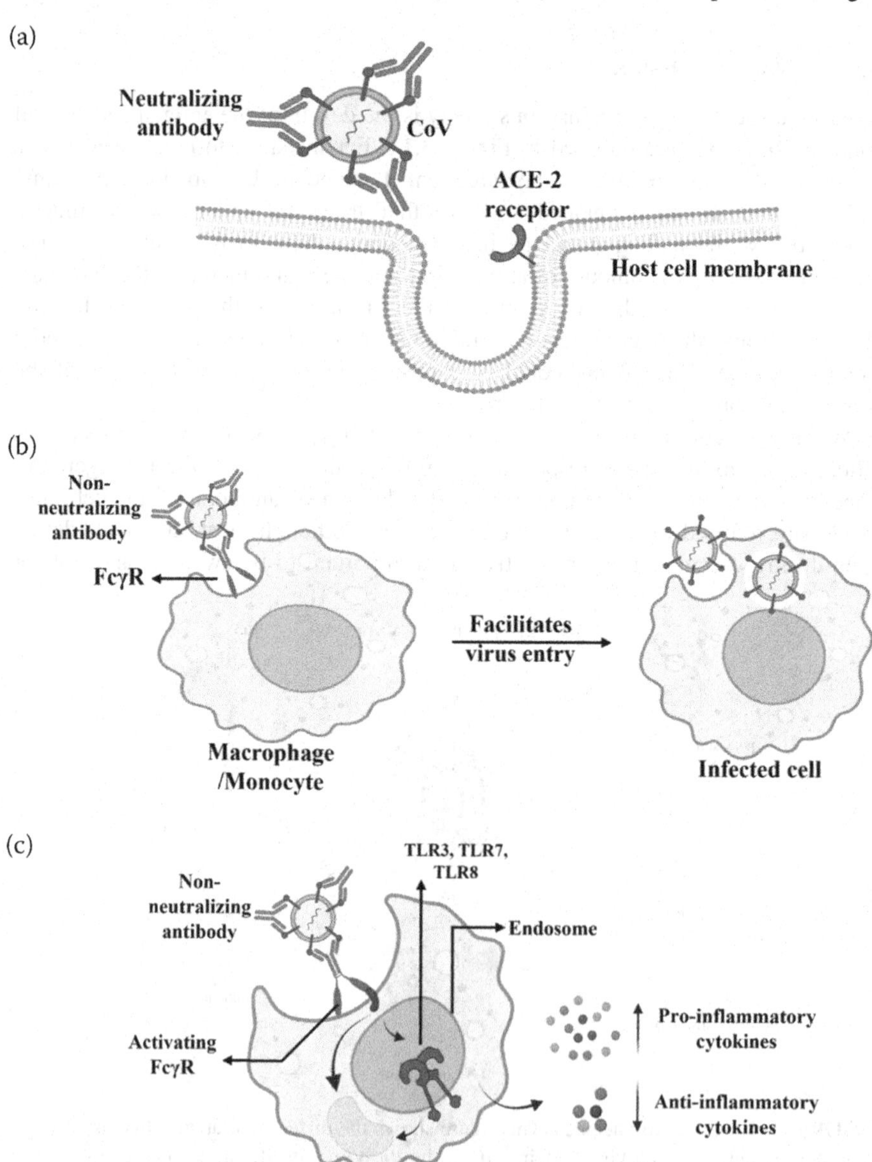

FIGURE 3.15 Antibody-dependent enhancement. (a) Virus bound by neutralizing antibodies cannot infect cell. (b) Once bound to a virus, the Fc domain of non-neutralizing antibodies can bind to cell surface Fc receptors thus facilitating the entry of virus into the cell. (c) resulting in an inflamatory response.

cell resulting in entry of the virus. Therefore, instead of neutralizing the virus, the antibody enhances the infection by increasing viral entry into target cells [9].

3.4.2 Live Attenuated Virus

There are several ways to attenuate a virus. One of these is to select a variant that cannot infect human cells as efficiently as the wild-type strain. This can be accomplished by passaging a pathogenic virus in a cell type with a different origin (such as monkey cells). At first, it will not be able to infect the monkey cells very well, but with increasing passages, variants with higher infectivity for the monkey cells, will be selected, such that, at one point, the variant that emerges will be weakened in its ability to infect human cells. Thus, this virus although still be able to replicate, cannot enter human cells as easily, and is therefore attenuated in its ability to cause disease in humans. Similarly, the virus can be adapted for survival at lower temperatures, making the human body an inhospitable host [10].

More recently, genetic strategies of inactivation have been developed. These include altering post-translational processing, the use of rare codon pairs or interfering with the virus's ability to suppress anti-viral pathways in the host [11].

A major advantage of using a live attenuated virus for vaccination is that it elicits the most robust immunity. However, since it can cause mild disease (because there is a replicating virus), it cannot be used for immunocompromised individuals, since they may suffer from more severe symptoms in the absence of an adequate immune response.

This method has been used successfully for yellow fever vaccines. In fact, the yellow fever virus has also been used as a vector developing vaccines for other viruses. For instance, if dengue virus genes introduced in yellow fever virus, an immune response will be generated to the Dengue virus also [12].

3.4.3 Inactivated Virus

To increase safety, pathogens can be inactivated so that they cannot replicate. This can be achieved by heat, by chemicals, or by radiation and although the virus is inactive, it maintains its immunogenicity, so that the immune system can still recognize it and develop an immune response.

Vaccine preparations of inactivated pathogens have several advantages: in addition to being more stable than those of attenuated viruses, these vaccines can also be administered to immunocompromised individuals. The disadvantage, however, is that the vaccine is not as strong and therefore needs a booster [10].

3.4.4 Subunit Vaccine

When a part of the pathogen is used as the immunogen, it is referred to as a subunit vaccine. Sub-unit vaccines are often proteins or their fragments, but can also be polysaccharides, such as those making up the bacterial capsule [10].

Given that only antigens corresponding to one specific component and not the whole virus, are presented to the immune system, sub-unit vaccines are not as efficient as inactivated or attenuated vaccines. Therefore, careful consideration has to be given to the choice of the antigen. Care also has to be taken to assure that the immunogenicity of the chosen candidate is maintained once isolated from its natural environment [13]. Several avenues of research are focused on identifying the most immunogenic viral sub-unit that could serve as an antigen and also how to maintain its immunogenicity during large-scale production and delivery [11].

When using a protein antigen, for instance, in the case of SARS-CoV-2, the Spike protein is an obvious choice since it is responsible for cell entry (Figure 3.16). The virus uses the spike protein to bind a cell surface protein, acetylcholine esterase 2 (ACE-2), to gain entry into the cell. Therefore, a neutralizing or protective antibody could be one that would disrupt this interaction reducing the ability of the virus to enter the target cell. While it is possible to use the entire spike protein, there are attempts to use just the part that interacts with the receptor, since in theory, it should be able to elicit the desired effect.

Regardless of the protein fragment that serves as the antigen, it can be delivered, either as a protein, or as a nucleic acid (DNA or RNA) fragment that can then be used as a template to produce the protein in the host [14].

3.4.5 DNA Vaccine

Once the antigen to be used in a vaccine has been identified, it is straightforward to determine the corresponding DNA sequence and use it to create a recombinant DNA molecule carrying an expression cassette for the protein in question. This DNA can be injected into the tissue, where it is taken up by the patient's cells that start producing the protein. The route of administration can vary. The DNA can be

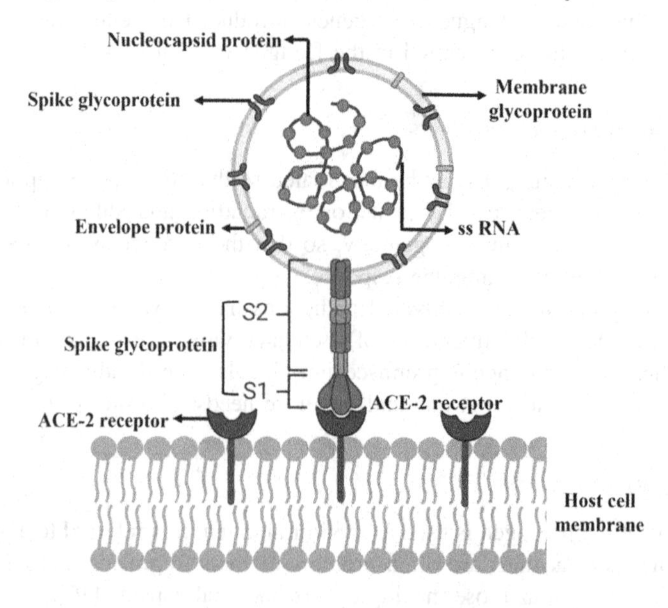

FIGURE 3.16 SARs-CoV-2 spike protein. The figure shows the interaction between the SARs-CoV-2 spike protein with the ACE-2 (acetylcholine esterase 2) receptor on a target cell.

directly injected into dendritic cells that can present the antigen to the adaptive immune system. Alternatively, an indirect route can be used, injecting the vaccine into the muscles or the keratinocytes, which will produce the protein. Antigen-presenting cells take up peptides or protein fragments released from muscle cells and present them to the adaptive immune system.

A major concern for using DNA as the immunogen is the possibility of chromosomal integration and consequently perturbation of the genome. However, careful evaluation of several veterinary DNA vaccines have not provided any evidence of genomic integration of the exogenous DNA [14]. Another cause for unease relates to the fact that autoimmunity is linked to the presence of antibodies against DNA and raises the possibility that administration of exogenous DNA could elicit an antibody response and hence generate or exacerbate autoimmunity. However, contrary to expectation, clinical trials with patients suffering from diabetes [15] and multiple sclerosis [16] were reassuring.

A big advantage of using genetic material as an immunogen is that the biomanufacturing process is a lot easier than that for proteins, given that all DNA molecules share similar physicochemical properties and therefore production does not have to be tailored to each vaccine. From the identification of the sequence to production of the DNA vaccine is a relatively rapid process in comparison to the production of a protein-based vaccine, which can involve purifying a protein subunit from a pathogen or producing recombinant protein to serve as the antigen. Despite these advantages, trials with DNA vaccines have not been very successful at eliciting a robust immune response and several approaches are being investigated to increase their immunogenicity. These include improvements in the design of the vector and the use of adjuvants [17].

3.4.6 RNA Vaccine

Similarly, instead of DNA, it is possible to use RNA. In fact, in theory, RNA vaccines have a distinct advantage in that RNA molecules need to be delivered just to the cytoplasm, avoiding the additional step of transport to the nucleus and there is no fear of genomic integration, as is the case with DNA. However, in practice, RNA vaccines were not favored for development due to the concerns of instability and their higher inflammatory potential as a result of recognition by several endosomal TLRs (TLR 3, 7 and 8) and cytoplasmic RNA sensors (RIG-1 and MDA5) [18]. Although efforts have been under way to address these concerns for several years, they came together in 2020 in the form of highly effective mRNA vaccines in the face of the SARS-CoV-2 pandemic. The identification of new nucleosides that could enhance stability, better understanding of the RNA-sensing mechanisms to control the inflammatory response and the development of lipid nanoparticles (LNP) as safe and effective delivery agents, have all contributed to this success [19].

3.5 CONCLUSION

Although we have learned a lot since the early attempts of vaccination against small pox over a 1,000 years ago, the search for strategies to increase immunogenicity without compromising safety continues. Various avenues of research including the application of artificial intelligence in the design of antigen, the use of new types of adjuvants to boost the immune response, and better delivery agents, are being pursued making this an exciting time in vaccine research.

REFERENCES

[1] O. Takeuchi and S. Akira, "Pattern recognition receptors and inflammation," *Cell*, vol. 140, pp. 805–820, 2010.

[2] T. Kawasaki and T. Kawai, "Toll-like receptor signaling pathways," *Front. Immunol.*, vol. 5, pp. 1–8, 2014.

[3] L. C. Borish and J. W. Steinke, "Cytokines and chemokines." *J. Allergy Clin. Immunol.*, vol. 111, pp. S460–S475, 2003.

[4] M. J. Kaplan and M. Radic, "Neutrophil extracellular traps (NETs): Double-edged swords of innate immunity," *J. Immunol*, vol. 189, pp. 2689–2695, 2012.

[5] H. W. Schroeder, Jr and L. Cavacini, "Structure and function of immunoglobulins," *J. Allergy Clin. Immunol.*, vol. 125, pp. S41–S52, 2010.

[6] M. Noris and G. Remuzzi, "Overview of complement activation and regulation," *Semin Nephrol.*, vol. 33, pp. 479–492, 2013.

[7] W. R. Heath and F. R. Carbone, "Cross-presentation in viral immunity and self-tolerance," *Nature Rev. Immunol.*, vol. 1, pp. 126–135, 2001.

[8] V. Votter, G. Denizer, L. R. Friedland, J. Krishnan and M. Shapiro, "Understanding modern-day vaccines: what you need to know," *Ann. Med.*, vol. 50, pp. 110–120, 2018.

[9] S. M. C. Tirado and K.-J. Yoon, "Antibody-dependent enhancement of virus infection and disease," *Viral Immunol.*, vol. 16, pp. 69–86, 2003.

[10] H. F. Maassab, "Adaptation and growth characteristics of influenza virus at 25 degrees c," *Nature*, vol. 213, pp. 612–614, 1967.

[11] M. Kanekiyo, D. Ellis and N. P. King, "New Vaccine Design and Delivery Technologies," *J. Infect Diseases*, vol. S1, pp. S88–S96, 2019.

[12] G. Bruno, F. Noriega, R. L. Ochiai et al., "A recombinant live attenuated tetravalent vaccine for the prevention of dengue," *Expert Rev. Vaccines*, vol. 16, pp. 1–13, 2017.

[13] C.-L. Hsieh, J. A. Goldsmith, J. M. Schaub et al., "Structure-based design of prefusion-stabilized SARS-CoV-2 spike," *Science*, vol. 369, pp. 1501–1505, 2020.

[14] M. A. Liu, "A comparison of plasmid DNA and mRNA as vaccine technologies," *Vaccines*, vol. 7, pp. 37–57, 2019.

[15] P. Gottlieb, P. J. Utz, W. Robinson and L. Steinman, "Clinical optimization of antigen specific modulation of type 1 diabetes with the plasmid DNA platform," *Clin. Immunol.*, vol. 149, pp. 297–306, 2013.

[16] H. Garren, W. H. Robinson, E. Krasulová, et al., "Phase 2 trial of a DNA vaccine encoding myelin basic protein for multiple sclerosis," *Ann. Neurol.* vol. 63, pp. 611–620, 2008.

[17] D. Hobernik and M. Bros "DNA vaccines—How far from clinical use?" *Int. J. Mol. Sci*, vol. 19, pp. 3605–3633, 2018.

[18] N. Pardi, M. J. Hogan, F. W. Porter and D. Weissman "mRNA vaccines—a new era in vaccinology," *Nat. Rev. Drug Discovery,* vol. 17, pp. 261–279, 2018.

[19] R. Verbeke, I. Lentacker, S. C. De Smedt, and H. Dewitte, "The dawn of mRNA vaccines: The COVID-19 case," *J. Controlled Release*, vol. 333, pp. 511–520, 2021.

LIST OF ABBREVIATIONS

PRR	Pattern Recognition Receptor
PAMP	Pathogen-associated Molecular Pattern
TLR	Toll-Like Receptors
CLR	C-Type Lectin Receptor
NLR	NOD-Like Receptor
RLR	RIG-Like Receptor
APC	Antigen-Presenting Cell
NK	Natural Killer
ADCC	Antigen-Dependent Cellular Cytotoxicity
ADCP	Antigen-Dependent Cellular Phagocytosis
ADE	Antibody-Dependent Enhancement
ACE	Acetylcholine Esterase

4 Cell lines for vaccine production

Isabelle Knott
GSK, Rixensart, Belgium

Jean-Philippe Matheise, Isabelle Ernest, and Jean-Pol Cassart
GSK, Wavre, Belgium

CONTENTS

DOI: 10.1201/9781003229797-4

4.1 INTRODUCTION

Viral vaccines are usually cell-culture–based medicinal products. As human viruses require a human host to replicate, vaccine researchers use the knowledge of animal-cell technology to develop existing and future vaccines.

Since the mid-twentieth century, numerous viral vaccines were developed to meet important medical needs such as poliomyelitis, measles, congenital disease related to rubella, or gastroenteritis due to rotavirus. Millions of lives have been saved, and the morbidity associated with those viral infections has been reduced substantially [1].

In the history of vaccines, animal-cell technology had been demonstrated as a key technology by the beginning of the twentieth century. The early progress made in tissue culture had led to the isolation of certain viruses, such as the yellow fever virus. In that period, tissue culture was based on the use of chicken eggs; in effect, the embryonic tissue of chickens. The first vaccines for yellow fever and influenza were developed with embryonated chicken eggs in 1930s. Later on, primary cells such as chicken-embryo fibroblasts were used. In the 1950s, cell cultures derived from monkey cells were developed to propagate poliomyelitis viruses; and this pioneering work led to Enders, Weller, and Robbins receiving the Nobel Prize in 1954. In the 1960s, diploid cell lines were isolated and considered as safe for use of human vaccine [2].

During the 1970s, the use of continuous cell lines was highly debated, because oncogenes had been discovered, and the DNA of continuous cell lines was perceived as a threat. Some continuous cell lines were approved for some vaccines, thanks to additional requirements being met.

Molecular biology opened the field to recombinant products. Some cell lines were quickly identified as expression systems for recombinant products, such as Chinese hamster ovary (CHO) cells.

During all these decades of viral-vaccine development and manufacturing, the safety of the products, including the cell lines, was a critical question. The safety of the viral vaccine remains the top priority, even though some risks associated with the use of viruses have been mitigated.

The aim of this chapter is to review which cell lines are useful for vaccine manufacturing, and how they are selected. The review includes information on viral safety and associated regulatory guidance; and on the analytical considerations for determining cell line quality with respect to vaccine safety and to its performance in general.

4.2 BASIC AND TECHNICAL CONSIDERATIONS

4.2.1 Basic Considerations (Including Vaccine Types and Cell Lines)

Vaccine development is based on a strong knowledge of the antigen and its immunogenicity. To identify antigen targets, immunologists are helped by an in-depth understanding of the natural disease the vaccine is intended to protect against. This knowledge drives the identification of key antigens, which are usually proteins in the case of viruses. Hence, their selection is also driven by the anticipated/desired immune response, which can be either humoral, or cell-mediated, or a combination of both. In the case of a humoral response, neutralizing antibodies against the pathogen are usually sought. For this purpose, conformational antigens must be produced. When a cell-mediated response is anticipated, antigens with sequential epitopes can be designed. These questions are more thoroughly addressed in Chapter 3.

In addition, the variability of the antigen should be considered to generate immunity to natural variants or strains of the virus, and hence to confer the broad or cross-strain protection.

The knowledge arising from immunology and specific viral diseases have led to three major types of marketed vaccines: live-attenuated vaccines, whole-inactivated vaccines, and sub-unit vaccines.

Live-attenuated vaccines consist of attenuated viruses reproducing aspects of natural infection but without the disease symptoms. They do not require adjuvant. Because the attenuated virus can replicate in the vaccine recipient, the dose amount is generally low. The immune response is generally broad because the attenuated virus contains numerous antigens and allows a better presentation to activated humoral and cellular immunity. One potential drawback of live-attenuated viruses is their potential reversion back to wild type. This is particularly the case for viruses with genomes consisting of positive RNA strands, such as the poliovirus. Reversion of OPV (oral poliomyelitis vaccine) has been well documented [3].

For live-attenuated vaccines, the selection of a cell line is highly dependent on its susceptibility to viral infection. Three main cell lines have now been used for several decades; CEF (chicken-embryo fibroblasts), MRC-5 and Vero (Table 4.1).

Whole-inactivated vaccines are made from viruses propagated in cells and then further purified to obtain antigens. The immunogenic dose (by virus-particle equivalent) is usually high in comparison to live-attenuated vaccines. This type of vaccine is usually adjuvanted, typically with aluminium salts. Because higher amounts of antigen are typically required, the capacity of the cell line to propagate

TABLE 4.1
Cell lines used for marketed live-attenuated vaccines (non-exhaustive list)

Cell Line	Marketed Vaccine	Disease
CEFs (chicken embryo fibroblasts)	Priorix (Me-Mu), Attenuvax (Me), Mumpsvax (Mu), Trimovax (Me), ProQuad (Me-Mu)	Measles and mumps
MRC-5	Priorix (Ru), Tresivac (MMR), Trimovax (Ru), Varilrix (Va), Biopox (Va), ProQuad (Va), Zostavax (Va), Varivax (Va), Priorix-Tetra (va)	Mumps, measles, rubella, varicella or chickenpox
Vero	Rotarix, Rotateq, ACAM2000 (smallpox), IMOJEV (JEV), OPV	Gastroenteritis due to rotavirus, smallpox, Japanese encephalitis, poliomyelitis
Wi-38	MeruvaxII (Ru), ProQuad (Ru), Adenovirus Type 4 and Type 7 Vaccine Live Oral	Rubella, adenoviruses

large quantities of virus is critical. Various cell lines have been used for this type of vaccine, of which Vero and MRC-5 are prominent (Table 4.2).

Sub-unit vaccines are based on purified antigens, either isolated from whole virions or produced as single component with recombinant technology. The antigens are highly purified and well characterized. There is no risk of reversion. Sub-unit vaccines usually require adjuvantation. Several doses and regular boosters are often needed to confer long-term protection. In the case of purified antigens from whole virions, cell-line selection is driven by the capacity of the cell line to propagate large quantities of virus, as is for whole inactivated vaccines. With regard to recombinant technology, other cell lines are considered. The main criteria for cell line selection are based on the ability to harbor the gene of interest (i.e., the gene coding for the sub-unit antigen) and express it appropriately. Several expression systems have been developed over the past 40 years (for a review, see [4]).

Cell cultures suitable for the development of vaccines can be categorized into three types: cultures based on (i) primary cells, (ii) diploid cell lines, and (iii) continuous cell lines (Table 4.3). Primary cells were the first substrates used for the development of vaccines. Primary cells consist of cells extracted from the tissue source and used without passage in tissue culture, in accordance with WHO guidance. Primary cells are not stored (in cell banks), and no longer accepted for the manufacture of new vaccines. Because the sources of the cells are not homogeneous, manufacturing consistency is difficult to guarantee; for example, CEFs are isolated from numerous different chicken flocks to manufacture each batch of vaccine. In addition, the

TABLE 4.2
Cell lines used for marketed inactivated vaccines (non-exhaustive list)

Cell Line	Marketed Vaccine	Disease
CEF (chicken-embryo fibroblasts)	Rabipur/RabAvert, Encepur (TBE), FMSE-Immun (TBE)	Rabies, tick-borne encephalitis
MRC-5	Havrix (HAV), Avaxim (HAV), Epaxal (HAV), VAQTA (HAV), Twinrix (HAV), Poliovax (discontinued), Imovax (rabies)	Hepatits A, poliomyelitis, rabies
Vero	Poliorix*, Boostrix*, Infanrix*, Pediarix*, Kinrix*, Quadracel*, Pentacel*, Pediacel*, IPOL (IPV), IMOVAX Polio, Adacel (IPV), Celvapan (p-flu), Preflucel (s-flu), Ixiaro (JEV), Jespect (JEV), Jenvac (JEV), Jeev (JEV), Encevac (JEV), VERORAB (Rab), Abhayrab (Rab), Speeda (Rab)	Poliomyelitis, pandemic and seasonal flu, Japanese encephalitis, rabies
Continuous cell lines MDCK	Optaflu, Flucelvax	Seasonal flu

Note
* IPV (inactivated polio virus) antigens are part of these vaccines.

TABLE 4.3
Types of cells used in the manufacture of marketed vaccines

Cell-type category	Cell line	Vaccine against
Primary cells	CEFs (Chicken Embryo Fibroblasts)	Measles, mumps, rabies, tick-borne encephalitis
Diploid cell lines	MRC-5	Measles, mumps, rubella
		Hepatitis A, poliomyelitis, rabies
		Varicella
	WI-38	Rubella, adenoviruses
Continuous cell lines	Vero	Poliomyelitis (inactivated)
		Rotavirus, pandemic and seasonal flu, Japanese encephalitis, rabies, smallpox
	CHO (Chinese Hamster Ovary)*	Zoster-shingles
	MDCK	Influenza
	Hi-5 with BEVS (Baculovirus Expression Vector System)*	HPV (human papilloma virus)

Note
* Recombinant systems

extensive testing protocols for identifying adventitious agents are not applied because those protocols are only used with cell banks. Nevertheless, European Pharmacopeia and CBER guidance amongst others, define strict testing programs for CEF-based vaccines to demonstrate that the appropriate safety criteria are satisfied (see also Section 4.3).

Diploid cell lines were developed in the 1960s, based on their capacity to withstand storage in a cell bank (see Section 4.2.2.1). These cells have a finite in vitro life span, are euploid (diploid) through the life span, and are structurally identical to those cells in the species from which they were derived. They also display contact inhibition and senescence. Diploid cell lines used for marketed human vaccines have been derived from human tissue. One advantage arising from the use of these cell lines is that any potential residual proteins are of human origin and therefore not susceptible to triggering an immune reaction. One disadvantage is the finite life span.

A continuous cell line is a cell line with an apparently unlimited capacity for population doubling [5]. The life span is indefinite allowing good cell amplification to large scale. However, many continuous cell lines display genetic aberrations manifested by aneuploidy. Therefore, the genetic stability needs to be well characterized: certain mutations could lead to tumor or oncogene activation [6]. The features of these cell lines led to define specific regulatory considerations.

4.2.2 Technical Considerations

This paragraph addresses the technical considerations to start development or manufacturing of a vaccine that requires cell lines for virus propagation or for recombinant-antigen expression. Other chapters will provide more in-depth review of key technologies.

4.2.2.1 Cell Banks

Creating a cell bank of the cell line represents the cornerstone of the process development and manufacturing. The cell bank provides the process with the same source for each production lot. It can be extensively defined and tested. In the case of a recombinant expression system such as CHO, the cell bank can be defined with respect to its genetic properties (including transgene copy number and integration site). In the case of a cell line for virus propagation, the capacity for virus infection and replication can be defined. All cell banks are extensively tested for safety. These include detailed testing programs to document the absence of adventitious agents (see Sections 4.3 and 4.4).

Cell banks are usually established as a two-tier system. This approach was first proposed in 1963, when the cryopreservation methods used were based on glycerol or DMSO [7]. The two-tier system consists of a master cell bank (MCB) and a working cell bank (WCB) derived from the MCB after few passages. The MCB is usually prepared from a clone or material obtained from a well-known bioresource archive such as ATCC or ECACC. Cell lines can also be licensed. The two-tier system generates a huge capacity for development and further commercial manufacturing. Indeed, each vial from the MCB can generate a new WCB. Hence manufacturing from the same MCB can continue for decades. Therefore, the long-term strategy with respect to the vaccine volumes required and other potential technical advances need to be considered by the manufacturer when the cell bank is established. A third cell bank, termed the end of production cells (EoPC), is also usually established. The purpose of this cell bank is to expand cells beyond the limit of the passage used in vaccine production.

MCB → WCB → Production Passage → EoPC

Hence, the three types of cell bank can be used to test the consistency of the biological system across the production framework; testing that is critical for the release of clinical lots or commercial vaccine batches.

4.2.2.2 Viral Seeds

For many viral vaccines, viral seeds are produced as well. Viral seeds are produced for live attenuated vaccines as well as inactivated vaccines. For certain recombinant vaccines, viral seeds are also produced for expression systems like the baculovirus expression vector system (BEVS), an adenovirus platform or a vaccinia recombinant system.

Viral seeds are also established as a two-tier system. This system consists of the master viral seed (MVS) and the working viral seed (WVS). A third viral seed is also prepared at the production passage or beyond, i.e., the post-production seed (PPS). The GMP production framework is similarly defined as follows:

MVS → WVS → Production Passage → PPS

The MVS and WVS are prepared from either the MCB or WCB, depending on the preference of the manufacturer. Some manufacturers derive an inoculum from WVS to manage seed capacity. Viral seeds are also tested to document consistency of the biological product across passages.

4.2.2.3 Technology Evolution

Several factors have shaped the evolution of cell-culture technology.

Viral safety associated with the selection of a cell line is a key element that will be discussed extensively in the next paragraph (see Section 4.3). In association with this topic, the raw materials required to prepare cell banks contribute to the safety of the cell line. This will also be reviewed.

The yield and cost-efficiency of the upstream processes are key aspects for consideration in the manufacture of vaccines, especially as many vaccines are recommended for universal or mass vaccination programs. To develop cost-efficient upstream processes, certain cells, including CHO and many continuous cell lines, have been developed or selected to be amenable for suspension (non-adherent) culturing. The shift to suspension allows high-cell density culturing and hence high yields. By contrast, some cell lines, such as MRC-5 and Vero cells, are adherent in culture, and hence usually display low to medium yields. For the manufacture of live-attenuated vaccines, a low virus yield may not be an issue because the content of a vaccine dose is low. For inactivated viral vaccines and recombinant products, high yields and hence high cell-culture densities are preferred.

To achieve high cell-culture densities, bioreactors are preferred over stationary technologies, such as cell factories. Several bioreactor technologies have been developed. Since 2000, the technologies have evolved from stainless-steel bioreactors to disposable bioreactors. Disposable bioreactors provide agility to GMP operations and can contribute to cost efficiency. Disposable bioreactors are also useful for establishing cell banks. The bioreactor technologies will be further discussed in Chapter 6.

4.3 VIRAL SAFETY

The risk of virus contamination in a manufacturing facility is low, but when it occurs, the impact is severe [8]. In a recent review published by the Consortium on Adventitious Agent Contamination in BioManufacturing (CAACB), 26 virus contaminations were reported over the past 36 years [9].

4.3.1 Risk and Impacts

The risks resulting from a viral contamination can be positioned at different levels:

- Risk to vaccine recipients from direct inoculation (safety risk when pathogenic for human)
- Risk to product quality
- Risk to other products in the facility
- Risk to product availability for vaccine recipients

Even in the absence of contamination, there is a risk of non-compliance with regulatory guidelines, if the plan to mitigate virus-contamination risk is inadequate. The risk of non-compliance can block license approval for specific markets with significant commercial impact.

The contamination of a manufacturing facility can lead to substantial economic losses during and after the incident, and can result in patient hardship as a consequence of the loss or interruption of the drug supply [8].

An adventitious virus contamination during cell culture manufacture of a biologic is incredibly disruptive [9]. Firstly, extensive efforts in time and resources need to be allocated for the identification of the contaminant, the source of the contamination and the extent of the contamination. These investigations can delay the development of new products. Secondly, corrective actions, such as changes of raw materials or decontamination of facilities, must be implemented. Finally, the reputation of the manufacturer can be tarnished, and its competitive advantage can be dented.

4.3.2 Sources of Viral Contaminations

Potential adventitious agents might be introduced into the process via three main sources, illustrated in Figure 4.1:

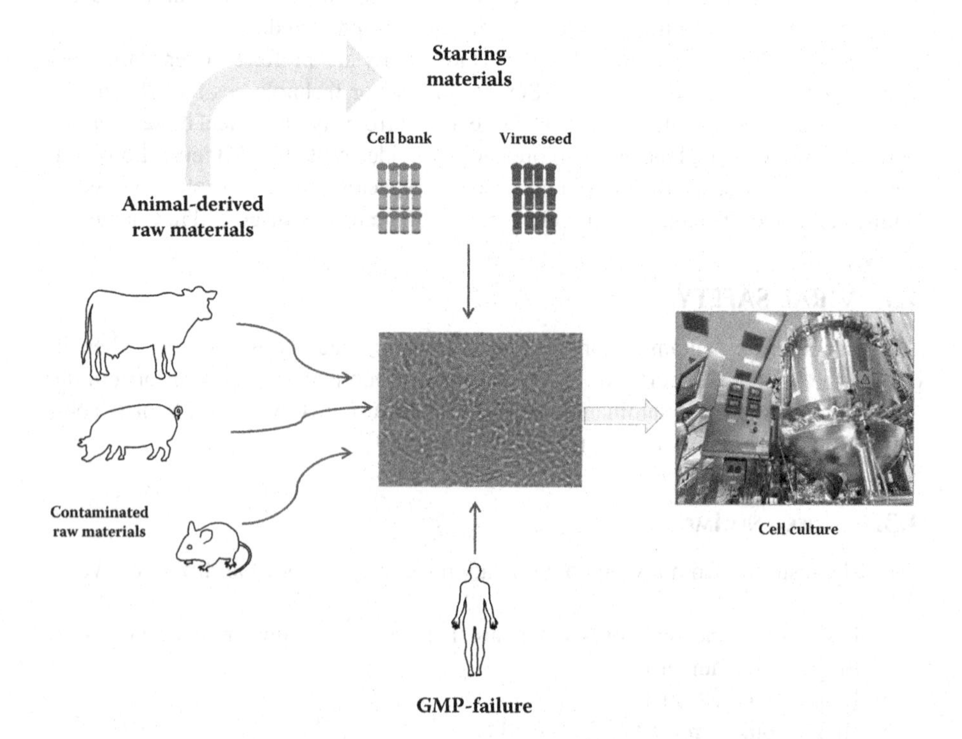

FIGURE 4.1 Three sources of viral contamination in biomanufacturing.

1. **Raw materials** of animal origin might carry extraneous agents of bovine, porcine, human or other origins (e.g. porcine trypsin, foetal bovine serum, or human serum albumin). Other raw materials can be also contaminated by contact with animals.
2. **Starting materials** including cell banks and virus seeds might contain extraneous agents introduced during their preparation and the amplification of such materials in different culture conditions (presence of animal derived material) or on different cell substrates (e.g., in the process of virus attenuation).
3. **Manufacturing operations** might bring viral contaminations from various sources:
   a. **Environmental contaminants** might be introduced via contaminated equipment and air. These contaminants are primarily agents that are very stable and have endured prolonged periods in a dry or liquid environments, which do not support growth of the contaminating agent.
   b. **Contamination by human operators** might occur if viruses/agents from the respiratory tract or from contaminated body surfaces are introduced into process material during open-air operations such as the preparation of culture media.

According to Onions et al. [10], the majority of contamination problems in the biotechnology and vaccine industry have a root cause associated with an adventitious agent in a raw material. In a review of viral contamination cases [9] three of the four viruses (blue tongue virus, CacheValley virus, and vesivirus 2117) that contaminated CHO cell cultures were suspected or identified as having come from serum. However, and by contrast, the minute virus of mice (MVM) was found to have contaminated a process in which no animal-derived material had been used [9].

4.3.3 Levels of Risk

Biologics, including vaccines, can be categorized according to their viral-risk level and the capacity to mitigate this risk. These risks are detailed below and described in Figure 4.2.

Risk Level 3, all three main sources of contamination are combined. Most vaccines produced from animal/human cell cultures are in this category. When the vaccines are from inactivated viruses, mitigation protocols in the manufacturing process can be used to reduce the risk. For the other live-attenuated vaccines, for which the purification is often limited or absent, the control is dependent on precautionary measures taken during the selection of the raw materials and during the entire manufacturing process. The risk also can be mitigated by implementing additional testing when a potential virus contaminant is highlighted by the risk assessment.

Risk Level 2, no primary animal-derived raw material is used during the manufacturing of the vaccine, which is also highly purified. However, the vaccines in this category are prepared from cell banks and seeds. Hence, the viral risk is still present.

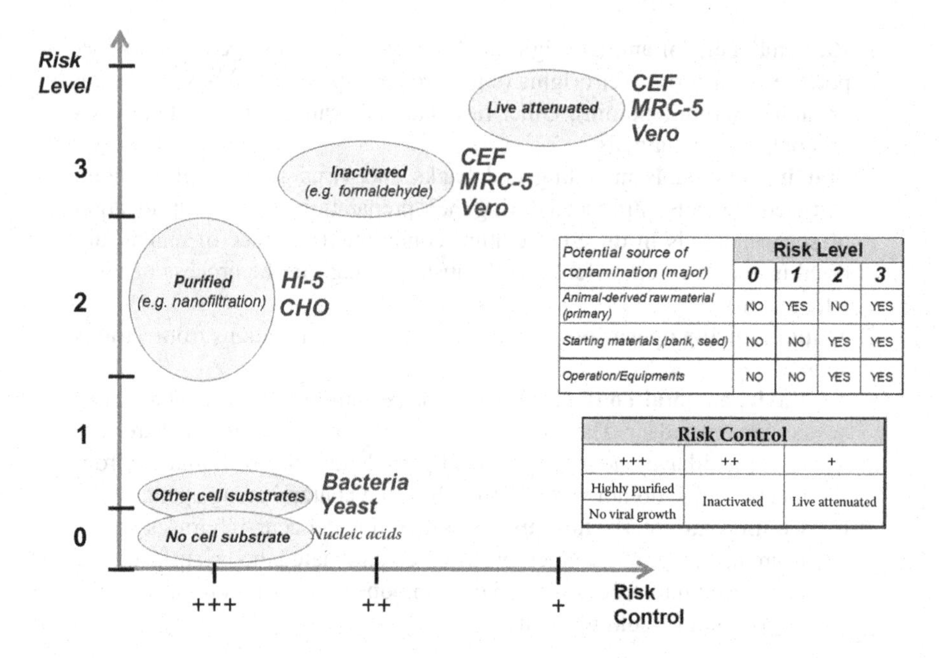

Potential source of contamination (major)	Risk Level 0	1	2	3
Animal-derived raw material (primary)	NO	YES	NO	YES
Starting materials (bank, seed)	NO	NO	YES	YES
Operation/Equipments	NO	NO	YES	YES

Risk Control		
+++	++	+
Highly purified No viral growth	Inactivated	Live attenuated

FIGURE 4.2 Viral safety risk and control for different vaccine types.

Risk Level 1, no animal/human cell-based starting materials are used in the manufacture of the vaccine. Vaccines in this category are prepared from bacteria and yeast. The major source of virus contamination comes from animal-derived raw material used to prepare the culture media. But any minor virus contamination cannot propagate in such a cell substrate.

Risk Level 0, no primary animal-derived materials and no starting materials are used in the manufacture of the vaccine. Any minor virus contamination present or occurring during manufacture cannot be amplified in the absence of cell substrate. Nucleic acids-based vaccines are in this category.

4.3.4 Case Study: The PCV1/Rotarix Incident

The discovery of a previously undetected contaminant in a licensed biological medicinal product occurs typically through the application of new analytical technologies, often with improved sensitivities or capabilities not previously available [11]. The porcine circovirus example is illustrative. Porcine circovirus (PCV1) sequences were detected in Rotarix by the group of Eric Delwart in 2010 using next-generation sequencing [12]. GlaxoSmithKline (GSK) initiated extensive investigation to identify that the source of the PCV1 contamination was most likely the porcine trypsin, used in 1983 to prepare the Vero MCB [13]. PCV1 had not been detected in conventional tests for more than 20 years, and therefore was able to propagate in the Vero cell substrate without being detected [14].

One week after the detection of PCV1 in Rotarix (15 March 2010), the FDA recommended that clinicians and public health professionals in the United States temporarily suspend the use of Rotarix while the agency and manufacturer investigated the finding of DNA from PCV1 in the vaccine [11]. On 14 May 2010, the FDA recommended resumption of the use of Rotarix. The EMA and FDA both concluded that the benefits of the vaccines were substantial and included the prevention of death and hospitalization for severe rotavirus disease. These benefits were considered to outweigh the risk, which is theoretical when considering that Rotarix is an orally administrated vaccine.

New sequencing technologies are now able to detect unknown viruses present in a biological sample (e.g., cell, seed, or vaccine) without prior knowledge of their presence and genetic sequences [15]. A direct consequence is the increasing number of new viruses to be included into the risk-assessment exercise.

In light of this greater level of virus diversity, NGS technology is revealing the limitations of those current risk-control strategies mainly based on infectivity assays and PCR testing. The poor performance of *in vitro* assays is not a problem of assay sensitivity but rather because the adventitious viral agents are not able to replicate in the cells used in the assay, or when viral replication is not detectable using classic read-outs, such as CPE or hemadsorption/hemagglutination. For example, the in-vitro 9CFR assays was identified as having limited capabilities in detecting a wide range of viruses, even within the same families, in bovine serum [16]. These viruses could infect human cells and represented potential contaminant risks to biological products. When viruses are incapable of propagating in cell culture, specific PCR assays are typically included in the testing plan. However, the potentially high genetic variability of the virus may need to be considered in the design of the PCR assay, or particular virus variants may pass undetected. For example, Gagnieur et al. [17] reported mismatches between the sequence of the PCR primers used to detect BVDV and the BVDV sequences present in the batches of bovine serum.

Thus, the viral-risk assessments and the associated risk-mitigation strategies based on the testing for specific viral contaminants can become rapidly obsolete. New assays cannot be systemically developed and implemented after each new virus/variant discovery. This gap in viral coverage of existing testing is increasingly recognized in the scientific community, particularly in the case of novel cell substrates and biologically derived raw materials [14].

4.3.5 Viral-Risk Assessment

From the lessons learned during the last decade, regulatory guidance has adapted quickly. Biopharmaceutical manufacturers are now advised to implement additional assays to detect adventitious agents as part of the manufacturing processes, to ensure the quality and safety of the biological products.

In the conventional viral-risk model, viral-risk assessments represent the first key stage to identify and reduce the risk.

According to the "Quality Risk Management ICH Q9" [18], a viral-risk assessment consists of the identification of hazards plus analysis and evaluation of risks associated with exposure to those hazards.

- **Risk identification** is a systematic use of information to identify adventitious viruses that can be potentially introduced at each step of the vaccine manufacturing process and derived from animal-derived raw materials.
- **Risk analysis** is the estimation of the risk associated with the identified viruses. This analysis can consist of interrogating biological information about the identified potential viral contaminants and the mitigation processes to deal with starting materials and raw materials. The result of such analysis can lead to the exclusion of some viruses from the risk.
- **Risk evaluation** is mainly based on an FMEA (failure mode effect analysis), which evaluates potential extraneous agents that might represent a potential risk for the vaccine. The risk evaluation takes into consideration three categories of criteria: occurrence (O), detectability (D), and severity (S). The composite risk score of each potential contaminant is calculated as the product of its three individual component ratings: O/D/S. This composite risk is called a risk priority number (RPN), which is used to categorize the risks.
- **Risk control** includes decision making to reduce the risk to a low or medium level. Decision makers have to put in place some action to mitigate the severity and the occurrence of the risk.
- **Risk acceptance** can be a formal decision to accept the residual risk based on an extensive justification.

The risk is globally controlled by the three approaches described in ICH, FDA, and EMEA regulatory guidance:

1. selecting and testing cell lines and other raw materials, including media components, for the absence of undesirable viruses which may be infectious and/or pathogenic for humans
2. assessing the capacity of the production processes to clear infectious viruses
3. testing the product at appropriate steps of production for absence of contaminating infectious viruses

A few months after the PCV1/Rotarix contamination alert in 2010, licensed vaccine manufacturers received a letter from CBER requesting information about the risk assessment and implementation of additional adventitious agent testing methods for viral vaccines and/or combination vaccines containing viral antigens.

4.3.6 Regulatory Considerations

The development of viral-risk assessment is increasingly mentioned in regulatory guidance as a prerequisite in the definition of a testing program.

With the 9th edition of the *European Pharmacopoeia* [19], two key chapters for vaccines were fundamentally revised (§5.2.3 and §2.6.16) in terms of viral-risk assessment and one new chapter (§5.2.14) was created with the new concept of *in vivo* test substitution.

- **Ph. Eur. Chapter 2.6.16** ["Tests for extraneous agents in viral vaccines for human use"] was revised in the 9th edition of the *European Pharmacopoeia* (9.3), which became effective on 01 January 2018. A new paragraph requests the introduction of a risk assessment to build the testing strategy that must be based on a full package of suitable tests with reference to Chapter 5.1.7. In the paragraph regarding the testing for extraneous agents in cell cultures, it is requested that the testing for extraneous agents must be carried out based on a risk assessment for each virus-seed lot, each virus harvest, and each production cell culture (control cells or control eggs).
- **Ph. Eur. Chapter 5.2.3:** "Cell Substrates for the production of vaccines for human use," Version 9:0 and updated Version 9.3 in July 2017. A greater flexibility is proposed when testing for infectious extraneous agents between master cell bank (MCB) and working cell bank (WCB), based on a risk assessment. Table 5.2.3.-1 is updated accordingly, and molecular methods are now considered.
- **The new Ph. Eur. Chapter 5.2.14**: ["Substitution of in vivo method(s) by in vitro method(s) for the quality control of vaccines"], Version 9.3 published in July 2017, was created to facilitate the introduction of in vitro methods.
- Finally, the **Ph. Eur. Chapter 5.1.7** provides general requirements concerning the viral safety of medicinal products whose manufacture has involved the use of materials of human or animal origin. The relevant factors to be considered in a risk assessment are listed.

Other general regulatory requirements in terms of viral risk assessments are also described in the following guidance:

- ICH Q5A 'Viral Safety Evaluation of Biotechnology Products Derived from Cell Lines of Human or Animal Origin' [20].
- ICH Q5D 'Derivation and Characterization of Cell Substrates Used for Production of Biotechnological/Biological Products' [21].
- Guidance for Industry: Characterization and Qualification of Cell Substrates and Other Biological Materials Used in the Production of Viral Vaccines for Infectious Disease [22].
- USP 1050 – Viral Safety Information / General information [23].
- EMA Guideline on Virus Safety Evaluation of Biotechnological Investigational Medicinal Products [24].
- WHO Recommendations for The Evaluation of Animal Cell Cultures as Substrates for The Manufacture of Biological Medicinal Products and for the Characterization of Cell Banks [5].

The development of risk assessment is also requested in product-specific guidance.

For example, the European "Guideline on the use of porcine trypsin used in the manufacture of human biological medicinal products" [25] provides general quality specifications for porcine trypsin, especially with respect to viral safety. It is requested that manufacturers specifically consider widely distributed viruses, which are difficult to inactivate (e.g., PCV and PPV) and with zoonotic potential (e.g. HEV); and that manufacturers conduct viral-risk assessment according to the general principles outlined in Ph. Eur. 5.1.7 (Viral Safety). The same recommendation is also made for the analysis of serum in the EMA guideline on viral-risk analysis [26]. This guidance strongly recommended that manufacturers of biological medicinal products conduct risk analyses, taking into account the quality and properties of the serum batches and the impact of these serum batches on the quality of the finished product, prior to use.

4.3.7 Raw Materials: From Biological Materials to Chemically Defined Materials for Cell Lines

Serum and trypsin are biological materials with potentially significant safety risks. Following the strategic imperative of removing non-essential animal-derived materials, manufacturers of biologics including vaccines progressively have been developing serum-free cell-culture media. Some plant components have been used such as rice extract. However, this was not the ideal solution because many new plant viruses have also been identified over the last 20 years [27]. Chemically defined cell-culture media is also being developed. Hence, over the years, upstream manufacturing processes for new vaccines are becoming free of animal-derived components. Cell and virus banks have benefited from this approach as well. It is clear that the preferred option of risk management for cell culture media is to prevent and remove the risks rather than to mitigate through detection.

4.4 ANALYTICAL CONSIDERATIONS AND CELL-BASED ASSAYS

4.4.1 Safety Tests and Future

A panel of tests are requested to be performed by manufacturers at relevant stages of the production process, to confirm that the biological products, and the biological starting materials from which they are manufactured, are free of adventitious agents. This extensive testing for adventitious agents prevents major contamination events and potential adverse clinical consequences.

4.4.1.1 In Vitro and in Vivo Safety Tests

The current in vitro and in vivo technologies for the detection of specific adventitious contaminants of vaccines were developed more than 50 years ago.

The current in vitro virus-detection assays use different cell lines in accordance with the relevant regulatory guidance (e.g., WHO, FDA guidance, Eur P, and USP, etc.). The assays work by incubating the test sample in the cell culture and then observing subsequent cytopathic effects (CPEs), or hemadsorption (HAD) or hemagglutination (HA), by introducing red blood cells. In vivo virus-detection assays require the inoculation of the sample in an animal susceptible to infection by the virus under consideration. The presence of virus is demonstrated by animal mortality or the presence of hemagglutinins in tissue at sites of potential infection.

Such virus-detection assays are also used to evaluate specific contaminants, such as retroviruses (endogenous to the cell line), bovine, and porcine viruses.

In the context of the 3Rs program (to Replace, Refine and Reduce [3Rs] the use of animal in the safety evaluations of products) the sensitivity and scope of current assays to detect adventitious agents have been reviewed [28] and suggests that in vitro assays are more limited than in vivo assays. Several proposals were made to ameliorate the current in vitro assays, including the potential inclusion of additional cell lines and the integration of new molecular methods.

4.4.1.2 New Technologies

New assays for detecting adventitious agents are being developed to have higher sensitivity or to detect new adventitious agents.

New technologies, based on nucleic-acid methods such as PCR, NGS, virus microarrays, and broad-range PCR combined with mass spectrometry are powerful tools for the detection and identification of viruses.

4.4.1.2.1 PCR-Based Methods

Today, the PCR-based assays are already widely used in the GMP environment, for routine activities in the detection of specific adventitious agents, such as viruses (PCV, MVM, Hep E) and bacteria (mycoplasma, mycobacterium).

In contrast to other nucleic-acid–based methods, such as quantitative PCR (qPCR), or hybridization-based assays, NGS provides a sensitive and quick platform assay for virus detection and identification in various types of biological sample, without prior sequence knowledge of potential viral contaminants [29]. Several NGS platforms are currently available, and these platforms have the capacity to generate a huge amount of sequence data. Two main NGS platforms are available today, primarily based on the read length; either short or long. Each system shows strengths and limitations, related to the number of reads, read length, error rate, run duration, and the complexity of the library construction. Compared with short-read platforms, which show a limited capacity to detect new viruses due to the length of the read, the long-read platforms are more able to identify new adventitious agents. Bioinformatics tools including tools to manage the pipeline, data quality, and data trimming/cleaning and data assembly, help establish the key parameters for the breadth and accuracy of virus detection.

In contrast to conventional adventitious agent detection methods, nucleic-acid–based methods report the genome copies/reaction or per volume present in the test article. Because viral preparations contain noninfectious or defective viral particles, in addition to infectious particles, the quantity of nucleic acid may not match with the titer of infectious particles. The same considerations have been largely discussed in the comparison of compendial tests for mycoplasma with PCR-based assays.

4.4.1.2.2 Microarrays

Microarray technology uses oligonucleotide probes for the detection and identification of known viral sequences. In contrast to NGS, the bioinformatic analysis is performed during the probe-design phase, meaning that the analysis mainly consists of calculating probabilities of virus detection, based on factors such as signal intensity, coverage across multiple targets within the same virus, and signals from closely related viruses.

4.4.1.2.3 *Mass Spectrometry*

Protein-mass spectrometry by MALDI TOF is also considered as an attractive alternative for the detection of viruses, but its application is very limited by the sensitivity required and by the complexity of the samples typically used. Coupled separation techniques like 2-D gel electrophoresis, in addition to peptide analysis by LC-MS/MS, are powerful methods used for detection of unknown viruses in complex samples.

These new methods using mass-spectrometry still need to be validated for their use in the detection of adventitious viruses.

4.4.2 Cell Line–Based Assays and the Future

4.4.2.1 Potency Tests

The efficacy of a vaccine is generally demonstrated by the ability of the immune system to protect against disease. The potency is a measurable characteristic of a vaccine that was correlated with the protection in the clinical study.

In practice, potency is used to demonstrate consistency (discriminating potent and subpotent) between vaccine lots, on the assumption that that the accepted range of potency correlates with the safety and the efficacy of the product. This range is part of the submission file approved by the regulatory authorities.

A broad panel of methods is currently used for potency tests and is dependent on the properties of the product and other analytical considerations.

4.4.2.1.1 *In Vivo Potency*

The most direct nonclinical method for testing the efficacy of a vaccine is a challenge test in an animal model (typically rodents). The objective is to demonstrate the prevention of mortality or pathogenesis in vaccinated animals.

If a challenge model is not available or feasible, the immune response to the vaccine is measured. A proxy for efficacy is evaluated by the detection of antibody titers, either by neutralizing activity or by ELISA, or by hemagglutination assays. In vivo potency tests for vaccine release are mainly required on specific inactivated-virus vaccines (Polio Salk) or purified products including antigen combinations.

Because animal-based tests are becoming ethically unacceptable, as well as being expensive, time-consuming, and often inaccurate, these tests are tending to be superseded by in vitro potency tests [30].

4.4.2.1.2 *In Vitro Potency*

The selection of the in vitro test is primarily shaped by the mechanism of action of the vaccine.

The in vitro potency tests for recombinant or purified vaccine are often immunoassays (e.g., single radial immunodiffusion potency test for the influenza vaccine and the enzyme-linked immunosorbent assay for recombinant Hepatitis B vaccines). These tests are generally based on the binding of key protective antigens in the vaccine to specific antibodies.

The progress in biosensor technology (e.g., surface plasmon resonance and biolayer interferometry) has now enabled the evaluation of the quantity and quality of vaccine antigen binding to antibody.

4.4.2.1.3 Live-Attenuated Vaccines

For live-attenuated vaccines (measles, mumps, rubella, varicella, oral polio vaccines), the efficacy of each vaccine lot is linked to the number of live viral particles that can be detected by cell-based assays.

Techniques directly quantifying the number of viral particles (e.g., transmission electron microscopy, ViroCyt, and Nanosight) have also been developed, but are still limited to specific applications and are only utilized in the release testing of a few vaccines. Moreover, these methods make no assessment of the viability of the virus particles.

Thus, the most common methods used to quantify viruses are based on cell infectivity and detection of cytopathic effects (e.g., for measles, mumps, and varicella vaccines). Serial-dilution methods in cell cultures (generally in microplates) may involve identifying the end-point dilution—the minimum titer necessary to cause infection—and the calculation of the dilution, leading to 50% cell infection (TCID50) [31]–[33]. However, these methods may result in a significant variability.

Plaque assays are based on the assumption that a single infectious virus particle infects one cell and its propagation leads to the surrounding cells also being infected, producing a delimited zone of cytopathology (plaque). The number of plaques is hence considered to represent the number of viable virus particles in the sample dilution.

The appropriate design of the plaque assay [34] reduces the assay variability [35] but the assay can be compromised if the virus is poorly lytic or the culture conditions are not correctly optimized (as with a semi-solid culture layer), potentially resulting in limited virus propagation and plaque identification. Alternative readouts based on the detection of single infected cells [36], [37] or the detection of nucleic acid resulting from virus replication [38] are increasingly being used to improve precision and increase throughput.

Finally, more refined methods [39], [40] combining the quantification and precise identification of the vaccine components are being developed, and may have the potential to identify the potency of the viral particles.

4.4.2.1.4 Live-Vectored Recombinant Virus

General methods described previously can also be applied for the detection of vaccines based on live recombinant virus platforms (e.g., lentivirus and adenovirus platforms). However, a combination of a potency test with an evaluation of the expression of the inserted sequence coding for the antigen is preferred.

4.4.2.1.5 Validation

Although during assay development, an evaluation of the assay performance and suitability is sufficient, the potency test has to be well defined and qualified for Phase III clinical studies and fully validated for consistency and release lots.

These potency tests should comply to strict validation criteria in terms of accuracy, precision (repeatability, intermediate precision), specificity, linearity, limit of detection and limit of quantitation, system suitability and robustness, in accordance with the ICH Q5A(R1) guideline [20].

4.4.3 Molecular Assays and Control of the Viral Risk in the Future

A major complication in the control of the viral risk is the continued discovery of new virus species/genera and variants, especially given the power of a new technology like NGS. This technology enables sequencing (i.e., identifying) millions of nucleic acid molecules with known and unknown preliminary knowledge of their sequences with unprecedented speed, quality, coverage, and depth.

The increasing number of newly identified viruses is illustrated in Figure 4.1. This graph is based on the taxonomy releases between 1971 and 2020 in the ICTV (International Committee on Taxonomy of Viruses) depository, in which the numbers of new virus species, genera, and families were counted. The number of virus species increased 2.5-fold between 2015 and 2020, from 3,704 to 9,110 (Figure 4.3).

The number of virus families, genera, and species are reported in historical ICTV (International Committee on Taxonomy of Viruses) taxonomy releases.

The viral risk assessments and the associated mitigation strategies based on the specific testing, such as PCRs, can become rapidly obsolete. New assays cannot be systemically developed and implemented after each new virus/variant discovery.

In addition, like other molecular-based assays, PCR and NGS cannot discriminate between nucleic acid from non-viable and viable/replicative viruses. Indeed, both methods can detect small DNA or RNA molecules which could be fragmented by inactivation treatments such as gamma-irradiation. For example, in a study by Toohey-Kurth et al. [41], NGS was used to analyze 26 bovine serum samples from 12 manufacturers. Twenty viruses belonging to nine virus families were detected in this study. Four viruses were taxonomically unassigned. The authors recognized that the presence of viral nucleic acids did not indicate contamination with infectious virus.

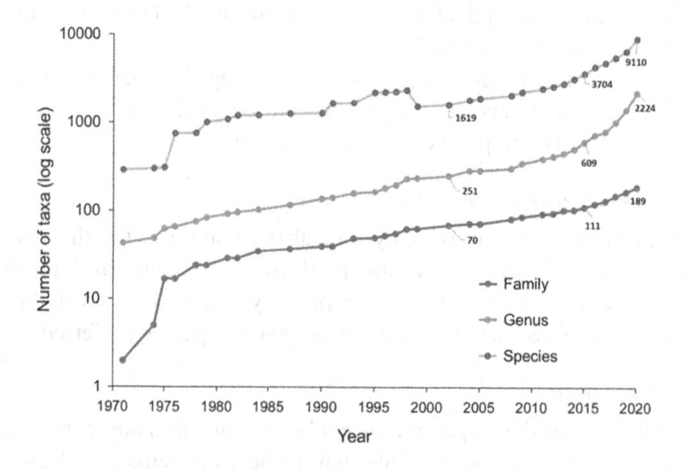

FIGURE 4.3 Illustration of the increasing number of new viruses.

Thus, the central question is the relevance of an NGS signal. Using a different approach, NGS transcriptome analysis is used as a readout in infectivity assays to detect viral replication in infected cells, indicating the presence of live virus in the tested material [42]. Transcriptomics analysis identified viral RNAs only synthesized during cell infection. Metabolic labeling of newly synthesized RNA has also been used to unambiguously and specifically detect virus replication in permissive cell lines [43].

According to these approaches, the transcriptome generated by NGS can not only be used to detect and identify virus sequences, but also to be used assign them to non-viable or viable/replicative, eliminating false positive signals.

Alternatively, new genetic-engineering technologies can be also used to modify or disrupt the activity of endogenous and extraneous viruses in cell lines. A genome editing technology has been used to disrupt in CHO cells, an active retrovirus C element, and completely suppress the release of RNA-loaded viral particles [44]. CHO cells have been made resistant to MVM infection by altering genes involved in cell-surface sialylation [45].

4.5 CONCLUSIONS

Cell lines are essential for vaccine manufacturing. For live-attenuated or inactivated viral vaccines, cell lines have been developed to allow viral replication.

Because cell cultures and raw materials can harbor adventitious agents, the viral safety assessment of cell lines and eukaryotic expression systems represents a key step in the choice of cell line for vaccine manufacture. Measures to prevent contamination by adventitious agents include the use of cell-culture media devoid of animal-derived components.

Numerous analytical methods are also part of control strategy to mitigate contamination by adventitious agents. Novel technologies are complementing or replacing existing conventional methods, such as those based on NGS, and have enabled the improved characterization of cell lines. The genetic engineering of cell lines to prevent potential contamination by adventitious agents is also a promising approach. These improvements will further contribute to safety of vaccines.

ACKNOWLEDGMENTS

All authors conceptualized the manuscript and contributed equally to the book chapter. All authors have read and agreed to the published version of the chapter. Competing financial interests: all authors are employees of the GSK group of companies. Trademarks used in the article are the property of their respective owners.

REFERENCES

[1] J. Ehreth, "The global value of vaccination," *Vaccine*, vol. 21, no. 7–8, pp. 596–600, Jan. 2003.

[2] I. Jordan and V. Sandig, "Matrix and backstage: cellular substrates for viral vaccines," *Viruses*, vol. 6, no. 4, pp. 1672–1700, Apr. 2014.

[3] A. J. Cann *et al.*, "Reversion to neurovirulence of the live-attenuated sabin type 3 oral poliovirus vaccine," *Nucleic Acids Res.*, vol. 12, no. 20, pp. 7787–7792, Oct. 1984.
[4] K. H. Khan, "Gene expression in Mammalian cells and its applications," *Adv. Pharm. Bull.*, vol. 3, no. 2, pp. 257–263, 2013.
[5] WHO, "Annex 3 Recommendations for the evaluation of animal cell cultures as substrates for the manufacture of biological medicinal products and for the characterization of cell banks Replacement of Annex 1 of WHO Technical Report Series, No. 878," 2013. https://www.who.int/biologicals/vaccines/TRS_978_Annex_3.pdf (accessed Oct-2021).
[6] F. Aubrit *et al.*, "Cell substrates for the production of viral vaccines," *Vaccine*, vol. 33, no. 44, pp. 5905–5912, Nov. 2015.
[7] R. I. Freshney, *Culture of Animal Cells: A Manual of Basic Technique*, 5th Edition. John Wiley & Sons, Inc.
[8] A. Stokes, "Managing potential virus and TSE contamination | Pharmaceutical engineering March/April,"2018. (https://ispe.org/pharmaceutical-engineering/march-april-2018/managing-potential-virus-and-tse-contamination)
[9] P. W. Barone *et al.*, "Viral contamination in biologic manufacture and implications for emerging therapies," *Nat. Biotechnol. 2020 385*, vol. 38, no. 5, pp. 563–572, Apr. 2020.
[10] D. Onions, C. Côté, B. Love, and J. Kolman, "Deep Sequencing Applications for Vaccine Development and Safety," in *Vaccine Analysis: Strategies, Principles, and Control*, B. K. Nunnally, V. E. Turula, and R. D. Sitrin, Eds. Berlin, Heidelberg: Springer-Verlag, 2015, pp. 445–477.
[11] J. Petricciani, R. Sheets, E. Griffiths, and I. Knezevic, "Adventitious agents in viral vaccines: lessons learned from 4 case studies," *Biologicals*, vol. 42, no. 5, pp. 223–236, Sep. 2014.
[12] J. Victoria *et al.*, "Viral nucleic acids in live-attenuated vaccines: detection of minority variants and an adventitious virus," *J. Virol.*, vol. 84, no. 12, pp. 6033–6040, Jun. 2010.
[13] G. Dubin *et al.*, "Investigation of a regulatory agency enquiry into potential porcine circovirus type 1 contamination of the human rotavirus vaccine, Rotarix: approach and outcome," *Hum. Vaccine Immunother.*, vol. 9, no. 11, pp. 2398–2408, Nov. 2013.
[14] R. L. Sheets and P. A. Duncan, "Role of Analytics in Viral Safety," in *Vaccine Analysis: Strategies, Principles, and Control,* B. Nunnally, V. E. Turula, and R.D. Sitrin, Eds. Berlin, Heidelberg: Springer, 2015.
[15] A. Khan *et al.*, "Advanced Virus Detection Technologies Interest Group (AVDTIG): Efforts on High Throughput Sequencing (HTS) for virus detection," *PDA J. Pharm. Sci. Technol.*, vol. 70, no. 6, pp. 591–595, Nov. 2016.
[16] C. Marcus-Sekura, J. Richardson, R. Harston, N. Sane, and R. Sheets, "Evaluation of the human host range of bovine and porcine viruses that may contaminate bovine serum and porcine trypsin used in the manufacture of biological products," *Biologicals*, vol. 39, no. 6, pp. 359–369, Nov. 2011.
[17] L. Gagnieur *et al.*, "Unbiased analysis by high throughput sequencing of the viral diversity in fetal bovine serum and trypsin used in cell culture," *Biologicals*, vol. 42, no. 3, pp. 145–152, 2014.
[18] ICH, "International conference on harmonisation of technical requirements for registration of pharmaceuticals for human use ich harmonised tripartite guideline quality risk management Q9," 2005. https://database.ich.org/sites/default/files/Q9%20Guideline.pdf (accessed Oct-2021).
[19] "EDQM (European Directorate for the Quality of Medicines – Council of Europe). European Pharmacopeia, 9th edition, Strasbourg (France)," 2016.

[20] ICH, "International conference on harmonisation of technical requirements for registration of pharmaceuticals for human use viral safety evaluation of biotechnology products derived from cell lines of human or animal origin Q5A(R1)," 1999. https://database.ich.org/sites/default/files/Q5A%28R1%29%20Guideline_0.pdf (accessed Oct-2021).
[21] ICH, "International conference on harmonisation of technical requirements for registration of pharmaceuticals for human use ich harmonised tripartite guideline derivation and characterisation of cell substrates used for production of biotechnological/biological," 1997. https://database.ich.org/sites/default/files/Q5D%20Guideline.pdf (accessed Oct-2021).
[22] FDA (U. S. Food and Drug Administration) "Guidance for Industry: Characterization and qualification of cell substrates and other biological materials used in the production of viral vaccines for infectious disease indications," 2010. https://www.fda.gov/regulatory-information/search-fda-guidance-documents/characterization-and-qualification-cell-substrates-and-other-biological-materials-used-production (accessed Oct-2021).
[23] U. S. Pharmacopeia, "General Chapters: <1050> VIRAL SAFETY EVALUATION OF BIOTECHNOLOGY PRODUCTS DERIVED FROM CELL LINES OF HUMAN OR ANIMAL ORIGIN," 2011. http://www.pharmacopeia.cn/v29240/usp29nf24s0_c1050_viewall.html (accessed Oct-2021).
[24] EMA, "Guideline on virus safety evaluation of biotechnological investigational medicinal products (EMEA/CHMP/BWP/398498/2005)," 2008. https://www.ema.europa.eu/en/documents/scientific-guideline/guideline-virus-safety-evaluation-biotechnological-investigational-medicinal-products_en.pdf (accessed Oct-2021).
[25] EMA, "Guideline on the use of porcine trypsin used in the manufacture of human biological medicinal products (EMA/CHMP/BWP/814397/2011)," 2014. https://www.ema.europa.eu/en/documents/scientific-guideline/guideline-use-porcine-trypsin-used-manufacture-human-biological-medicinal-products_en.pdf (accessed Oct-2021).
[26] EMA, "Guideline on the use of bovine serum in the manufacture of human biological medicinal products (CPMP/BWP/1793/02)," 2013. https://www.ema.europa.eu/en/documents/scientific-guideline/guideline-use-bovine-serum-manufacture-human-biological-medicinal-products_en.pdf (accessed Oct-2021).
[27] M. McLeish, A. Fraile, and F. García-Arenal, "Ecological complexity in plant virus host range evolution," *Adv. Virus Res.*, vol. 101, pp. 293–339, Jan. 2018.
[28] J. Gombold *et al.*, "Systematic evaluation of in vitro and in vivo adventitious virus assays for the detection of viral contamination of cell banks and biological products," *Vaccine*, vol. 32, no. 24, pp. 2916–2926, May 2014.
[29] C. Lambert *et al.*, "Considerations for optimization of high-throughput sequencing bioinformatics pipelines for virus detection," *Viruses*, vol. 10, no. 10, 1–28, Oct. 2018.
[30] B. Metz, C. Hendriksen, W. Jiskoot, and G. Kersten, "Reduction of animal use in human vaccine quality control: opportunities and problems," *Vaccine*, vol. 20, no. 19–20, pp. 2411–2430, Jun. 2002.
[31] G. Kärber, "Beitrag zur kollektiven Behandlung pharmakologischer Reihenversuche," *Naunyn-Schmiedebergs Arch. für Exp. Pathol. und Pharmakologie 1931 1624*, vol. 162, no. 4, pp. 480–483, Jul. 1931.
[32] L. J. Reed and H. Muench, "A simple method of estimating fifty per cent endpoints," *Am. J. Epidemiol.*, vol. 27, no. 3, pp. 493–497, May 1938.
[33] C. Spearman, "The method of 'Right and wrong cases' ('Constant stimuli') without Gauss's Formulae," *Br. J. Psychol. 1904–1920*, vol. 2, no. 3, pp. 227–242, 1908.

[34] D. LaBarre and R. Lowy, "Improvements in methods for calculating virus titer estimates from TCID50 and plaque assays," *J. Virol. Methods*, vol. 96, no. 2, pp. 107–126, 2001.

[35] A. Roldão, R. Oliveira, M. Carrondo, and P. Alves, "Error assessment in recombinant baculovirus titration: evaluation of different methods," *J. Virol. Methods*, vol. 159, no. 1, pp. 69–80, Jul. 2009.

[36] I. V. Gates, Y. Zhang, C. Shambaugh, M. A. Bauman, C. Tan, and J. L. Bodmer, "Quantitative measurement of varicella-zoster virus infection by semiautomated flow cytometry," *Appl. Environ. Microbiol.*, vol. 75, no. 7, pp. 2027–2036, Apr. 2009.

[37] B. Grigorov, J. Rabilloud, P. Lawrence, and D. Gerlier, "Rapid titration of measles and other viruses: optimization with determination of replication cycle length," *PLoS One*, vol. 6, no. 9, p. e24135, 2011.

[38] R. Gustafsson, E. Engdahl, and A. Fogdell-Hahn, "Development and validation of a Q-PCR based TCID50 method for human herpesvirus 6," *Virol. J.*, vol. 9, 311, 2012. https://doi.org/10.1186/1743-422X-9-311

[39] A. Calderaro, M. Arcangeletti, I. Rodighiero et al., "Identification of different respiratory viruses, after a cell culture step, by matrix assisted laser desorption/ionization time of flight mass spectrometry (MALDI-TOF MS)," *Sci. Rep.*, vol. 6, 36082, Oct. 2016.

[40] T. Santiago-Rodriguez, "Identification and quantification of DNA viral populations in human urine using next-generation sequencing approaches," *Methods Mol. Biol.*, vol. 1838, pp. 191–200, 2018.

[41] K. Toohey-Kurth, S. Sibley, and T. Goldberg, "Metagenomic assessment of adventitious viruses in commercial bovine sera," *Biologicals*, vol. 47, pp. 64–68, May 2017.

[42] A. Brussel *et al.*, "Use of a new RNA next generation sequencing approach for the specific detection of virus infection in cells," *Biologicals*, vol. 59, pp. 29–36, May 2019.

[43] J. Cheval, E. Muth, G. Gonzalez , M. Coulpier, P. Beurdeley, S. Cruveiller, and M. Eloit, "Adventitious virus detection in cells by high-throughput sequencing of newly synthesized RNAs: unambiguous differentiation of cell infection from carryover of viral nucleic acids," *mSphere*, vol. 4, no. 3, e00298-19, Jun. 2019.

[44] P. Duroy et al., "Characterization and mutagenesis of Chinese hamster ovary cells endogenous retroviruses to inactivate viral particle release," *Biotechnol. Bioeng.*, vol. 117, no. 2, pp. 466–485, Feb. 2020.

[45] J. Mascarenhas et al., "Genetic engineering of CHO cells for viral resistance to minute virus of mice," *Biotechnol. Bioeng.*, vol. 114, no. 3, pp. 576–588, Mar. 2017.

5 Upstream processing for viral vaccines–General aspects

Lars Pelz and Sven Göbel
Bioprocess Engineering, Max Planck Institute for Dynamics of Complex Technical Systems, Magdeburg, Germany

Karim Jaen
Department of Internal Medicine II, Klinikum rechts der Isar, Technical University of Munich, Munich, Germany and Bioprocess Engineering, Max Planck Institute for Dynamics of Complex Technical Systems, Magdeburg, Germany

Udo Reichl
Bioprocess Engineering, Max Planck Institute for Dynamics of Complex Technical Systems, Magdeburg, Germany and Chair for Bioprocess Engineering, Otto-von-Guericke-University Magdeburg, Magdeburg, Germany

Yvonne Genzel
Bioprocess Engineering, Max Planck Institute for Dynamics of Complex Technical Systems, Magdeburg, Germany

CONTENTS

DOI: 10.1201/9781003229797-5

SYMBOLS

CSVY	Cell-specific virus yield	virions/cell
C_{vir}	Virus particle concentration	virions/mL
$C^*_{O_2}$	Oxygen saturation concentration	mM
C_{O_2}	Actual oxygen concentration	mM
C_s	Substrate concentration	mM
C_x	Cell concentration	cells/mL
η	Dynamic viscosity	Pa s
k_La	Volumetric mass transfer coefficient	1/h
MOI	Multiplicity of infection	infectious units/cell
OTR	Oxygen transfer rate	mM/h
PFU	Plaque forming units	PFU/mL
pO_2	Partial pressure of oxygen	%
rpm	Rounds per minute	rpm
τ	Shear stress	Pa
$TCID_{50}$	Tissue culture infectious dose 50	$TCID_{50}$/mL
TOH	Time of harvest	h
TOI	Time of infection	h

VVP	Volumetric virus productivity	virions/L/d
wv	Working volume	L
$\frac{\partial u_x}{\partial y}$	Velocity gradient	s^{-1}

5.1 INTRODUCTION

This chapter focuses on general aspects of virus production in small-scale vessels and bioreactors. It provides experimental, practical, and theoretical aspects that are essential for cell-culture–based upstream processing (USP) of viral vaccines. It is not the intention of the authors, however, to cover all basics of animal cell culture. Moreover, the literature and citations are not exhaustive as the authors wanted to highlight only selected aspects of viral vaccine manufacturing. The examples shown mainly concern studies that were performed at laboratory scale. Nevertheless, most of the covered aspects certainly should also hold for large-scale manufacturing. Regarding details of large-scale processes, see the case studies described in chapters 9 to 12.

5.2 BASIC DESIGN OF A VIRUS PRODUCTION PROCESS

Cell-culture–based production of viral vaccines is a complex process that involves several steps until the final product is obtained (Figure 5.1). First, a vaccine candidate is considered. Then a set of cell lines permissive to infection is screened to identify the most promising host for high-yield virus production. In a next step, the generation and testing of master and working banks of both virus and cell lines is pivotal. To initiate production, cryopreserved cells from the working cell bank are thawed to inoculate the first of a series of precultures of increasing volume (inoculum train). Cells successfully maintained in exponential growth are subsequently transferred to a stirred tank reactor (STR), followed by cultivation in successive STR runs at increasing volumes (seed train) until the required amount of cells is generated to start the production run. Finally, in the production bioreactor, once the target cell concentration is reached (cell growth phase), the seed virus is added to initiate the virus infection phase. Often, this involves a dilution step or a complete medium exchange to provide the required substrates and to reduce levels of potential inhibitor concentrations that might interfere with virus production. Virus replication and release into the extracellular medium often involves cell death and cell lysis. In a next step, the virus harvest is collected and cells and debris are removed by clarification. For subsequent virus purification, host cell nucleic acids (mainly deoxyribonucleic acid [DNA]) contaminating the virus harvest are degraded enzymatically through addition of a nuclease. In downstream processing (DSP) the level of particulate contaminants, residual nucleic acids, host cell proteins and other compounds of the virus harvest is reduced by a purification train typically comprising several filtration, concentration, and chromatography steps. In a few cases, when virus particles accumulate within cells or aggregates are formed, a homogenization step is required that may involve freeze-thawing or use of

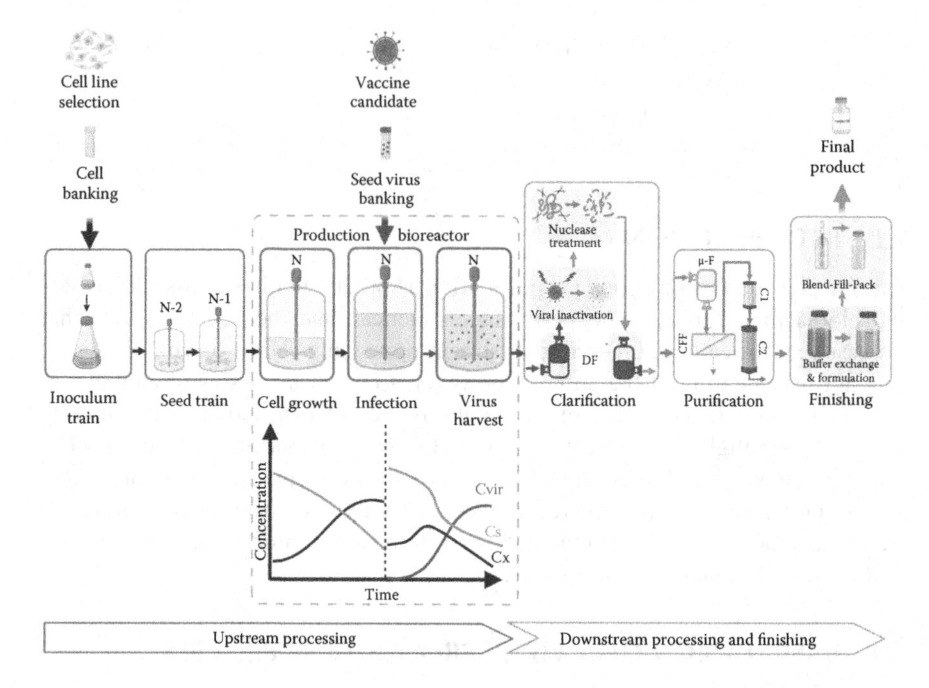

FIGURE 5.1 Overview on steps of a cell-culture–based virus production process. Cells undergo expansion through a series of precultures until production scale in a bioreactor. Compared to other biopharmaceutical products produced with animal cell culture, upstream processing of viral vaccines is typically a two-phase process: 1) Cell growth to the target cell concentration; 2) infection, virus replication, virus release together with cell death and cell lysis, and finally virus harvest (upstream processing). After inactivation, the virus harvest is subjected to a purification train (downstream processing), followed by formulation and filling (finishing). In some cases, virus inactivation is done after downstream processing (not shown). Abbreviations: N, number of cell expansions for production; DF, depth filtration; μ-F, microfiltration; CFF, cross-flow filtration; C1: chromatography step 1; C2 chromatography step 2. Arrows color indicates no virus (black), active virus (red), or inactive virus (green) flows. Arrow size indicates internal (small) and external process flows (large). Discontinuous arrows represent a waste flow. C_{vir}: virus concentration; C_S: substrate concentration; C_X: cell concentration over process time for cell growth and virus production phase. Created with Biorender (www.biorender.com).

high-pressure cell lysis (not shown). For inactivated vaccines, an inactivation step is introduced either before or after DSP. Next, the buffer used to recover the virus in the last step of purification is replaced by a buffer suitable for the intended administration, and additional compounds, i.e., adjuvants, are added for formulation of the vaccine bulk (finishing). Subsequently, the primary containers are filled with the vaccine and, in case a solid form is required, a lyophilization step may be performed. Lastly, the containers are capped and sealed. For the final product, further steps such as labeling, packaging, and storage, among others are required.

The following sections describe in detail the individual steps of USP, resulting in a virus harvest ready for purification.

5.3 VACCINE CANDIDATES

Viral vaccine formulations that contain or are derived from whole virus particles are mainly produced in animal cell cultures. The choice of a suitable host cell system depends on the vaccine type (see Table 5.1): Live-attenuated, inactivated, split, or sub-unit vaccines are mainly produced in mammalian, avian, or fish cell lines. For manufacturing of virus-like particles (VLPs) and recombinant protein vaccines, apart from mammalian cells, insect cells, but also yeast or bacteria, are used. Alternatively, viral antigens can be introduced to the immune system with mRNA or DNA technology or via viral vectors. Therefore, the choice of production system depends on the vaccine type, but is also strongly influenced by the virus, the biosafety level (BSL) of the virus, the required quantity, as well as the target group of vaccinees (age, place, human or veterinary, costs). Moreover, the disease itself is a further decision criterium: depending on severity and prevalence of the disease, the need for a vaccine is more evident. Thus, higher costs are accepted and more companies will take an effort to develop a manufacturing process. Regulatory hurdles, subsidization from governments or other funding agencies (e.g.,, priority review vouchers, orphan incentives, and expedited programs), and "time to market" will further decide on production innovations or if manufacturers will go back to what historically works and is safe.

Some viruses, like Hepatitis C virus (HCV), are only poorly or not replicating at all in available cell lines or in primary cells. Producing such viruses for inactivated vaccines, where a high virus concentration per dose is needed, will be difficult. Here, often the lack of vaccines indicates such production issues. For HCV, virus replication typically results in less than five infectious virions/cell (cell-specific virus yield: CSVY) [4].

Regarding efficacy and duration of immunity, live-attenuated vaccines are often preferred over other vaccine types. However, developing attenuated virus strains and demonstrating their safety is challenging. This is reflected by the relatively low number of such vaccines in the market: yellow fever virus (YFV), polio virus, some influenza virus strains, modified vaccinia virus ankara (MVA) virus, varicella zoster virus, rotavirus, pox virus, and the vaccine against measles, mumps, and rubella (MMR). Reverse genetics uses of non-human viruses as vectors (e.g., vesicular stomatitis virus [VSV]) is a very promising alternative for the development of vaccines against some highly pathogenic viruses. As before, the replication of these vectors in cell cultures is often poor. Examples here are the production of measles viruses for a Chikungunya vaccine [5], VSV for an Ebola vaccine [6], or YFV for a dengue vaccine [7].

Some vaccines are administered as a multivalent vaccine. For instance, recent influenza vaccines may contain up to four influenza A and B subtypes (quadrivalent influenza vaccines). Other vaccines are combining even different viruses, such as the MMR vaccine. For dengue vaccines, four virus subtypes need to be in the vaccine in the correct ratio to each other to not cause complications [8]. Obviously,

TABLE 5.1
Vaccine types and their specificities

Vaccine type	Antigen	Immunogenicity/ booster?/duration of protection/ adjuvants	Principle	Risks and safety	Dosage/ storage	Production platform	Examples for vaccines
Live-attenuated	Whole virus	Very high/no/long protection/none	Attenuation of the virus under non-physiological conditions, multiple passages. In vaccinee low level virus replication.	Risk of reversion, higher risk for immunocompromised vaccinees	e.g., 1E05-1E07 inf. units per dose/cold chain	Mammalian & avian cells	YFV, measles, polio, influenza
Inactivated	Whole virus	High/yes/short protection/yes	Virus is chemically inactivated, but stays intact	Risk of incomplete inactivation, required biosafety level in production determined by pathogenicity of the virus, inactivation might influence antigen structure	e.g., 1E07-1E10 units/dose, difficult for low CSVY	Mammalian & avian cells	influenza, polio
Split	Mix of disrupted virus particles/ antigens	Low/yes/short protection/yes	Whole viruses are disrupted by detergents (split)	No risk of infection as virus is disrupted and split, required biosafety level in production determined by pathogenicity of the virus, inactivation and detergents might influence antigen structure	ng protein/dose	Mammalian & avian cells	influenza

Subunit	Purified sub-units from virus material	Low/yes/short protection/yes	Whole viruses are disrupted by detergents and purified into sub-units	No risk of infection as virus is disrupted and split, required biosafety level in production determined by pathogenicity of the virus, inactivation and detergents might influence antigen structure	ng protein/dose	Mammalian & avian cells	influenza
Recombinant sub-unit	Viral antigen (protein)	Medium/yes/short protection/yes	Recombinant expression of viral genes to produce virus proteins	Low risk as antigens have no viral genome, lower immune response as some virus antigens are missing	ng protein/dose	Mostly insect cells, HEK293 and CHO	HA (hemagglu-tinin) from influenza virus
Recombinant virus-like particle	Virus-like particle (VLP)	Medium/yes/short protection/yes	Recombinant expression of viral genes to produce VLPs	Low risk as VLPs have no viral genome, lower immune response as some virus antigens are missing	ng protein/dose	Mostly insect cells, HEK293 and CHO	HA from influenza virus
Recombinant vector/ chimeric virus	Viral antigen (protein) plus vector whole virus	High/yes/short protection/yes	Attenuated or recombinant viral vector express viral antigen in the vaccinated person, chimeric virus displays recombinant antigen on its surface	Antigens from vector can serve as adjuvant, non-replicating vectors makes this interesting for viruses of high biosafety level	e.g., ng protein/ dosee.g., 1E08 vector units/ dose/cold chain	Mammalian, avian and insect cells	VSV for Ebola, YFV, AdV (adenovirus) for SARS-CoV-2 (severe acute respiratory syndrome corona virus type 2)
mRNA	Viral antigen (protein)	High/yes/not enough data/no	Viral antigen is expressed in the vaccinated person from the introduced mRNA, mRNA is packed in lipid particle to be	No risk of infection, no risk of integration of mRNA into genome, relatively new technology, possible concerns not fully clear	0.1-100 µg/ dosecold chain–50 to −15°C, 2–8°C 30 days and up	Enzymatic, lipid chemistry	SARS-CoV-2

(Continued)

TABLE 5.1 (Continued)
Vaccine types and their specificities

Vaccine type	Antigen	Immunogenicity/ booster?/duration of protection/ adjuvants	Principle	Risks and safety	Dosage/ storage	Production platform	Examples for vaccines
			introduced into cells, RNA can be self-amplifying, then contains alphaviral replication machinery		to 25°C for 24 h.		
pDNA	Viral antigen (protein)	High/yes/not enough data/ no	Viral antigen is expressed in the vaccinated person from pDNA molecules delivered intradermally through a needle-free-injector. The pDNA is not replicative in recipient cells [1].	No risk of infection, no risk of integration into genome.	1-3 mg/dose cold chain2–8°C and up to 25°C for 3 months.	*Escherichia coli.*	SARS-CoV-2

Adapted from [2].
mRNA vaccines adapted from [3].

these factors might equally play a role for the design of the corresponding manufacturing process.

5.4 CELL LINES FOR VIRUS PRODUCTION

5.4.1 Selection of Host Cell Lines

To produce a new potential vaccine candidate based on whole viruses or on one of the available or even new viral vectors, virologists typically use small-scale vessels, not too much focused on the host cell line as long as titers are high enough for the studies required. Thus, often adherent "standard cell lines" such as Vero, A549, or HEK293 cells as well as diploid cell lines like MRC5 or even primary cells like chicken embryo fibroblasts (CEFs) are used (see Table 5.2). Human cell lines are often used as diagnostic tools, but a problem for their use in vaccine production is the presence of adventitious agents with direct link to humans, which would require very costly and comprehensive testing and is often considered not to be acceptable for manufacturing of vaccines for human use.

Thus, latest, once the viral vaccine candidate is found, it should be considered, which host cell line is suitable for production. For a more detailed discussion on cell lines see chapter 4. Typically, a vaccine is administered to a healthy person who does not accept any severe side effects of vaccination. Thus, in contrast to a patient suffering from a disease, the hurdle of acceptance is much more stringent and vaccine production has to comply to extremely high-quality safety and efficacy standards. This also results in a very conservative approach towards the adaptation of new technologies in cell culture-based vaccine production. For example, conventional batch processing is still performed for the majority of vaccine manufacturing processes despite the clearly demonstrated advantages of advanced production technologies including fed-batch or perfusion cultivations. Also, cell line selection seems unnecessary restricted for historical reasons. This applies, in particular for baby hamster kidney (BHK21) or Chinese hamster ovary (CHO) cells that are still considered as a "no go" as early evaluations and risk assessments have identified these cell lines as possible tumorigenic/cancerogenic. These cell lines could be used in veterinary vaccine manufacturing but not for human vaccines covered in this chapter. However, this circumstance is actually strongly discussed in the research community, as the tumorigenicity/cancerogenicity potential of cell preparations tested is only one of many factors that need to be considered to evaluate the safety of a vaccine. Perhaps more important is the extent to which USP/DSP eliminate cellular factors of concern [9]. Today, not only powerful analytical tools, but also highly efficient DSP methods are available for characterization and purification of cell culture-derived vaccines. Accordingly, a comprehensive re-evaluation of safety aspects for use of these cell lines including process data might be considered. Especially, suspension BHK21 cells would be very interesting candidates for human vaccine production as for many viruses and viral vectors very high titers can be achieved in bioreactors. However, such a re-evaluation of cell substrates is very costly and time will show if commercial aspects and society needs are strong enough to attract the required investments.

TABLE 5.2
Selected host cell lines for virus production

Name	Cell type and tissue origin	Morphology	Adherent / suspension / provider	GMO / BSL/ cancerogenic	Used in vaccine production	Comments	Selected viruses that replicate well
A549	Human lung carcinoma	Epithelial	Both / culture collection & iBET	No / 1 / yes	No	Used for many assays	IAV
AGE1.CR.pIX	Duck retinoblast		Both / ProBioGen	Yes / 1 / no	?	Growth in CDM	MVA, IAV, VSV
BHK21	Baby hamster kidney	Fibroblast	Both / culture collection & Ceva	No / 1 / yes	Veterinary only	Safe?	Rabies, YFV, VSV
CEF	Chicken embryo	Fibroblast	Adherent / primary	No / 1 / no	Yes	Needs to be prepared new each time, needs serum	MVA
CHO	Chinese hamster ovary	Epithelial	Both / culture collection	No / 1 / yes	No	Can grow to very high concentrations (2E08 cells/ mL), growth in CDM	–
EB66	Duck embryonic stem cell	Stem cell	Suspension / Valneva	No / 1 / no	?	Can grow to very high concentrations (1.6E08 cells/ mL), growth in CDM	MVA, YFV, ZIKV, IAV
HepG2	Human hepatocellular carcinoma	Epithelial	Adherent / culture collection	No / 1 / yes	No	Needs serum	HCV, Hepatitis E virus
HEK293	Human embryonic kidney	Fibroblast	Both / culture collection	Yes / 1 / no	Yes	Often used for recombinant sub-unit vaccines and vector vaccines	Adenovirus, AAV, IAV
High five	Insect		Both / culture collection	No / 1 / no	Yes	Growth at 28°C	Baculovirus
Huh7.5	Human hepatoma	Epithelial	Adherent / culture collection	No / 1 / yes	No	Needs serum	HCV

MDBK	Porcine kidney	Epithelial	Both / culture collection	No / 1 / no	Veterinary only?	Veterinary vaccines	IAV
MDCK	Dog kidney	Epithelial	Both / culture collection	No / 1 / no	Yes	Polarized, growth in CDM	IAV
MRC-5	Human lung	Fibroblast	Adherent / culture collection	No / 1 / no	Yes	Needs serum, diploid	Polio, measles
PER.C6	Human retinoblast		Suspension / Johnson & Johnson	Yes / 1 / no	No	Can grow to very high concentrations (2E08 cells/mL?)	IAV, adenovirus,
PK15	Porcine kidney	Epithelial	Both / culture collection & ProBioGen	No / 2 / no	Veterinary only	Growth in CDM	IAV
SF9	Insect	Epithelial	Both / culture collection	No / 1 / no	Yes	Grows at 28°C	Baculovirus
Vero	African green monkey kidney	Epithelial	Both / culture collection & NRC	No / 1 / no	Yes	Used for many assays, available suspension cells still show very low doubling times	YFV, SARS, polio, rabies, HSV1

Please note that there are several subpopulations of cell lines with a different history available, often all running under the same name. Some of these subpopulations may differ in possible adaptations to growth in suspension, display lower or higher cell-specific virus yields, etc. The list of selected viruses is certainly not complete.

The transition from vaccine research and development to manufacturing requires a good manufacturing practice (GMP)–qualified cell line growing in a GMP-compliant medium under GMP conditions. The cell line should be fully characterized and free of adventitious agents (mycoplasms, prions, endogenous retroviruses) and should not be tumorigenic/cancerogenic. Furthermore, cell substrates need to be accepted by the EMA (European Medicines Agency) or FDA (Food and Drug Administration) [10,11]. Therefore, cell lines are often used that have already been approved for other vaccines and have otherwise proven to be safe. This also greatly enhances the possibility to be "first on the market" or to establish a vaccine manufacturing in a country for self-supply. Together with the safety aspects discussed previously, this also explains the limited availability of new cell substrates and obtaining a license covering manufacturing for such fully characterized new cell lines is equally very expensive. Nevertheless, a license for a certain cell line that covers the use for manufacturing is essential unless a company's own cell line has been developed. More details are found in Chapter 5.9.

5.4.2 Cell Banking

One crucial step in the design of a vaccine manufacturing process is cell banking to ensure the cell line availability. Preferably, cell banking is done reproducible and under GMP conditions. Cryo tubes of cells are kept in liquid nitrogen for long-term storage. Freezing and thawing of cells are very critical steps to guarantee high cell survival and rapid growth. Therefore, appropriate protocols should be followed or developed. First, a master cell bank (MCB) is created consisting of several cryo tubes with cells of a low passage number. Based on the MCB, a working cell bank (WCB) is prepared mostly with up to four more passages. The cells from the WCB are then used for scale up with the final number of passages defined after comprehensive testing. Typically, cells are frozen in cryo tubes at 1.5 mL. By using disposable cryobags to freeze the cells in a higher volume (and higher cell concentration) a high cell count for direct inoculation of bioreactors is possible and with that the time to production scale can be reduced to allow for more efficient and flexible manufacturing (see next chapter). Comprehensive characterization of the cell banks is required for sterility (bacteria, viruses), absence of mycoplasm and retroviruses, cell identity, and genetic stability [10,11]. Furthermore, cell banks should be large enough to provide consistent access to cells for production over many years. For establishing the cell banks, cell freezing is conducted during exponential growth phase of the respective cells. To avoid the formation of ice crystals and osmotic stress, dimethyl sulfoxide (DMSO) (about 6–10% (v/v)) is added as cryoprotectant to the cell suspension. Typically, adherent cells are frozen at 1E06 cells/cryo tube and suspension cells at 1E07 cells/cryo tube. Freezing is conducted either stepwise (−20°C, −80°C) or gradually (−1°C/min, −80°C) by using a freezing container or dedicated equipment before aliquots are placed into the gas or liquid phase of nitrogen. For cell thawing, a cryo tube is thawed until only a small ice pellet is visible and the cell suspension is carefully transferred into a T-flask or spin tube/shake flask. A medium exchange after thawing is highly recommended to outdilute any cell toxic DMSO. Before release of a MCB and WCB, successful thawing and good cell growth of the thawed cells needs to be demonstrated for short and longer storage times. To avoid losses of cell banks, storage at two locations should be considered.

5.4.3 Preculture and Maintenance of Cells

To allow a continuous growth/maintenance, passaging of cells is necessary due to depletion of nutrients, accumulation of toxic by-products, or limitation of the growth surface (adherent cells). During cell passaging, old medium is withdrawn and cells are seeded in fresh medium at a lower cell concentration. The passage number describes the number of these subcultivations. Although the growth of continuous cell lines is not limited per se, a regulatory limitation exists. For cells that pose a potential tumorigenicity hazard, including Vero and MDCK cells, the maximum allowed passage number for production is often limited to 20. In contrast, new designer cell lines (e.g., PER.C6) grown in suspension are often very well characterized and can be used for up to 100 passages [12]. The cell line, starting from the working bank, undergoes several passages with increasing volume during seed train expansion until inoculation at production scale. Furthermore, cells can be maintained at a small volume, e.g., laboratory scale experiments or vaccine development studies. Adherent cells are typically passaged once or twice a week, shortly before they reach confluence. For passaging, the attached cells have to be detached by proteases (e.g., trypsin). Use of trypsin (activity/concentration, incubation time, pH) and inactivation of protease activity by addition of either serum or medium needs to be optimized for the respective cell line, medium and the cultivation vessel (e.g., T-flask, roller bottle, cell factory, microcarrier culture). A typical split ratio for adherent cells is 1:4-1:10. For suspension cells, typical cell concentrations in batch mode reach 2E06-10E06 cells/mL with doubling times of 20–30 h. Suspension cells will be subcultured every 3 and 4 or 2 and 3 days at the late end of the exponential growth phase and are seeded usually at a cell concentration of 0.5-0.8E06 cells/mL, while a small fraction of the cell suspension is filled up with fresh medium (split ratio: 1:4-1:20). Preferably, no antibiotics are added to the cell culture medium to avoid masking of contaminations. Some of these possible contaminations only become evident when switching to active aeration and use of antibiotics can potentially create resistances. Therefore, when using antibiotics, a clear strategy of changing antibiotics in a regular mode needs to be followed.

Offline measurement of cell concentration is either done via trypan blue dyes to discriminate between viable and non-viable cells (integrity of the membrane) or via crystal violet by staining of cell nuclei. Cells are either counted manually using a microscope or with dedicated cell counters. Cristal violet has the problem that some cells have multiple nuclei, which can result in overestimation of cell concentrations. The methods of cell counters need to be carefully set, so that cell aggregates or cell debris are correctly determined. The typical standard deviation in measuring cell concentration is 6–10% (but can easily increase up to 30% depending on the method and the experience in case of manual cell counting). With increasing cell concentrations, the needed dilution increases and with that the potential dilution error. For bioreactors, various online options are available. Turbidity or optical density measurements from optical sensors are used to follow cellular growth online. Although seemingly simple, these measurements have many limitations. First, they are extremely sensitive to the presence of bubbles, cellular debris, and other particles in the medium. Thus, their performance decreases over the process time. Another disadvantage is that changes in cellular size and morphology can strongly influence cell counts, limiting the validity of the results. Additionally, both methods do not allow to discriminate viable cells. In contrast, dielectric spectroscopy takes advantage of the unique electrical properties cells exhibit upon exposure to an electrical field. Cells behave as microscopic charge containers or capacitors. The changes in cell permittivity at different frequencies describe a sigmoidal curve designated as β-dispersion. At the inflection point of the curve, the permittivity measurement correlates with the viable cell concentration. This method is practically insensitive to the presence of non-cellular material or non-viable cells [13]. Capacitance probes consist of platinum electrodes embedded in a resin and housed in a stainless steel body connected to a pre-amplifier. The additional option to perform conductivity measurements is another useful feature that could be exploited to track changes in the culture medium that arise from by-product excretion (lactate, ammonium) or pH control such as addition of $NaHCO_3$ or NaOH. For their routine verification, signal simulators and electrolyte conductivity standard solutions are used, but calibration requires specialized service. Another option for cell growth monitoring is the use of online imaging systems that allow to carry out cell counts (total and viable cell concentration, cell diameter and viability) through differential digital holographic microscopy. For example, the iLineF systems consist of a single-use closed loop tube mounted in the bioreactor, where an integrated pump recirculates continuously the cell culture through the measurement chamber for image acquisition [14]. Finally, flow cytometry is increasingly used to obtain a deeper insight into the composition of cells populations. For instance, the use of this system allows online monitoring of cell size, apoptosis, and cell cycle [15].

Precise determination of the cell concentration and cell viability are key for process monitoring, optimization and control. It directly influences medium feeding to provide a sufficient supply of nutrients, and allows to determine the optimum time of infection (TOI) and the required ratio of the number of infectious viruses to the number of viable cells (MOI). Furthermore, many derived parameters calculated to evaluate process performance directly relate to cell concentration measurements. This concerns, in particular, the cell-specific virus yield (CSVY) or the cell-specific substrate feeding rate in perfusion cultures.

5.4.4 Adherent Versus Suspension Cells

As most animal cells used in research laboratories and for biologics manufacturing are tissue-derived, they will primarily grow as adherent cells and, thus, require a growth surface for proliferation. This surface can be plastic materials as for T-flasks (coated or uncoated, charged or uncharged) or carriers (porous or non-porous). Many viral vaccine production processes still rely on adherent cells due to ease of handling, high CSVY, well-established production facilities, and years of experience concerning regulatory approval. Moreover, recent improvements regarding packed-bed or fluidized-bed reactors or hollow fiber bioreactors (HFBR) allow cultivation of adherent cells with low shear stress at very high cell concentrations including options for medium replacement or feeding. Nevertheless, there is a shift towards suspension cells as scale-up and passaging is significantly easier, cell growth is rapid, and very high cell concentrations by perfusion cultivations can be achieved. Furthermore, suspension cell based processes have many advantages regarding online monitoring and control. Typical cell lines used for viral vaccine and vector production, including MDCK [16] and HEK293 cells [17], but also designer cell lines (AGE1.CR.pIX [18], EB66 [19], CAP [20]) were recently adapted to suspension growth under serum-free conditions for high titer production of viral vaccines.

Adaptation to suspension growth can be very tedious and lengthy. Moreover, it is not clear which of the various approaches proposed in literature will eventually achieve the desired result. Currently, it seems reasonable to start parallel adaptations with cell lines from different sources (i.e., cell culture collections), with different passage histories or in different media to increase the rate of success. One adaptation approach that is often used follows a two-step adaptation. Starting with adherent cells that are cultivated in serum-containing media, the serum content in the medium is first stepwise reduced (serum wheaning) by diluting with serum-free or chemically defined medium. Next, confluent cells are maintained in T-flasks by continuous refreshment of the medium over several weeks. By reaching a super-confluency state, cells start to form aggregates above the confluent layer and in the supernatant. These cell spheroids are then cultivated under agitation for several passages in spinner flasks aiming for single-cell growth in suspension culture (separation of small and large aggregates) [21,22]. Moreover, adaptation by a direct transfer of cells to a new medium might be successful [23]. Alternatively, suspension growth can be triggered by targeted transfection as shown for HEK293, AGE1.CR, and PER.C6 cells (Ad5 genes E1A and E1B, [24]) or MDCK cells (siat7e gene [25]). Finally, whatever approach was chosen, the stability of single-cell growth over several runs should be tested and the doubling time should remain between 20–30 hours.

For commercial application, suspension cells have been mainly used for the production of recombinant proteins or veterinary vaccines (BHK21, against foot and mouth disease [26] and rabies [27]). A major concern regarding the use of suspension cell lines for vaccine production for human use is traceability, risk of adventitious agents and tumorigenicity/cancerogenicity. Due to enormous progress made in methods to allow for rigorous cell line characterization, suspension cells have been established for human influenza virus production, e.g., Optaflu® (MDCK cells, Novartis) licensed in 2007 [28] or Flucelvax Tetra (MDCK cells, Seqirus), both currently available in Europe [29] and the United States [30]. Nevertheless, drawbacks of cultivation with suspension cells are the risk of cell aggregation and the requirement for cell retention devices for perfusion cultivation or the medium exchange prior to infection (minor challenge for adherent cells). However, scaling up of adherent cells is significantly more complicated and labor intensive. For adherent cells, the maximum cell concentration is restricted by the provided surface area, which must be increased during scale-up. Therefore, microcarrier systems were established to increase the surface/volume ratio. Here, cells grow on microcarriers, which are suspended by agitation in stirred tank bioreactors (STRs) or are used in packed or fluidized bed reactors. Microcarrier systems have been scaled up to 6,000 L for Vero cells [31]. For suspension cells, the concentration of metabolites is the restricting factor, thus, the volume of the vessel is of interest for scaling up. STRs using suspension cells could be scaled up to 10,000 L, but 2,000 L is the preferred size. Another drawback of adherent cell cultivation is the necessity for cell detachment when scaling up or during subcultivation, which can be realized via proteases such as trypsin. Moreover, cell counting for cell growth monitoring requires cell detachment from the carriers, which is difficult during a production run, while measurement for suspension cells is conducted offline by taking a sample via a dip tube. For adherent as well as suspension cells today online measurement of the cell volume/concentration via the permittivity of polarized cells is now possible, overcoming this drawback.

5.4.4.1 Cell Attachment and Use of Microcarriers

The adhesion of adherent cells to a surface is divided into two main phases as shown in Figure 5.2A [32]. After cell seeding, cells attach to the surface and are characterized by a round cell shape. To provide a suitable growth surface, flasks as well as microcarriers are often coated, as cell attachment is dependent on a

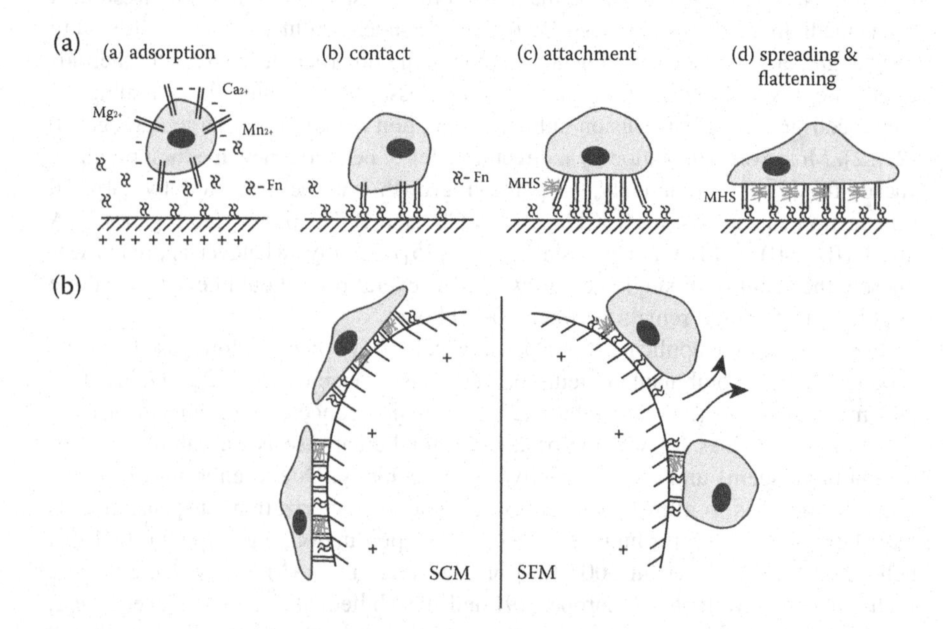

FIGURE 5.2 Schematic representation of cell adhesion (a) without shear stress (T-flasks), and (b) with shear stress on microcarriers either in serum containing medium (SCM) or in serum-free medium (SFM). Cells that do not flatten will be easily detached from the carrier surface due to the shear forces from stirring (indicated by arrows). Figure A adapted from [33]. MHS: multivalent heparan sulfate; Fn: fibronectin.

hydrophilic and charged surface [33,34]. Typical microcarrier materials are dextran, polyester and collagen and the surface is modified with DEAE (diethylethanolamine) or cations. Moreover, attachment factors are crucial to ensure the attachment, which are either provided by the serum or addition of exogenously fibronectin under serum-free conditions [33]. For optimal growth, homogenous cell attachment is required, thus, the ratio of cells to microcarriers needs optimization. As second phase, cell flattening occurs to enlarge the contact area for stable binding, a process called cell spreading [32]. By using serum-free medium, the cells tend not to flatten, resulting in easier cell detachment due to shear stress during agitation (Figure 5.2B). To help cell attachment, intermediate stirring can be tried, in which a short stirring for homogenization is followed by 30 minutes without stirring, repeated over about 2 hours in total. Moreover, using macroporous microcarriers, adherent cells are attached within the carriers reducing the shear stress by agitation and sparging [33,35]. Furthermore, volume can be reduced after cell seeding to increase the chance of cell surface contact.

5.5 CULTIVATION MEDIA AND METABOLISM

5.5.1 Media and Additives

The choice of cell growth medium is essential for a successful cultivation of animal cells and, consequently, the virus production. Media developed over many years now allow growth of suspension cells in batch mode up to 2E07 cells/mL and in perfusion mode up to 2E08 cells/mL. Traditionally, cell-culture–based viral vaccines are produced in serum-containing media (e.g., fetal bovine serum, FBS). This may be required even today as many adherent cells, and in particular primary cells, are the only host cells that can be used to propagate some viruses at high yield. However, serum-containing media pose numerous disadvantages for efficient viral vaccine production by complicating the purification process and therefore increasing the costs. Moreover, serum can be contaminated with adventitious agents and unwanted by-products (i.e., antibodies against the virus to be produced). In addition, the quality of serum is difficult to assess and batch-to-batch variation can cause significant inconsistencies in cell growth and virus yields. Thus, serum-free media were developed, where only particular serum components, including hydrolysates, amino acids, growth factors, hormones, lipids, adhesion factors, and other compounds are added at a defined concentration (see Table 5.3). This trend is moving towards the use of animal component-free (AFM) or even chemically defined media (CDM), where all ingredients are defined and, thus, contribute to process robustness. To enable the formulation of AFM variants of media, non-animal substitutes have been developed for the various supplements derived from serum (see Table 5.5). Today, the preferred media for production are chemically defined, free of animal components, and certified for GMP production. The adaptation to serum-free variants of medium typically follows two different protocols, the sequential adaptation (changing the ratio of serum-containing and serum-free medium) or the adaptation by using conditioned medium (also see "Adherent versus suspension cells").

TABLE 5.3
Categories of media used in animal cell culture

Abbrev.	Definition	Serum[1]	Hydrolysate[2]	Human-derived components[3]	Animal-derived components[4]	Peptides[5]
SCM	Serum-containing	Yes	Yes	Yes	Yes	Yes
SFM	Serum-free	No	Yes	Yes	Yes	Yes
AFM	Animal component-free	No	Yes	Yes	No	Yes
XFM	Xeno-free	No	Yes	Yes/No[6]	No	Yes
PFM	Protein-free	No	No[7]	No	No	Yes[8]
CDM	Chemically defined	No	No	Yes	Yes	Yes

Notes

[1] Serum: mainly fetal bovine or fetal calf serum (FBS or FCS) from traceable sources without the risk of prion contamination, batch-to-batch variations, human or equine sera may also be used.

[2] Hydrolysate from mechanical, chemical or enzymatic treatment of soy, yeast, or animal extracts, batch to batch variations, might still contain proteins.

[3] Typically recombinant human serum albumin

[4] Enzymes and/or growth factors that are not produced recombinant.

[5] Protein hydrolysates.

[6] Depends e.g., on cell line origin; human components may be required for a human cell line.

[7] May contain hydrolysates only in special cases, then hydrolysates should be protein-free.

[8] Includes amino acids or di-/tripeptides, but not polypeptides or proteins [36].

Yes: is allowed to be part of the medium, No: is not allowed; some suppliers differ in definition of composition, especially regarding peptides.

Media development can be considered an art of itself. Besides medium composition, a major requirement for their use is sterility, meaning the absence of microorganisms (bacteria, yeasts, fungi), which is ensured mainly by sterile filtration. Moreover, to allow optimal cell growth and virus production, media should be buffered to provide a physiological pH value. Often a bicarbonate buffer system is applied, where sodium bicarbonate ($NaHCO_3$) is added to the medium, while carbon dioxide (CO_2) can be provided via the inlet air of the incubator or cultivation vessels. Another key requirement of the medium is the adequate supply of nutrients, such as glucose or glutamine, which are necessary for cell survival and proliferation. However, each cell type, along with the desired product, has specific component and concentration requirements. Hence, the medium formulation needs to be adjusted and optimized very carefully. Moreover, care should be taken during media development to restrict the accumulation of unwanted, metabolic by-products including ammonium and lactate. Therefore, design of experiment approaches are frequently used to maximize the specific growth rate and virus yields. As media formulations often comprise more than 40 compounds, such as inorganic salts, amino acids, vitamins, growth factors, sugars, co-factors, and metals and some of these components are soluble at different pH, the order of addition of some compounds is important (see Table 5.4). In a few cases, certain components are not stable (e.g., glutamine, only 1–2 months of shelf life, degradation with side-product ammonium which may inhibit cell growth and virus production) and are, thus,

TABLE 5.4
Specific media components, typical concentrations, characteristics, and functions

Name	Concentration	Function	Characteristics
A			
Glucose (gluc)	10–40 mM	Substrate, growth and energy metabolism	Conversion to lactate: growth inhibitor & pH decrease
Glutamine (gln)	2–8 mM	Substrate, growth and energy metabolism; decomposes to pyrrolidone-carboxylic acid and ammonia with time	Conversion to ammonia: growth inhibitor, inhibition of virus replication
Amino acids	0.2–2 mM	Protein synthesis, growth	Some: pH value decrease; some: ammonia release
Keto-acids (pyruvate)	1–4 mM	Growth	Additional carbon source
Dipeptides, ala-gln (GlutaMAX)	2–6 mM	E.g., more stable supply of glutamine, reduced ammonia release	Using the di-peptide ala-gln gln monitoring to follow substrate uptake is not straightforward
Glutathione	0.5–1 mg/L	Protects from oxidative stress, redox environment control, facilitates protein secretion	Tripeptide
Lipids (e.g., cholesterol)	0–10 mg/L	Membrane synthesis including viral membranes	Plant or animal components
Insulin	1–20 mg/L	Promotes glucose and amino acid uptake	Rec. or animal component
Vitamins (folic acid, niacinamide, i-inositol, thiamine, ...)	2–7 mg/L	E.g., activates pyruvate metabolism, precursors of various cofactors, antioxidant effects	Present in most of the basal media but adjusted to cell types
B			
Hypoxanthine	1–5 mg/L	Intermediate in DNA synthesis, energy metabolism	Purine derivative
Thymidine	0.2–0.7 mg/L	Intermediate in DNA synthesis	Deoxy nucleoside

(Continued)

TABLE 5.4 (Continued)
Specific media components, typical concentrations, characteristics, and functions

Name	Concentration	Function	Characteristics
Albumin		Antioxidant and growth factor carrier	Rec. or animal component
Fetuin		Cell attachment	Plant or animal component
Fibronectin		Cell attachment	Animal component
Transferrin	7.5 mg/L	Antioxidant, binds iron	Rec. or animal component
Alpha2-macroglobulin		Trypsin inhibitor	Rec. or animal component
Trypsin		Cell detachment, reduces cell aggregation, supports virus entry	Rec. or animal component
C			
Endothelial cell growth factor (ECGF)		Mitogen	Rec. or animal component
Epidermal growth factor (EGF)		Mitogen	Rec. or animal component
Fibroblast growth factor		Mitogen	Rec. or animal component
Insulin-like growth factors (IGF-1, IGF-2)		Mitogen	Rec. or animal component
Platelet-derived growth factor (PDGF)		Mitogen & major growth factor	Rec. or animal component
Interleukin-1 (IL-1)		Induces IL-2 release	Rec. or animal component
Interleukin-6 (IL-6)		Promotes differentiation	Rec. or animal component
Hydrocortisone		Promotes attachment & proliferation	Steroid
D			
Iron, copper, zinc, selenium	Traces	Enzyme cofactors	Metals, minerals
E			
Chlortetracycline	5 mg/L	Gram-pos. and Gram-neg. bacteria	Inhibitor of protein synthesis
Gentamicin sulfate	50 mg/L	Gram-pos. and Gram-neg. bacteria, mycoplasmas	Inhibitor of protein synthesis
Nystatin	50 mg/L	Fungi and yeast	Inhibitor of cell membrane function
Penicillin G	100 U/mL	Gram-pos. bacteria	Inhibitor of cell wall synthesis
Spectinomycin	20 mg/L	Gram-pos. and Gram-neg. bacteria	Inhibitor of protein synthesis
F			
2-Mercaptoethanol	0–2 mg/L	Reduce disulfide bonds, antioxidant	
Inorganic salts	0.1-6 g/L	Buffer and osmolality	

TABLE 5.4 (Continued)
Specific media components, typical concentrations, characteristics, and functions

Name	Concentration	Function	Characteristics
G			
Pluronic F68	0.1 %	Shear protectant, anti foam	
FoamAway	3%	Anti foam	Sterile, ready-to-use, pre-diluted to 3%
Poloxamer 188 and others	0.5-2 g/L	Shear protectant	

A: compounds involved in cellular metabolism.
B: proteins.
C: growth factors.
D: metals, minerals.
E: antibiotics (antibiotics should, if used at all, be changed on a regular interval to avoid developing antibiotic resistance).
F: others,
empty field: no data found.
G: shear protectants, some examples are given.
Data provided based on homepages of Thermo Fisher Scientific, Sigma-Aldrich, Gibco, Xell, and other media suppliers [38,39].

solubilized and added by the customer only shortly before the start of the cultivation. Moreover, some media are very fragile to degradation by light. Hence, containers used for storage should be light-protected. For CHO cells, many media supporting efficient growth and high-product yields are available today. This is in strong contrast to media available for many cell lines used in vaccine production. For the latter, there is still much room for optimization. This concerns not only growth properties but also metabolic pathways, which modulate virus production. Many effects of media supplementation have been identified including changes in lipid-, energy-, nucleic acid-, oxidative stress-, and polyamine metabolism, protein processing and post-translational modification as well as control of cell cycle and apoptosis. For instance, an increase in infectious virus titers of up to sixfold through additions of various supplements has been reported by Rodrigues et al. [37].

Media formulation is additionally constrained to osmolality. The osmolality of the medium refers to "the concentration of osmotically active particles in that solution" [40]. Human blood plasma has an osmolality of approximately 290 mOsmol/kg. To mimic physiological conditions, cell culture media are adjusted to values of 260–330 mOsmol/kg. However, most continuously growing mammalian cells can show a very wide tolerance up to 500 mOsmol/kg [41]. Yet, cells will react to the respective osmolality of the medium by either shrinking (hyperosmotic, >330 mOsmol/kg) or swelling (hypoosmotic, <260 mOsmol/kg). During the cultivation, osmolality is changed by addition of salts, release of metabolites (lactate), feeding, and addition of buffers. Furthermore, monitoring of the osmolality can be

TABLE 5.5
Non-animal substitutes of media components

Constituent	Animal source	Non-animal source
Insulin	Bovine/porcine pancreas	Bovine or human rec. from *E. coli* or yeast
Transferrin	Bovine, porcine, or human plasma fraction	Inorganic iron carriers/chelates
Serum protein fractions (e.g., albumin, fetuin, lipoproteins)	Bovine or animal serum	Lipid-delivery alternativesPlant-derived hydrolysates
Protein hydrolysates	Lactalbumin, peptone, casein	Plant-derived hydrolysates
Lipids/sterols	Ovine/human cholesterol;piscine lipids, porcine liver	Plant-derived sterols;Synthetic and plant-derived fatty acids
Growth and attachment factors	Murine/bovine organ digests	Rec. factorsCollagen precursors
Amino acids (e.g., tyrosine, cyst(e)ine, hydroxyproline)	Human hair; avian feathers, bovine collagen; bovine/porcine bone gelatin	Synthetic (fermentation) or plant-derived amino acids
Surfactants (e.g., Tween™ 80)	Bovine tallow	Plant-derived polysorbate
Dissociating enzymes (e.g., trypsin)	Porcine pancreas	Plant-derived enzymesMicrobial enzymes
Albumin	Bovine	Rec. from bacteria and yeast

Table from [33].

very useful as quality control of the medium or to verify consistency between lots. Moreover, determination of cell size/cell volume via cell counter or capacitance probes should consider that cells do not only change in diameter due to osmolality, but also due to cell cycle phase, metabolic state or cell lysis (virus production).

5.5.2 Monitoring Cell Metabolism

In process development, cell growth and metabolism needs to be captured closely. High cell concentrations can only be achieved, when nutrient supply is ample and neither metabolites nor other toxic by-products are approaching inhibiting levels. Furthermore, the pH of the medium should be controlled in a range of pH 7.0–7.6, and oxygen supply should exceed 40% pO_2. Moreover, agitation speed and stirrer type affects cell growth. In small-scale cultivations offline measurement of the most important metabolites (glucose, glutamine, lactate, ammonia), pH, and cell concentration is often enough to optimize cell growth. When envisioning large-scale cultivation or cultivation at very high cell concentrations, the concentration of amino acids or other media compounds may also be monitored. In Figure 5.3, main metabolite concentrations of a typical batch cultivation are given together with typical stoichiometric ratios.

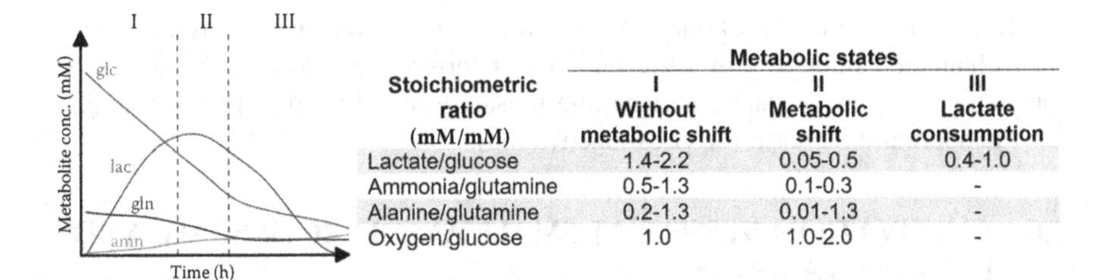

Stoichiometric ratio (mM/mM)	Metabolic states		
	I Without metabolic shift	II Metabolic shift	III Lactate consumption
Lactate/glucose	1.4-2.2	0.05-0.5	0.4-1.0
Ammonia/glutamine	0.5-1.3	0.1-0.3	-
Alanine/glutamine	0.2-1.3	0.01-1.3	-
Oxygen/glucose	1.0	1.0-2.0	-

FIGURE 5.3 Typical time course for metabolites of animal cells in batch culture (left) and typical stoichiometric ratios for different metabolic states (right) (adapted from [39]). When the availability of glucose or glutamine is limited, some cell lines shift their metabolism towards lactate consumption [42]. In state I, a high uptake rate of glucose and glutamine is necessary for rapid cells proliferation. In states II and III, along with lactate, glucose, and glutamine are consumed, but at lower uptake rates than in state I, mainly for cell maintenance [43]. Glc: glucose; gln: glutamine; amn: ammonia; lac: lactate.

Online measurements of glucose, lactate, glutamine, and glutamate are possible through miniaturized versions of the well established electrochemical enzyme-based sensors [44]. In addition, it is common that stand-alone analysis systems are coupled to sampling units [45]. The measurements are often based on the hydrogen peroxide concomitantly produced upon oxidation of the metabolites catalyzed by membrane immobilized specific oxidases. The oxidation of the hydrogen peroxide on a platinum anode generates a nanoamperometric signal proportional to the metabolite present in the sample. Potassium, sodium, and ammonium ions are measured by ion selective sensors. Examples of current commercially available systems for animal cell culture are the YSI 2940-2980 Multichannel Online Monitor, the Sartorius BioPAT® Trace (currently limited to glucose and lactate) or the Bioprofile Flex2 online autosampler from Nova Biomedical. These systems are also capable to monitor multiple bioreactors. These measurements have enabled process control in high cell density cultures to adjust perfusion rates to maintain constant glucose and lactate concentrations [46]. In addition, online chromatography systems are available that are mainly used in research laboratories. For example, using anion exchange chromatography and amperometric detection, glucose and 19 amino acids were measured in CHO cell cultures [47]. In comparison to chromatography methods, enzymatic methods for the relevant metabolites in cell culture require less time to generate specific results, do not require exhaustive sample preparation and are less prone to interferences. Clearly, all the methods addressed above require either manual sampling or automatic sampling systems. These consist of pumps and a valve system that withdraws a cell-free sterile sample (through sampling probes suited either with filter or dialysis elements) from the bioreactor and dilutes it before injection either into a sampling port or onto a column. In addition, there is a growing number of non-invasive technologies that are based on spectroscopic methods (near-infrared spectroscopy [NIR], mid-infrared spectroscopy [MIR], fluorescence), but still need considerable effort to fully exploit their potential, especially regarding data mining and data analysis [48,49]. Nevertheless, these techniques seem to constitute the future in bioprocess monitoring and control. For example, online NADH and NADPH cofactor measurements have been obtained, although still suffering from interferences. When analyzing virus containing samples, precautions regarding biosafety might need to be taken (e.g., a short temperature increase for virus inactivation).

5.6 CULTIVATION VESSELS AT SMALL AND LARGER SCALE AND PROCESS VARIABLES

5.6.1 Cultivation Vessels

Cultivation vessels, from well-plates and T-flasks to STR, are systems where the substrates in the culture medium are converted into the desired product by biochemical processes that take place inside the cells. They must provide the appropriate conditions for cell growth and product synthesis [50]. Different choices of static and dynamic cultivation systems including bioreactors are available for adherent and suspension growth adapted cells. This section presents an overview of the cultivation vessels used for virus production, their major features, the impact of some process variables and some considerations for their scale-up.

For adherent and suspension growth adapted cells different vessels at different sizes are available (Figure 5.4). Static bioreactors or vessels provide a surface pretreated to confer specific surface properties for attachment of adherent cells through some coatings over e.g., a polystyrene matrix [51]. Their scale-up is based on the available cell culture surface [52]. Aeration takes place through the gas-liquid interface by diffusion from the air contained in the vessel. Initially, in screw-capped vessels, partial twist of the cap allowed some gas exchange to the outside of the vessel. Today, this is done via membrane-filtered vent caps. Furthermore, notches for dishes, gas-permeable membranes for multi-plates, and multitray stacked systems

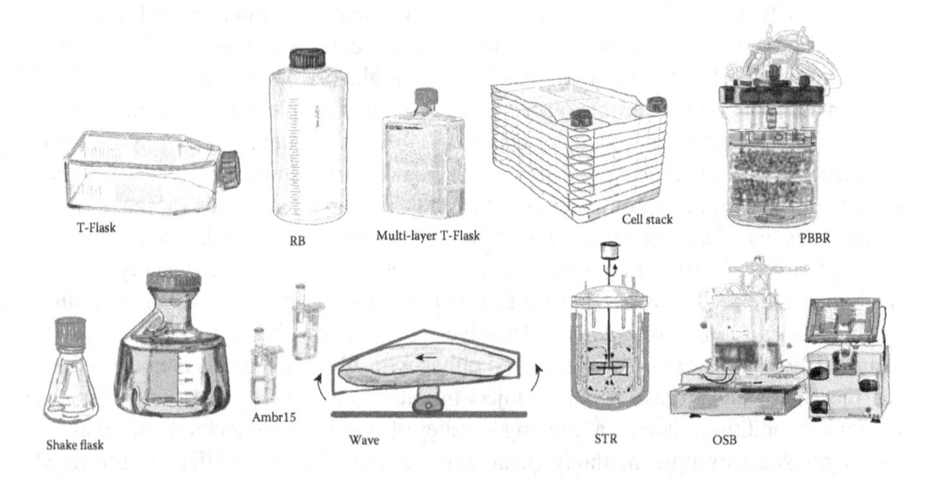

FIGURE 5.4 Overview of cultivation vessels for adherent (top) and suspension (bottom) cells. RB: roller bottle; PBBR: packed bed bioreactor; STR: stirred tank bioreactor; OSB: orbital shaken bioreactor.

were introduced. Thus, additional air exchange with the outside of the vessel is realized and with CO_2 incubators the composition of this air can be controlled. These vessels often require accessory equipment such as incubators or roller bottle racks to accomplish e.g., temperature and CO_2 control.

Multiwell-plates and dishes are used mainly for cell screening and discovery stages, but also for cell-based assays such as virus titrations. Dishes consist of a flat bottom and a cap commonly round with surface areas from 8.8 cm^2 up to 145 cm^2. T-flasks are the workhorse for cell culture expansion in laboratories. They come in different sizes from 25 cm^2 up to 175 cm^2 surface. Roller bottles and stacked multitrays achieve further expansion at low scale (>850 cm^2; Table 5.6). Although some systems, like roller bottles, can sustain medium mixing and gas diffusion, the lack of control options for pH and pO_2 can limit their use. Systems with perfusion options circumvent these limitations. Beyond tens to hundreds m^2 scales, adherent cell culture is carried out in packed-bed reactors, either in fixed or flow mode. In CelCradle and TideXcell (Esco VacciXell), aeration takes place by the direct intermittent exposition of the carriers in the vessel´s headspace that is replenished with fresh air, created by the alternating up and down media flow. In iCellis (Pall), CelliGen/BioBLU (Eppendorf), and scale-X (Univercells technologies) vessels the medium is moved and distributed through the packed bed or sheets. Clearly, at larger size and volume, manual handling of these vessels becomes difficult and automation is often available, like harvesting systems that supply reagents and shake vigorously the vessel for in-situ cell detachment.

For suspension cells, preculture is mainly done in shake flasks. Thus, passaging and scale-up is promoting the growth of cells that are perfectly adapted to shaken mode and head space aeration. These cells are then transferred to a STR where they need to adapt to stirring and (conventionally) additional aeration through spargers. The gas supplied is either air, a mixture of nitrogen, oxygen and carbon dioxide, or pure oxygen. In addition, the pH is controlled by adjustment of gases (in particular CO_2) and addition of buffers, acids (HCl) or bases (NaOH or Na_2CO_3). Thus, for the cells many parameters change and often this results in a decreased growth rate after transfer into STRs. Since many years, the impact of this switch is widely discussed. As alternatives, small-scale Ambr15 cultivation system (15 mL working volume (wv)) could be used for cell line development and screening under stirred conditions. Likewise, orbital shaken bioreactors (OSBs) allow to continue cultivation in shaken mode up to a 2,500 L scale.

For adherent cells, e.g., growing on Cytodex 1 microcarriers, very low stirrer speeds are recommended. Especially, if serum-free medium is used that does not support cell attachment as good as serum containing medium, selection of the proper stirring speed might be crucial for process performance (see also Figure 5.2). Typically, 60 to 100 rpm are used to achieve a balance of shear stress resulting in detachment of cells from the carriers and a high enough agitation to keep the microcarriers in motion and prevent settling to the bottom. During the initial phase of cultivations (first 1–3 h), intermittent stirring (stirrer 5 min on, 30 min off) might be useful to support cell attachment. For suspension cultures, stirring speeds can be higher depending on the cell line and the cell concentration. Here, 100–200 rpm are often described. For all cells, together with the aeration mode (head space,

TABLE 5.6

Overview on static cultivation vessels for adherent (and suspension) cells

Vessel	Supplier	Area (cm^2)	Average cell yield	Working volume (L)	Cells/volume (normalized to T175)	Growth support matrix	Control options	Perfusion options
T175	e.g., Corning	1.8E02	1.8E07	0.05	1.0	Bottom inner wall	Offline	No
RB	e.g., Corning	8.5E02	8.5E07	0.26	0.9	Inner wall	Offline	No
		1.8E03	1.8E08	0.53	0.9			
CellStack	Corning	6.4E02	6.4E07	0.2	0.9	Trays	Offline	No
		2.5E04	2.5E09	7.6	0.9			
HYPERFlask	Corning	1.7E03	1.9E08	0.6	0.8	Trays	Offline	No
HYPERStack	Corning	6.0E03	4.0E09	1.2	9.3	Trays	Offline	No
		6.0E04	4.0E10	12.0	9.3			
Cell Factory	Nunc	6.3E02		0.2		Trays	Offline	No
		2.5E04		8.0				
CellCube	Corning	8.5E03	8.5E08	0.6	3.9	Plates	Offline	Yes
		8.5E04	8.5E09	6.0	3.9			
Tide Motion CelCradle	Esco VacciXell	1.6E04	5.5E09	0.5	30.6	Fluidized bed macrocarriers	Offline	Alternate submerging
TideXcell	Esco VacciXell	3.2E05	7.0E10	2.0	97.2	Fluidized bed with matrix cassettes macrocarriers	Off- and online control	Yes
		1.5E06	3.5E11	20.0	97.2			
		1.6E07	3.5E12	100.0	97.2			
CelliGen BioBLU 5p	Eppendorf	3.0E04	2.5E10	5.0	13.9	Fixed bedmacrocarriers	Off- and online control	Yes
		6.0E05	5.0E11	50.0	27.8			
iCellis	Pall	5.3E03	1.5E09	1.0	4.2	Fixed bedmacrocarriers	Off- and online control	Yes
		5.0E06	1.5E12	70.0	59.5			
Scale-X	Univercells technologies					Fixed bed sheets of non-woven PET fabric	Off- and online control	Yes
Hydro		2.4E04		0.8				
Carbo		3.0E05		3.3				
Nitro		6.0E06		60.0				

Adapted from [2]. Additional information was included from the websites from Univercells technologies and Esco VacciXell.

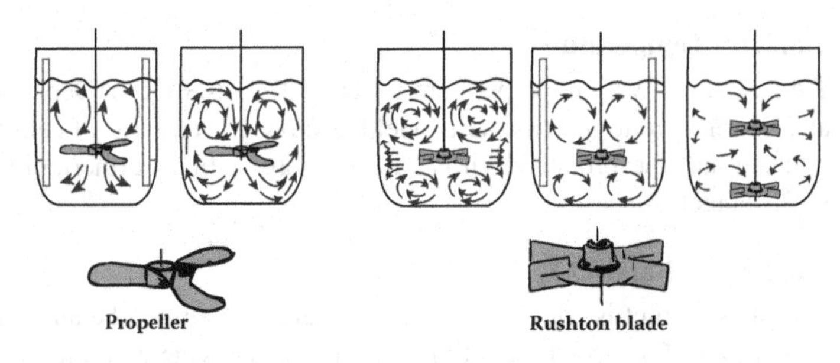

Attributes	Axial	Radial
Gassing	Less suitable	Highly suitable
Dispersing	Less suitable	Highly suitable
Suspending	Highly suitable	Less suitable
Blending	Highly suitable	Suitable
Baffles	Highly suitable	Less vortex, less swirl
Shear	More shear	More shear
Other types	Marine impeller, pitched blade	Scada, Rushton 4-blade, pitched blade

FIGURE 5.5 Possible flow of impellers: axial or radial flow. Adapted from: [53].

spargers), gas flow rate, stirrer configuration, and stirrer type and use of pumps (perfusion mode) optimal mixing conditions need to be found.

When multiple impellers are mounted on the stirrer shaft of larger bioreactors the respective types can be combined and the direction of mounting the impeller also needs to be considered. For bioreactors with baffles stirring will likewise need to be adapted. In small bioreactors, some sensors or tubes might even function like baffles.

As shown in Figure 5.5, flat blade or rushton turbine, pitched blade, marine impeller, axial flow blade, or propeller three blade can be used. Depending on the blade type either axial or radial flow will develop.

In spinner flasks (non-controlled small-scale STRs), especially for microcarrier cultivation, additional stirrers are used, such as pendular stirrers or stirrer units that are covered by silicone tubings for bubble-free aeration. Also paddle impellers and plough impellers can be obtained, particularly in disposable bioreactors. In all STRs, in particular glass and stainless steel bioreactors, the position of the impellers and stirring direction needs to be adapted to the filling height and cultivation need. Disposable stirred tank bioreactors are typically preconfigured and cannot be modified as easy. Furthermore, it seems not always trivial to position the stirrer shaft correctly in larger bags (the largest disposable STR currently available is 5,000 L (Thermo Fischer Scientific S.U.B.)). This issue could be avoided by using OSBs that are available up to 2,500 L.

5.6.2 Process Variables

Physicochemical variables play an important role in cell culture performance and virus production. Options for monitoring and control depend strongly on the built-in capacities of the bioreactor systems.

5.6.2.1 Temperature

Typically, cultivation temperature is set to 37°C (<±1°C)). Decreasing to 32–34°C at TOI can help to increase virus yield as enzyme activities of cellular proteases are reduced and the produced virus can be more stable. Furthermore, oxygen supply will change.

5.6.2.2 pH and CO_2

The pH set point is equally important. There will always be an optimal pH for cell growth, one for virus production and one for enzyme activities such as trypsin or other important enzymes for virus binding for example. Thus, different pH regimes during the different process phases might be accepted. Typically, the pH value is controlled in the interval of 7.2–7.4. Lactate and ammonia produced by the cells exert the major changes in cell culture supernatant, which are buffered preferably by a CO_2/bicarbonate system. For this buffer system, there is a very fragile balance between aeration rate, CO_2 addition, choice of sparger, stirring speed and head space aeration that needs to be fine-tuned for each filling height of the STRs. However, also for static bioreactors in CO_2 incubators, the CO_2 and $NaHCO_3$ buffering has to be considered, as pH decreases and increases occur when a T-flask is taken in or out of the incubator, respectively. Equilibration of fresh medium in the incubator might take as long as 1 hour Increasing CO_2 set point of the incubator from e.g., 5 to 10% can reduce the pH by up to 0.5.

To increase the pH value often either CO_2 supply is decreased or a base is added. Typically, this base is $NaHCO_3$. Addition of more $NaHCO_3$ to a medium that already contains up to 12 mM will increase the pH sharply (up to 1 log unit). In contrast, with $NaHCO_3$ concentrations exceeding 12 mM, further addition of $NaHCO_3$ will not increase pH significantly anymore. Often, NaOH is used as an alternative for pH control, but it is not well tolerated by many cell lines. Thus, both options should be evaluated at small scale. Intracellular pH is often overlooked but of great relevance for cell physiology. As long as the extracellular pH is above 7.2, the intracellular pH value is typically lower than the extracellular value [54].

5.6.2.3 Shear Stress

Cells, but equally virus particles, are sensitive to shear stress. Shear stress can be understood as a force applied tangentially on a fluid element at rest over a static surface that deforms the fluid element parallelly through planes slippage. Newton's first law relates the velocity gradient and the shear stress as follows [55]:

$$\tau = \eta \cdot \frac{\partial u_x}{\partial y} \tag{5.1}$$

where τ [Pa] is the shear tress, η [Pa.s] the dynamic viscosity and $\frac{\partial u_x}{\partial y}$ [s^{-1}] the velocity gradient. The resulting flow has a characteristic velocity profile. Two general flow regimes are distinguished depending on the hydrodynamic properties of the fluid and the flow conditions. Low shear stress conditions are found in laminar flow regimes, where the fluid flows in layers in one direction. Increasing shear stress turns this flow into a turbulent flow; the fluid moves randomly in all directions but yields a flow in one direction with an average velocity. Turbulent flow regimes are typically associated with higher shear stress.

Shear stress occurs in a bioreactor by stirring or when air bubbles disrupt. Likewise, high shear stress conditions can occur, when cells or virus particles are pumped through tubings. Peristaltic pumps are often used for such pumping actions for example for inoculum or virus addition or harvesting. But peristaltic pumps can cause very high shear stress that often results in cell damage or cell death. An alternative option is, e.g., the use of magnetic levitating pumps from Levitronix. Furthermore, for the transfer of all liquids containing cells or virus particles, the flow rate and inner tube diameter should be well chosen. In particular, switching between diameters can cause additional shear (equally the transfer through connectors). For vessel-to-vessel transfers, tubes should end as close as possible to the liquid level to avoid further cell damage.

5.6.2.4 Oxygen Supply

Lastly and of great relevance is the oxygen supply. Oxygen is an essential substrate for the cell, due to its crucial role as last electron acceptor within the respiratory chain. However, for their cultivation, cells are submerged in an aqueous liquid phase. Since O_2 is a gaseous non-polar molecule, its solubility is extremely low, around 7 mg/L at 37°C and 1 atm [56]. The biomass, salts, sugars, and other complex components in conventional culture media lower O_2 solubility in an additive way as a function of their concentration, so it can be up to 30% lower in comparison to its solubility in water at the same temperature and pressure [57,58]. To make O_2 available to the cells in culture, it has to be transferred from the gaseous to the liquid phase, which is the limiting step in this process. The overall oxygen transfer rate (OTR) is described by the following expression:

$$OTR = k_L a \left(C^*_{O_2} - C_{O_2} \right) \tag{5.2}$$

where $k_L a$ [h^{-1}] is the volumetric mass transfer coefficient, $C^*_{O_2}$ [mM] is the oxygen saturation concentration and C_{O_2} [mM] the actual oxygen concentration.

As long as the O_2 demand increases during the cell growth phase or early virus production phase (before virus-induced apoptosis and cell death), pO_2 decreases progressively [56]. Below a threshold value (about 5% pO_2), animal cells reshape their metabolism to cope with their energy and reducing power needs. Under this scenario, it is said that the cell culture is O_2 limited. In industrial scale bioreactors, O_2 limitation not only is a function of time, but also of space in case of the inefficient mixing. The latter results in a poor dispersion of air bubbles, so that regions with different O_2 concentrations and OTRs can appear [59]. The heterogeneities to which the cells are exposed to and that are magnified with the increase in scale, are often responsible of their low process performance.

Maximizing the OTR in bioreactor design and operation might prolong cells' productive time, from the high-throughput screening systems to the tenths m^3 production bioreactors. The fundamental guidelines in the strategies to increase

OTR in bioreactors rely on improving mixing and gas dispersion as well as in increasing O_2 solubility.

To increase the oxygen transfer, the following strategies can be used:

a. *Increase the partial pressure of O_2 in the gas fed to the STR*
According to Henry´s law, O_2 concentration in liquid phase is directly proportional to the O_2 partial pressure in the gaseous phase. Therefore, increasing the O_2 content in the aeration gas stream will allow more O_2 to be dissolved. Thus, aeration can be carried out either with air (21% pO_2) and CO_2 or with N_2, pure O_2 and pure CO_2. The use of enriched O_2 mixtures has allowed increasing cell density [60,61]. However, high pO_2 concentration might trigger oxidative stress responses [62,63].
b. *Pressurization*
Increasing the pressure in the bioreactor increases pO_2 and oxygen solubility in the culture medium. This allows a more efficient use of the volumetric power input that is reflected at high OTR (Knoll et al., 2005). A disadvantage is that dissolved CO_2 can reach inhibitory concentrations, since it is five times more soluble than O_2 [58].
c. *Decreasing culture temperature*
Although mammalian cells are often cultivated at 37°C, decreasing temperature to 30°C increases O_2 availability in the culture medium up to 10%.

Increasing the k_La value can be achieved as follows:

The volumetric mass transfer coefficient k_La in equation 5.2 is a key factor that determines the OTR. It is strongly influenced by aeration rate, sparger type, properties of the culture medium, mixing and vessel geometry including baffles, vessel pressure, and gas composition.

a. *Gas phase dispersion*
Most spargers are tubular structures, generally ring-shaped with drilled holes through which the gas stream is dispersed. The greater the number and the smaller the diameter of the holes, the smaller the size of the bubbles and consequently the larger the gas-liquid contact surface area. The latter applies, in particular, for the use of sintered steel or ceramic microspargers that allow for very small bubble sizes and contribute to an increased gas hold-up and bubble residence time [64]. However, this extent of gas dispersion can be critical for shear stress sensitive cells [65,66]. The geometry, size, and position of the impellers, as well as the sparging rate, influence importantly gas dispersion in terms of hold-up [67]. The composition and the flow rate of the gas input controls the amount of oxygen supplied, but can also have the disadvantage that it pushes CO_2 out of the liquid and with that increases the pH value [68]. Gas sparging results undesirably in foaming. Foam formed from medium compounds or proteins released into the extracellular medium can trap cells and could eventually clog vent filters resulting in an unwanted termination of a

production run. Foam buildup can be counteracted mechanically through devices attached to the stirrer´s shaft or by the addition of surfactants (antifoam agents). The latter, however, can impact negatively the pO_2 control as well as subsequent clarification and DSP steps. For high cell density (HCD) cultivations exceeding 4E07 cells/mL it might even be required to use two spargers to provide appropriate aeration for cell growth. Using pure oxygen instead of air again can help to reduce foaming as the flow rate can be lowered. However, aeration with pure oxygen results in a CO_2-free atmosphere, thus, favoring pH increase. In contrast, if pH decreases too fast it might be favorable to aerate with pure oxygen [54].

b. *Surface aeration and non-contact systems*
Bubble-free aeration systems aim to supply oxygen to shear sensitive cell cultures. In surface aeration, oxygen transfers from the headspace to the culture through the gas-liquid interface area in either OSB or STR, where vessel geometry, agitation speed, and working volume influence its rate [69]. In addition, O_2 supply can be enhanced through immersed silicon tubing aerators, owing to the relatively high O_2 permeability in this material. Unlike surface aeration, the transfer rate is independent of the stirring speed due to the fact that the main mass transfer resistance relies on the silicon tubing wall. By selection of an adequate length and wall thickness of the silicon tubing as well as the volumetric flow rate and the O_2 content in the flow stream, these aerators can achieve oxygen transfer rates similar to those of gas spargers [63].

c. *Mixing*
Short mixing times for dispersion of either gas or nutrients is especially challenging at larger size of the bioreactor vessels. As this often cannot be achieved by faster stirring, increase of turbulence and reduction of stagnant zones is often obtained through vessel and impeller design [70]. In general, the use of radial flow impellers, such as the Rushton type, results in a better dispersion of the gaseous phase at the expense of increased power consumption and lower mixing efficiency [71]. The use of two or more impellers increases mixing efficiency depending on their spacing along the shaft [72,73]. When radial and axial flow impellers are combined, lower power consumption and mixing times are obtained. However, a configuration of three 6-blade Rushton impellers spaced with a distance equal to the impeller diameter, can deliver the highest k_La per delivered power unit [74]. The considerable progress and refinement in computational fluid dynamics (CFD) offers options for an a *priori* evaluation of the mixing system. For example, Gelves and coworkers compared the performance of a typical Rushton impeller system against novel pitched blade impellers with rotatory microspargers through CFD mass transfer and hydrodynamic modeling. Experimental and computational results confirmed an increase a 34-fold increase in k_La with a 50% saving in power in comparison to the conventional system used as a reference [75].

5.6.3 Choosing a Production Vessel

Once the decision on the host cell line is made and the number of vaccine doses required is known, the final volume of the production vessel can be chosen. Adherent cells will need an attachment surface and large-scale production in STR requires the use of microcarriers. In the case that packed-bed reactors are considered, macrocarriers can be chosen. Nevertheless, some adherent cell lines are difficult to expand in such systems, especially when serum-free or chemically defined media are required. Here, the use of roller bottles or cell stacks should be considered. Handling of such systems with all their drawbacks can be cumbersome, but for some vaccine production processes, especially where multiple harvesting allows to dramatically increase virus yields, the cultivation of adherent cells can actually be advantageous. Finally, many factors decide on the selection of production vessels: previous experience with similar manufacturing challenges, size of the facility and equipment available, investment and running costs for equipment, utilities (electricity, water, steam, cooling, etc.) and media, regulatory requirements, prizing of a vaccine, production needs, location, speed to market, and qualification of staff.

With the change of using classical stainless-steel STR with up to 20,000 L to different disposable cultivation vessels in static or mixing mode, additional decision criteria have come up. If parts of the production stream are disposable these parts will be part of a supply chain. To be able to provide enough vaccine doses, this supply chain needs to be guaranteed and a back-up solution needs to be available. All suppliers must be audited and certified and should be able to guarantee a supply for at least the next 10 years. Delivery to all production facilities needs to be guaranteed and quality needs to be constant and consistent throughout different batches. Just imagining that the disposable bioreactor vessels of choice are not available, because they have been taken from the market or the company selling them is not existing anymore. Or a change of supplier for the production of the vessel is needed (plastic foil) and then the quality cannot be met anymore. Such risks need to be well-considered and corresponding concerns might, of course, slow down innovation at some point. Alternatively, such technology should then be integrated into the manufacturing company or large batches of these vessels need to be bought and for this, storage capacity needs to be considered. Nevertheless, this explains also why currently many of these single-use technologies and new bioreactor designs are collected together at only a few companies.

5.7 VIRUS PRODUCTION AND PROCESS DEVELOPMENT

5.7.1 Virus Production

This part now focuses on the virus production phase as the final stage of USP in cell culture-based vaccine production (Figure 5.1). As discussed previously, viral vaccines can be produced in adherent or suspension cell lines. Nowadays, production processes with suspension cells constitute the preferred choice due to an easy scalability, improved monitoring and process control and the option for high-cell density cultivations (see also next chapter).

In general, the cell-culture–based production of viral vaccines involves a biphasic process divided into the cell growth phase (3–5 days) and the subsequent

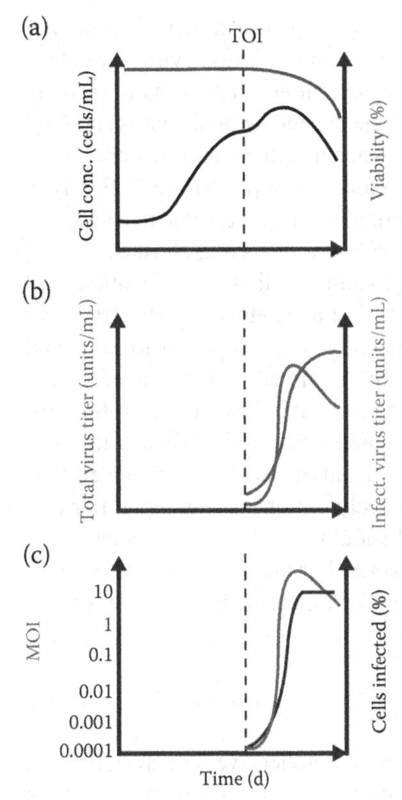

FIGURE 5.6 Typical time courses for a virus production process. Typically, this process is a two phase process with first cell growth (3–5 days) and second virus replication phase (2–4 days) (here indicated by the vertical dashed line at time of infection (TOI)). (a): cell concentration and viability, (b): total and infectious virus titer, (c): multiplicity of infection (MOI) and percentage of infected cells. As a reference, cells are often infected at 2E06 cells/mL with a MOI below 0.01. Due to the release of progeny virions, the MOI in the vessel increases (>10) and the cytopathic effect reduces cell concentration and viability. Conditions should be selected to allow productive infection of all cells to achieve the maximum virus yield. Ratios between infectious and non-infectious virus particles can be 1:10 and lower than 1:10,000, depending on the virus strain. Non-infectious viruses can be defective and interfering (defective interfering particles, DIPs). These DIPs need co-infections with standard virus particles for their replication; a high MOI scenario within the vessel increases DIP production and may reduce the infectious virus titer and the total virus titer [76].

virus production phase (2–4 days) (Figure 5.6). First, cells are seeded in the cultivation vessel (<1E06 cells/mL) and optimal process parameters are set (temperature, pH, pO_2, rpm). In a batch process, cells grow up to a certain cell concentration in the range of 1E06–1E07 cells/mL prior to infection. Taken into account that virus production is inhibited by the limitation of substrates and by accumulating by-products, a dilution or medium-exchange before time of infection (TOI) is beneficial for some processes.

Typically, infection takes place at the end of the exponential growth phase (e.g., for lytic viruses such as IAV). However, for slow replicating lytic viruses (e.g., YFV, MVA), an infection in the middle of the exponential growth phase should be considered, as cells continue to grow and the additional increasing cell concentrations might result in higher virus yields. For some viruses, like mink enteritis virus (MEV), the infection is even carried out directly after cell seeding as MEV replication only takes place in mitotic cells [77].

After infection, the viruses utilize the host cells to replicate their genome and synthesize viral proteins. The assembly and release of progeny virions completes a replication cycle. Infectious viruses released then infect uninfected cells until preferably the whole cell population is infected. Virus particles accumulate in the vessel and highest virus titers are reached, depending on the replication time of the virus, between 2–4 days after TOI (Figure 5.6). Typically, the infectious virus titer peaks earlier than the total virus titer (infectious plus non-infectious virus particles), and the total virus titer remains more or less constant after reaching its maximum, whereas the infectious titer often decreases with time depending on the virus stability. This is important for generation of seed virus and when infectious virus material is the product (e.g., live-attenuated vaccines, viral vectors, oncolytic viruses). For lytic viruses, the cytopathic effect leads to the termination of the process due to cell death. Depending on titers and the level of contaminating by-products (e.g., host cell DNA, proteins), the virus harvest is collected, clarified (depth filtration, centrifugation) and inactivated by (e.g., formaldehyde, β-propiolactone or binary ethyleneimine (BEI) for manufacturing of inactivated vaccines). Subsequently, it is subjected to DSP and formulation (Figure 5.1). After sterilization of the equipment or exchange of the single-use equipment, a new batch cycle can be conducted. Monitoring of the production process is carried out by measuring the concentrations of cells and metabolites, pH value, total and infectious virus titer, DNA and (host cell) protein levels (see also chapter 8). Most of these measurements still rely on manual sampling. For off-line analytics, samples are stored at −80°C and should only be thawed once for titrations as viruses are sensitive to freeze-thaw cycles. For other assays, heat or other inactivation of samples should be considered with respect to biosafety and virus contaminations of equipment.

5.7.2 Process Development and Optimization

Ideally, process development should be based on comprehensive data available for cultivations that were performed at various scales and under different conditions with at least three replicates using a design of experiments (DoE) approach. To obtain maximum virus titers, experiments should include the adaptation of viruses to the specific host cell line selected and variations in TOI and MOI. Moreover, the replicates should be performed from different precultures and with different aliquots from the virus seed. As animal cell culture is time consuming and many of the assays to determine virus titers require several days before read-outs are available, the total time for a set of experiments including analytics can require several weeks (typically 4–6 weeks). In addition, contamination risks and handling issues with small bioreactors can be challenging.

Furthermore, regarding data evaluation and the selection of optimum process conditions, the error of the respective assays has to be taken into account. This concerns in particular, virus titrations based on dilution series which often display errors in the ±0.3 log range. Together with relatively high batch-to-batch variations of cultivations performed in triplicates, the final selection of optimum process conditions is not trivial. Depending on resources (staff, equipment, analytics), it's generally advisable to first perform scouting experiments (possibly in triplicates) at very small scales (well plates, T-flasks, shake flasks) and then plan for DoE approaches at the bioreactor scale for a low number of (pre-)selected parameters only. Another option is the use of small-scale or micro-bioreactors such as Ambr® (Sartorius) or BioLector (m2plabs). However, working volumes and sampling have to be selected carefully as for some analytics sample volumes of 1 mL are required.

For cultivations where samples are taken twice a day and cultivation times exceed a week, the rather low working volume might not be ideal. Furthermore, when planning experiments, care should be taken that appropriate controls run in parallel, for instance mock cultivations where seed virus is not added. Especially, when switching to bioreactor cultivations at larger scale, sterility controls of samples after important process steps and a parallel run in small scale might help to identify possible contamination sources. For cell passaging it is equally helpful to use separate media bottles and sometimes even separate batches per person to have adequate backup solutions in case of contaminations.

Overall, optimization of cell growth of the host cell will take a considerable part of process optimization (see Table 5.7). It will be a continuous process as new medium formulations, sensors, or cultivation vessels are constantly developed and then a re-evaluation of process options might be needed. Taken together, relatively high costs of equipment and media together with time-consuming cultivations and assays make process development and optimization a demanding task.

5.7.2.1 Seed Virus Generation

A crucial part is the selection and production of a proper seed virus. The virus can either be isolated from an infected person or be obtained from a biological reference material bank such as the National Institute for Biological Standards and Control (NIBSC). As some viruses only enter and replicate in a small subset of GMP-relevant cell lines, the choice of the appropriate host cell line is of utmost importance. A cell line screening should be performed, where different cell lines are infected with the virus at different MOI (and other infection conditions) to identify the cell line with the highest virus yield. Such scouting experiments can be conducted in small scale in well plates, T-flasks, or shake flasks. From there, the virus is passaged serially in a cell line, while the sample with the highest infectious titer is used for the next passage to try to adapt the virus to the cell line. This might result in a more efficient production due to an accelerated virus propagation and increased maximum virus titer. This adaptation can also be used to adapt to lower MOI for instance.

5.7.2.2 Process Optimization Options

To obtain maximum virus yields and a high product quality, various parameters need to be considered and optimized in process development (see Table 5.7). Depending on vaccine type, the definition of product quality is variable. For live-attenuated vaccines the infectious titer is important and if possible the ratio of non-infectious to infectious particles should be low. However, if the non-infectious particles function as an adjuvant by increasing the immune response, this might not be too important. For sub-unit or split vaccines, the amount of functional antigens is of interest with possible low contaminations with host-cell proteins and DNA. Viral vectors or chimeric viruses generate the correct immune response by providing the antigen in the correct formation with the correct structure and if important glycosylation. Product quality is thus not only one attribute, but is an overall result of several virus/antigen properties. Therefore, optimization of the production process will need to consider on the one hand, how to obtain maximum yield and on the other

TABLE 5.7
Points to consider for process optimization either for adherent cells or for suspension cell lines

Evaluation/ optimization	Adherent cells	Suspension cells
Cell growth, small-scale	T-flasks, roller bottles, spinner flasks with microcarriers in SCM or SFM up to 2E06 cells/mL (good cell attachment rate and cell flattening); split ratio, split timing, inoculation concentration, maximum number of passages	Shake flasks in SCM, SFM or CDM at different rpm and pH up to 1E07 cells/mL (single cells/aggregates, cell size); split ratio, split timing, inoculation concentration, maximum number of passages
Cell growth, bioreactor	Growth on microcarriers or macrocarriers up to at least 2E06 cells/mL; inoculation concentration, trypsinization	Inoculation concentration, rpm, pH up to 1E07 cells/mL
Cell counting	Complete cell detachment	No aggregates
Cell banking	Check best medium, cell conc. and successful thawing	Check best medium, cell conc. and successful thawing, check freezing at high cell concentration
Process intensification	Check packed bed or HFBR	Check semi-perfusion in shake flasks and perfusion devices up to 2-5E07 cells/mL
Cell characterization	Check for sterility, mycoplasmas (PCR), etc. depending on requirements (FDA, EP, WHO)	Check for sterility, mycoplasmas (PCR), etc. depending on requirements (FDA, EP, WHO)
Seed virus	Adapt virus strain to host cells, 3–5 passages should be fine Check for sterility, mycoplasmas (PCR), identity, etc. depending on requirements (EP, FDA, WHO)	Adapt virus strain to host cells, 3–5 passages should be fine Check for sterility, mycoplasmas (PCR), identity, etc. depending on requirements (EP, FDA, WHO)
Virus production	Check virus replication at 2E06 cells/mL Check TOI and MOI Check yield at 37°C and reduced temperature (32-34°C)	Check virus replication at 2E06 cells/mL Check virus replication at >2E06 cells/mL for high cell density effect Check TOI and MOI Check yield at 37°C and reduced temperature (34°C)

SCM: serum containing medium; SFM: serum free medium; CDM: chemically defined medium; EP: *European Pharmacopeia*; FDA: Food and Drug Administration; WHO: World Health Organization; TOI: time of infection; MOI: multiplicity of infection.

hand efficiency, stability, ratio of infectious to non-infectious particles, and glycosylation. Furthermore, all these product quality parameters need to remain throughout purification and formulation as well as with storage, transport and final application as only few side effects will be tolerated by the vaccinated person. To be able to consider this already in upstream processing and process development is not always easy. Therefore, process optimization focuses on maximum yield. Nevertheless, as early as possible produced material should be evaluated in its efficacy. The selection of the optimum process conditions (e.g., temperature, pH, pO_2, rpm) is key (see Table 5.8). The pH typically ranges between 7.0–7.6, as a more acidic pH often results in virus inactivation or degradation. Optimal temperature for cell growth and virus production is mostly found at the physiological temperature of 37°C. However, a lower temperature (32–34°C) in the virus production phase is sometimes beneficial. Here, cell growth and virus propagation is inhibited, hence, maximum cell concentrations and virus titers peak delayed. Nevertheless, lower temperature might boost the maximum virus yield and prevent temperature-induced inactivation. Moreover, cold-adapted virus strains (attenuated virus strains for live vaccine manufacturing) also require a cultivation temperature lower than body temperature for efficient propagation in cell culture.

To optimize infection conditions, different TOIs and MOIs should be tested. Usually, virus infection takes place at low MOI in the range of 1E-02 to 1E-05, at which high virus yields are obtained and the risk of DIP propagation by unwanted co-infections with standard virus is low (see below). However, this optimum must be identified for every cell/virus combination and production process. At low MOI, it takes longer before maximum titers are reached. This poses the risk of early media depletion, which could have a negative impact on further virus propagation. Thus, special care for sufficient nutrient supply has to be taken. In contrast, infection at high MOI (>1E-01) is not cost-efficient, as a high volume of seed virus is required. Moreover, early virus-induced cell death might occur due to the high virus input, further limiting yields. Moreover, infection at very high MOI leads for almost all viruses to an accumulation of defective particles, that might even interfere with virus replication (defective interfering particles: DIPs). Overall, defective particles and, in particular, DIPs can significantly reduce yields or lead to non-optimal seed virus. Due to an internal deletion in their genome, DIPs cannot replicate on their own. They need a co-infection with standard virus (STV), where they rely on the complementing genome of the STV for their replication. But DIPs replicate faster and interfere with STV replication and, thus, may limit the overall virus yield. (However, DIPs might work as adjuvant in the final vaccine product by stimulating the immune response after vaccination.) One approach to reduce the level of DIPs in the virus seed is a serial passaging at low MOI [79].

In general, different attributes are determined by the host cell line, virus, and the process (see Table 5.9). Cell growth and virus production is heavily dependent on nutrients. During the cell growth phase, the main substrates glucose and glutamine are consumed for cell growth. The metabolization of these substrates leads to the accumulation of the metabolic by-products lactate, ammonia, and glutamate (Figure 5.3). At TOI a medium exchange is favored (fully or at least 50%) to improve substrate supply and dilute by-products released into the supernatant (ammonia is toxic at higher concentrations, lactate lowers the extracellular pH). Overall, cell growth and virus production continues until substrates are depleted, by-products reach inhibiting levels or cell death occurs by virus-induced apoptosis and lysis. During cultivation, a metabolic shift may occur, after which lactate is no

TABLE 5.8
Parameters for optimization of cell culture-based virus production processes

Parameter	Examples: Stirred Tank, Batch	Comment
Choice of cell line, engineering cell line	Adherent (continuous, primary)/ suspension; e.g., improve metabolism or reduce apoptosis	Adherent Vero cells often allow to be fast on the market; e.g., AGE1.pIX cells designed for better MVA replication
pH value	Optimal 7.0–7.6	Below pH 7: risk of virus inactivation
pO_2	Above 40%	Hypoxia might be beneficial for virus replication of some viruses [78]
rpm	100–250	Impacts cell growth
Temperature shift	Reduce to 32–34°C for virus infection phase	Can increase virus stability and yield
TOI	Often at 2E06 cells/mL to avoid any cell density effect	Cell concentration, cell cycle phase?
MOI	Between 1E-02 and 1E-05	As low as possible
Seed virus (DIPs, titer, engineering?)	Patient isolate?, egg or cell culture-derived, chimeric viruses (dengue/YFV)	Generate at low MOI, adaptation needed? aim for high titers, low DIP contamination level, engineering could have risk of creating a more dangerous virus?
Virus adaptation	3–5 passages in host cell	For IAV, switch from eggs to cell culture, adaptation to a new cell line
Infection mode	Washing step with PBS to remove serum at TOI, medium exchange, low volume infection, change to medium supporting virus propagation (e.g., by cell aggregation)	Complete medium exchange might be beneficial for virus replication
Time of harvest (TOH)	At max. virus titer (infectious or total virus particle concentration depending on vaccine type)	Amount of cell debris/DNA and protein contamination levels versus virus titer; cell disruption (e.g., freeze-thaw cycles) needed for virion release?
Multiple harvests	For non-lytic viruses and unstable viruses	Can reduce DIP production, limits virus degradation
Reduce inhibitors, medium selection	Lactate, ammonia, trypsin inhibitors, released interferons	Addition of so called "virus booster" (e.g., lipid cocktails) or virus medium to induce cell aggregation (e.g., MVA)
Additives	Trypsin, "virus booster," cholesterol, nucleosides, Pluronic F68	Needs to be optimized for medium, vessel and possible shear, multiple additions? stability, pH?

MOI: Multiplicity of infection; T: temperature; TOI: time of infection; DIPs: defective interfering particles.

TABLE 5.9
Parameters for virus replication determined either by cell line, virus, or process conditions

Determined by cell line	Determined by virus	Determined by process
Membrane composition of virus	Range of permissive cell lines	Volumetric virus productivity (VVP)
Glycosylation of viral proteins	Generation of DIPs	pH, temperature
Ratio infectious/non-infectious virions	Virus stability, virus degradation rate	Virus stability, virus degradation rate
Replication time	Replication time	
Cell-specific virus yield (CSVY)	Cell debris, cell lysis, apoptosis	Amount of cell debris

VVP: Volumetric virus productivity, virions produced per total volume medium and total process time (e.g., virions/L/d).

longer produced but consumed. Furthermore, both cell growth and virus replication depend strongly on availability and performance of certain commercially available media. Therefore, screening experiments are necessary to identify a medium that supports fast growth to high cell concentrations (first phase) and maximum virus yields. Furthermore, supplementation (e.g., concentration of glucose and glutamine or other additives) should be optimized.

Osmolality also impacts cell growth and virus production. For adenovirus production using HEK293 cells in suspension culture, hyperosmotic conditions during the cell growth phase were reported to lead to a reduced specific growth rate and a lower maximum cell concentration, but stimulated subsequent virus production (higher CSVY) [80]. A similar finding was already reported for antibody production [65,81,82]. Thus, osmolality should be screened during media development to determine the right balance between cell growth and virus production to ultimately optimize yields.

The harvest time point not only affects maximum virus yields, but also virus quality. As already mentioned, virus yields peak at a maximum in the range of 1–5 days post infection depending on the virus strain, host cell system and MOI. The optimal harvest time was reported for different viruses as: IAV (24 h post infection (hpi)) [16], YFV (48–96 hpi) [19], and MVA (72 hpi) [83]. However, infectious virus titer and total virus titer are typically not reaching their maximum at the same time (Figure 5.6). For IAV, the total virus titer typically further increases, while the infectious titer is stationary at its maximum or already decreases due to virus degradation [84]. This is due to the circumstance that at late time points, the proportion of non-infectious viruses, including DIPs, increases significantly. Moreover, depending on the harvest time point the level of contaminating DNA and protein might change. Typically, the total protein and host cell DNA level is most pronounced at later production time points due to cell lysis and the associated release of cell contents. This poses a major problem for DSP to achieve values lower than the maximum allowed levels of these contaminants in the final product. Host cell DNA might be very sticky and some viruses might attach to either host cell debris or host cell DNA complicating clarification steps (depth filtration, centrifugation) after virus harvesting. Salt concentrations/osmolality of media as well as choice of membrane material and cut-off of the respective filters will need to be screened thoroughly as each medium and cell line will give a different background and change of harvest time point will immediately change the cell broth composition.

In order to determine the optimal harvest time point, the ratio of total virus titer to contaminants should be considered. For attenuated vaccines and viral vectors, a high infectious titer is necessary to achieve high potency, hence, an early harvest time point should be targeted. Certain viruses show a low stability, which is characterized by a steep decrease in infectivity over time. One possible countermeasure is a multiple harvest strategy, in which the virus is harvested and stored at lower temperature and new medium is added to the bioreactor. This strategy is often used for adherent cells and slowly propagating lytic viruses. Moreover, continuous harvesting with subsequent cooling to prevent degradation could also be applied (see chapter 6).

Specific aspects of the intracellular virus replication cycle on process performance can also not be neglected. Some viruses bud from the (apical) cell membrane during virus release and, hence, carry a lipid bilayer as an envelope (see Table 5.9). Such enveloped viruses are often less stable at higher temperature, sensitive to lower pH values and fast degraded by contact with detergents; all resulting in infectivity losses. For a few viruses (e.g., MVA), a considerable number of virus particles remain within the cell. For reaching maximum virus yields, freeze-thaw cycles or the use of high-pressure homogenizers is recommended to disrupt the cell membrane and to release the virions.

Another factor to consider during viral vaccine production is the effect of shear stress (e.g., agitation, pumping) and aeration (e.g., O_2 and CO_2). High shear forces can hinder virus binding to the cells or can lead to early cell death and with that to lower virus titers. Cells go either into apoptosis due to virus infection or into necrosis due to shear stress. For some viruses, such as IAV, the right timing of apoptosis induction is important for virus release and with that virus yield. It is thus not as simple as just trying to avoid cell death for as long as possible, to keep cells productive. It will always be a combination of parameters and events that will result in higher virus yields. In principle, apoptosis is hallmarked by DNA fragmentation, plasma-membrane blebbing, and creation of apoptotic bodies (fragmentation) [85]. Different methods (e.g., imaging flow cytometry, NMR spectroscopy, proteomic approaches) are now established to further investigate cellular bottlenecks during viral vaccine production that might result in low virus yields (see also chapter 8). This understanding might help to identify high-producer cells and to optimize production processes.

Finally, the vaccine type (live-attenuated, inactivated, vector) also has a significant impact on design and optimization of virus production processes. Many attenuated vaccine strains show a lower replication rate and, thus, often reduced virus yields. In contrast, manufacturing processes for inactivated vaccines that comprise infectious and non-infectious virus particles often display very high titers. For production of pathogenic viruses without an option to vaccinate employees (e.g., early production campaigns of SARS-CoV-2 in adherent Vero cells) safety considerations should have a high priority and may require handling of virus-containing materials at least in a BSL3 environment. For manufacturing of inactivated vaccines, inactivation (heat, formaldehyde, β-propiolactone) is necessary and needs to be carefully validated and confirmed by innocuity assays. For recombinant sub-unit vaccines, higher concentrations are needed due to the lower immune response they induce. Taking all these points together explains that process optimization for viral vaccine manufacturing can be very lengthy and complicated.

5.8 SOMETIMES NOTHING MAKES SENSE—SOME IDEAS ON TROUBLE SHOOTING...

This part addresses some ideas on trouble shooting. It is a collection of experience of the authors, that is certainly not exhaustive.

Considering the high complexity of vaccine production processes with many, partially correlated, parameters, sometimes seemingly small changes can have a dramatic impact on process performance. It could happen that the medium supplier has a new provider for one compound of the cell culture medium. Now this compound has a slightly different quality or purity. It is prepared or stored in a different vessel and interacts differently with other medium components. It could equally result in differences regarding the medium filtrations performed. This could then have an impact on the performance of the cells, such as lower specific growth rate, reduced maximum cell concentration, or decreased cell viability. It could also display no direct impact on the cells, but when thawing the cells in this medium, they do not grow. Finally, and worst case, all cell growth–related properties are more or less the same, but virus yields are reduced. According to GMP requirements, the supplier will notify the customers about this change, as it could be a process relevant change. However, testing the impact of such small changes is not only time consuming and expensive, but also very challenging and flaws might not be identified before problems in large-scale productions arise. Due to these possibilities, the quality and consistency of all consumables that are used, should be verified on a regular interval when it comes to manufacturing. This includes regular audits of the respective suppliers. As the manufacturer is responsible for the quality of the product, everything needs to be documented very carefully following cGMP guidelines. At the end, the goal is to have a safe and potent vaccine produced with the same quality for many years.

In particular, cell culture media are critical. Quality can already change depending on where it is produced due to the water quality. Special care is needed when changing from liquid medium to powder medium that is prepared on-site to lower the costs and to extend in-house shelf life of lots. How the medium will be prepared will be different compared to what was done by the supplier (equipment, size, filters, water). Therefore, the medium might not show the same performance as the medium bought as liquid. Similar effects can be caused by the change of sterile filters for medium filtration, change of water quality (water cartridges of the Millipore systems, via reverse osmosis), change of flasks for cell growth (interaction of surface and medium components, leakage of unwanted compounds, etc.), change of supplier of bags for disposable bioreactors or change of hollow fiber membrane suppliers or some change for the carriers used. Finally, even chemically defined medium can have quite some variations.

Small changes in cell maintenance or preculture handling can equally have an impact on the performance of the cells in a bioreactor, for example. Cultivation in different vessels can lead to changes in aeration, mixing and shear conditions which will certainly influence metabolism and cell growth performance. After infection, either a very fast onset of cell lysis or almost no cell lysis can be observed or increased foaming, increased oxygen demand or significant changes in metabolism.

Transfer of cells through tubings also might need thorough analysis: The inner diameter of tubes, changes in the inner diameter of connectors, squeezing of tubes, tube materials that are air permeable or not, choice of pumps, length of tubing, etc.

Sometimes the CO_2 incubator might be the reason for troubles. For instance, someone placed additional equipment or flasks into the incubator that otherwise is not present. This results in temperature gradients or an additional heating from the equipment, and the cells do not grow at the optimal temperature anymore. Recalibrating the pH sensor during a bioreactor run, because of an off-set between online and off-line measurement can also cause trouble. For some products a shift of pH as low as 0.1–0.2 can result in changes in growth properties, antigen glycosylation, or reduce virus production.

Nevertheless, while all this seems to give the impression that animal cell culture and virus production is coping only with troubles, this is not the case. Strict rules regarding the quality control performed by the companies, detailed documentation and regulations, and the implementation of backup solutions have strongly improved vaccine quality and consistency.

5.9 CMC AND GMP CONSIDERATIONS RIGHT FROM THE START

Any development of a new vaccine starts at the research and discovery stage. Initially, scientists developing new vaccine candidates focus on mechanisms of the infectious organism causing disease. Laboratory research is conducted *in vitro* and *in vivo* until a feasible vaccine candidate for further development is found, often lasting 2–4 years. Additional research seeking options for establishment of laboratory-scale processes that allow to produce enough material for testing in animals to obtain first safety and mechanism of action data is performed. Finally, a process is established that enables large-scale manufacturing of the vaccine. Before a new vaccine can be administered to humans, the sponsor has to file an IND (investigational new drug application) in the United States of America or a CTA (clinical trial application) in the European Union (Figure 5.7). Latest at this point, academic research groups will discover that their production process proposed has to be GMP certified (good manufacturing practice or minimum "GMP ready") and provide all required "chemistry, manufacturing, and controls" information (CMC).

In contrast to chemical pharmaceuticals, biological products including vaccines are inherently more difficult to manufacture due to their derivation from cellular sources. Because manufacturing requires the propagation of the immunizing agents (e.g., viruses, VLPs, recombinant proteins, etc.) in cellular sources, the complexity to establish optimal conditions for growth, and

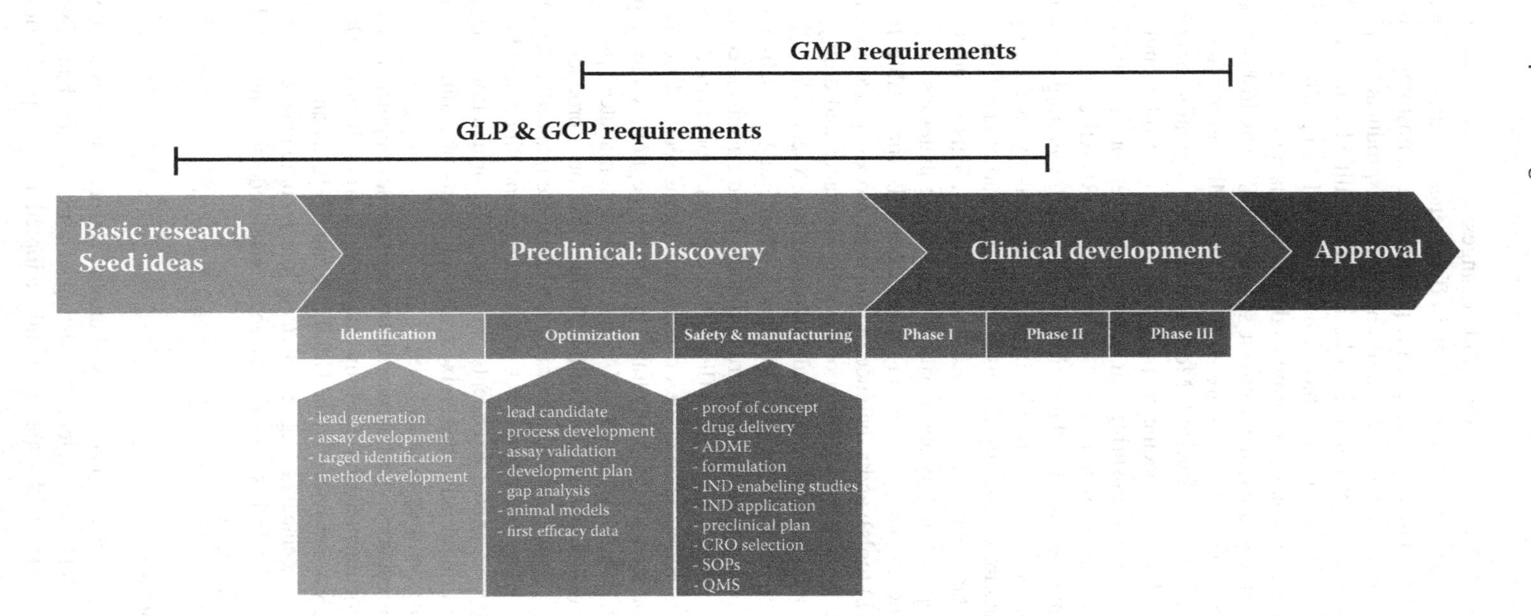

FIGURE 5.7 Typical stages of drug development from basic research to approval and market authorization. GLP: Good Laboratory Practice; GCP: Good Clinical Practice; GMP: Good Manufacturing Practice; ADME: Adsorption, Distribution, Metabolism, and Excretion; CRO: Contract Research Organization; IND: Investigational New Drug; SOP: Standard Operating Procedure; QMS: Quality Management System.

expression is increased. Moreover, even subtle changes in materials and process conditions can drastically affect the safety and effectiveness of the vaccine. For this reason, strict control and monitoring of the process is of utmost importance to produce consistent, reproducible, and well-characterized products [86]. As a general rule of thumb, the cell line has to be a GMP qualified cell line with a manufacturing license, the cultivation has to be performed using GMP certified media, and the virus material has to be produced in a GMP laboratory under GMP conditions. An adequate quality management system (QMS) has to be adopted, demonstrating that all documentations are according to GMP and the full history of materials and equipment is outlined. More specifically, regulatory authorities have to be assured that the vaccine is produced under controlled and stable conditions, with suitable tests ensuring that the vaccine meets prospective criteria (e.g., identity, purity, safety, potency, and stability) [87,88]. All equipment, assays and methods involved need to be validated. Standard operating procedures (SOPs) for all process steps are required. Obviously, quality and purity of a vaccine cannot be assured solely by downstream testing and assays, but depend on strict control of the manufacturing process as well [86]. This includes, but is not limited to, appropriate quality and purity of the starting materials (e.g., cells, viruses, reagents, media, etc.), use of in-process controls (preferable process analytical tools (PAT) tools), and adherence to validated process procedures [89].

This results in the regulatory definition of the drug substance and drug product for vaccine production as follows: "The drug substance of a vaccine is defined as the unformulated active (immunogenic) substance which may be subsequently formulated with excipients to produce the drug product. The drug substance may be whole bacterial cells, viruses, or parasites (live or killed); crude or purified antigens isolated from killed or living cells; crude or purified antigens secreted from living cells; recombinant or synthetic carbohydrate, protein or peptide antigens; polynucleotides (as in mRNA and plasmid DNA vaccines); or conjugates" [89]. "The drug product is the finished dosage form of the product. The drug product contains the drug substance(s) formulated with other ingredients in the finished dosage form ready for marketing. Other ingredients, active or inactive, may include adjuvants, preservatives, stabilizers, and/or excipients. For vaccine formulation, the drug substance(s) may be diluted, adsorbed, mixed with adjuvants or additives, and/or lyophilized to become the drug product" [89].

Thus, keeping all regulatory issues in mind throughout the development many little aspects need to be documented and carefully chosen right from the start. Detailed guidance's for industry are provided by the U.S. FDA and EU EMA and will slightly vary depending on the vaccine type (e.g., inactivated, live, recombinant, etc.) [89,90]. In the following, a brief overview on basic aspects considered for a successful regulatory submission filing is given.

5.9.1 Cellular Sources

Cellular sources typically refer to cell lines of animal (insects, humans, and other mammalians) origin. Cell bank systems consisting of a master cell bank (MCB) and a working cell bank (WCB) are frequently adopted for the production of vaccines (regardless if they are used as a substrate for virus propagation and recombinant DNA products or as vaccine components). A detailed description of the cell bank system including procedures used to avoid microbial/viral contamination, safety precautions in case of catastrophic events, and storage conditions must be provided [89]. Moreover, the cells of the MCB must be completely characterized and must be able to show a complete history. Characterization includes, but is not limited to, biochemistry (cell surface markers, mRNA, etc.), specific identifiers (morphology, serotype, etc.), karyology and tumorgenicity, virulence markers, genetic markers, and media (e.g., serum) [89]. The WCB is directly derived from the MCB and characterized for cell viability before use in the manufacturing process. In contrast to the MCB, documentation outlining procedures for storage and assays used for qualification, purity, and characterization has to be provided for each new WCB [89].

5.9.2 Virus Sources

Similar to cellular sources, seed lot systems including a master seed lot (MSL) and a working seed lot (WSL) are typical components of a vaccine production process. A full characterization of the WSL is required, but depends on several factors such as the nature of the virus strain/vector, genetic modifications, and the history of the cell source used in seed lot preparation [90]. Nevertheless, genetic and phenotypic properties of the virus including a side-to-side comparison with the parental virus must be included. Genetic characterization typically involves next-generation sequencing of the virus, PCR analysis, southern blotting, and restriction mapping. Moreover, verification of possible endogenous retroviruses present in the viral genome is critical to ensure that no additional virus material is introduced. Special attention should be paid regarding genetic stability ensuring no changes in regions involved in e.g., attenuation. For this, genetic stability studies of the vaccine seed of a passage level comparable to a production batch should be conducted. Phenotypic characterization focuses on *in vitro* and *in vivo* studies measuring markers for attenuation/modification and expression of heterologous antigens. Particularly for viral vaccines, infectious virus titers, total particle numbers, virus yields, and *in vivo* growth characteristics in a suitable animal model are a critical part of this characterization. WSL are usually prepared by passage of the MSL in the cellular source used in the production process. Here, special attention must be paid regarding passage numbers, ideally limiting the number of passages to minimize the possibility of genetic and phenotypic changes of the virus [90].

5.9.3 Cell Growth and Harvesting

Production processes for various vaccine types are likely to be similar. In many cases, there will be minimal downstream processing and the basic requirements for control and manufacturing are the same [90]. For large-scale production, the working cell bank as well as the seed virus may have to be expanded before inoculation of the production cell culture. As discussed above, the number of passages between WSL and production lot should be kept low. Whatever the final production process looks like, a clear definition of the method is required [90]. This includes descriptions of each stage of cell growth: inoculation and growth of initial and precultures (volumes, time, temperature, pH, etc.), main culture system including operating and control parameters (temperature, pO_2, pH, static vs. agitated, aerobic vs. anaerobic, volume of bioreactor, number of bioreactors, process mode, etc.) [89]. Moreover, in-process controls and testing (purity, viability, yields, etc.) as well as sterilization/preparation/cleaning procedures should be described. All used materials (e.g., water, media, serum, additives, antibiotics) must comply with the respective guidelines provided by the regulatory agencies. At a certain point during the production process, the drug substance has to be harvested. A brief description outlining the criteria for harvesting and the monitored process parameters should be described. Method(s) used for separation of unrefined drug substance from the cell broth (e.g., precipitation, filtration, centrifugation, etc.) must be provided. Each harvest has to be tested for extraneous agents and characterized regarding identity of the virus and infectious/total virus titers. Here, thresholds for acceptable titers should be established [89,90].

5.9.4 Purification and Downstream Processing

Detailed descriptions of purification and downstream methods can be found elsewhere. For chemistry, manufacturing and controls (CMC) submissions, the used material and methods for separation and concentration of the drug substance should be described. The final purification step can be prepared from a single virus harvest or several virus pools. Each step (e.g., inactivation, purification, stability processing, detoxification, etc.) must be outlined, including adopted or developed analytical tests to show identity, purity, concentration, and levels of impurities. If final formulation substances such as stabilizers or diluents are used, efficacy and safety of the drug product cannot be impaired and verification of the stability should be demonstrated. The use of antibiotics as antimicrobial preservatives is usually strictly prohibited. Potential titer loss during filling, freeze-drying, and shelf life, should be compensated by a higher titer in the final bulk vaccine. Final testing of the vaccine should be performed on samples from each batch, analyzing sterility, stability, potency, and identity. Quantitative and qualitative analyses must include thermal stability, endotoxin levels, residual host cell protein, residual agents/animal serum proteins, infectious virus titer, or immunogenicity *in vivo* [89,90].

Regardless of the vaccine type, they all have to fulfil strict regulatory requirements throughout their manufacturing process. An understanding of those requirements is critical for developers to manufacture a reproducible, consistent, safe, and effective vaccine. Even though regulatory processes do not directly impact the early development stage, choices for equipment, process steps, and *in vitro* and *in vivo* testing design could be reconsidered as they may not fulfill the needed qualifications. Professional consultation by knowledgeable qualified persons or regulatory consultants already at the research stage might streamline the vaccine development process and alleviate stress and unnecessary delays with regulatory submissions. With the SARS-CoV-2 pandemic this became very prominent as those vaccines that became available very fast, were developed by industrial cooperation's using existing infrastructures of similar vaccine production processes. In contrast, academic approaches could only provide proof-of-principle alternatives, but could not contribute to the establishment of manufacturing processes within the required time frame [91,92]. Nevertheless, by integrating regulatory guidelines at early stages of process development and establishment of a clear and focused regulatory strategy, even academic approaches may be fast tracked at least for animal testing and early clinical development.

5.10 OTHER POINTS TO CONSIDER

5.10.1 Manufacturing Options

If the vaccine type and host cell system is chosen, and dose input as well as booster regime and annual demand are known, manufacturing scale and bioreactor vessel(s) can be defined. Table 5.10 gives some examples of possible settings for such estimations for a scenario assuming a need of 1E07 infectious units/dose for a live-attenuated vaccine and 1E09 total virions/dose for an inactivated vaccine, no booster vaccination, a loss of 50% in downstream processing and the same CSVY for all production modes. For an inactivated vaccine produced at 2,000 L wv in a STR, this would require 2,667 bioreactor runs or handling of 1.3E08 roller bottles (see Table 5.10) to cover the world supply. Even with the current SARS-CoV-2 vaccine production, one of the next vaccines designated for Europe will be produced in roller bottles using adherent Vero cells (Valneva). This exercise might help to demonstrate that vaccine production is about producing many doses at low cost and many parameters need to be considered at all time.

5.10.2 Biosafety

Typically, viruses considered for production of vaccines will be wild type viruses provided by WHO, NIBSC, and others. At early stages of vaccine development even isolates from patients might be handled. If available, attenuated virus strains will be preferred as their biosafety level will be BSL1 or BSL2. This also applies to most of viral vectors. In the future, certainly more and more advanced methods for genetic engineering including reverse genetics or CRISPR-Cas technologies will be used to either modify virus strains, viral vectors or production cell lines. Then risk assessments need to be updated accordingly. One risk that always needs to be considered in this respect is the risk to generate a highly pathogenic virus instead of an attenuated strain. Equally, for attenuated strains, the risk of reversion of the attenuation will need to be addressed; likewise for inactivated vaccines, the inactivation procedure needs to be carefully validated.

Therefore, not only efficacy and quality of vaccines, but also safety needs to be controlled on a regular basis. Handling several viruses or subtypes in parallel could promote unwanted recombinations or generation of new subtypes (i.e., for IAV). Furthermore, cross-contaminations in cell culture as well as contaminations of media, cell lines, or virus seeds should be checked on a regular basis.

TABLE 5.10

Calculation exercise for possible bioreactor needs

Vaccine type	Dosage (infect. units/ dose) (total virions/dose)	Doses per year	Total viruses needed	CSVY (virus/cell)	Total cells needed	Vessel	Max. cells/ vessel[1]	Vessel wv (L)[1]	Vessel units needed[2]	Wv needed (L)[2]
Live-atten.	1E07	1E07	1E14	150	6.7E11	RB	8.5E07	0.26	1.6E04	4078
Live-atten.	1E07	1E07	1E14	150	6.7E11	cc	8.5E08	0.60	1.6E03	941
Live-atten.	1E07	1E07	1E14	150	6.7E11	ic	1.5E12	70	1	62
Live-atten.	1E07	1E07	1E14	150	6.7E11	STR	2.0E09	1	667	667
Live-atten.	1E07	1E07	1E14	150	6.7E11	pSTR	5.0E10	1	27	27
Live-atten.	1E07	8E10	8E17	150	5.3E15	RB	8.5E07	0.26	1.3E08	3.3E07
Live-atten.	1E07	8E10	8E17	150	5.3E15	cc	8.5E08	0.60	1.3E07	7.5E06
Live-atten.	1E07	8E10	8E17	150	5.3E15	ic	1.5E12	70	7.1E03	5.0E05
Live-atten.	1E07	8E10	8E17	150	5.3E15	STR	2.0E09	1	5.3E06	5.3E06
Live-atten.	1E07	8E10	8E17	150	5.3E15	pSTR	5.0E10	1	2.1E05	2.1E05
Live-atten.	1E07	8E10	8E17	20	4.0E16	RB	8.5E07	0.26	9.4E08	2.5E08
Live-atten.	1E07	8E10	8E17	20	4.0E16	cc	8.5E08	0.60	9.4E07	5.6E07
Live-atten.	1E07	8E10	8E17	20	4.0E16	ic	1.5E12	70	5.3E04	3.7E06
Live-atten.	1E07	8E10	8E17	20	4.0E16	STR	2.0E09	1	4.0E07	4.0E07
Live-atten.	1E07	8E10	8E17	20	4.0E16	pSTR	5.0E10	1	1.6E06	1.6E06
Inactivated	1E09	8E10	8E19	1.5E04	5.3E15	RB	8.5E07	0.26	1.3E08	3.3E07
Inactivated	1E09	8E10	8E19	1.5E04	5.3E15	cc	8.5E08	0.60	1.3E07	7.5E06
Inactivated	1E09	8E10	8E19	1.5E04	5.3E15	ic	1.5E12	70	7.1E03	5.0E05
Inactivated	1E09	8E10	8E19	1.5E04	5.3E15	STR	2.0E09	1	5.3E06	5.3E06
Inactivated	1E09	8E10	8E19	1.5E04	5.3E15	pSTR	5.0E10	1	2.1E05	2.1E05

Notes

[1] For each vessel, the maximum cells per vessel unit and vessel working volume (see Table 5.6). RB: roller bottle (850 cm^2); STR: stirred tank bioreactor (1 L wv); pSTR: perfusion STR (1 L wv); cc: cell cube; ic: icellis max. For pSTR additional medium for perfusion is needed. Cultivation either of suspension or of adherent cells is possible in the vessel.

[2] To consider 50% loss in DSP all values are multiplied by two.

(These calculations assume that cell specific virus yield (CSVY) is the same in each vessel. This is often not the case.)

TABLE 5.11
Basic recommendations in case of accidental spilling of virus material

Laboratory	Person
• Avoid aerosols	• Remove contaminated clothing
• Leave contaminated area, block area, wait for 30 min to allow aerosols to settle	• Clean, rinse thoroughly wounds or eyes
• Put on lab coat, goggles, masks, gloves	• Get medical help if needed
• Watch out for glass breakage, syringes	• Calm down, think first!
• Disinfect with liquid pouring or tissues	• Safety first, experiment last!
• No spraying!	• Reduce number of people in contaminated area
• Autoclave contaminated liquids and waste	• Get help if needed
• Make sure no GMOs, virus material gets into the environment	• Inform your supervisor, biosafety officer

GMO: genetically modified organism.

Some viruses that are handled are known to be spreading via aerosols and others do not. Handling viruses under laboratory conditions in much higher concentrations than under "real-life" conditions might also lead to a spreading of such non-aerosol viruses via aerosols.

Thus, in case of accidental spilling of larger volumes of high titer virus material some basic rules should be followed. In addition, for each laboratory specific rules need to be defined according to the safety regulations for handling of viruses and genetically modified material implemented in the respective country (see Table 5.11).

5.11 CONCLUSION

With the emergence of new viruses (such as Zika virus or SARS-CoV-2), it is clear that the development of viral vaccines including in-depth knowledge on immune response and virology is of great importance. Even if mRNA vaccines will complement or replace some of the currently licensed viral vaccines, the demand for production of viruses and viral vectors will still increase. This applies, in particular, to therapeutic viruses once the full potential is better understood. To meet all these production needs, certainly ideas on process intensification to increase yields beyond batch mode will be required.

REFERENCES

[1] T. Momin *et al.*, "Safety and Immunogenicity of a DNA SARS-CoV-2 vaccine (ZyCoV-D): Results of an open-label, non-randomized phase I part of phase I/II clinical study by intradermal route in healthy subjects in India," *EClinicalMedicine*, vol. 38, 2021. doi: 10.1016/j.eclinm.2021.101020

[2] L. E. Gallo-Ramirez, A. Nikolay, Y. Genzel, and U. Reichl, "Bioreactor concepts for cell culture-based viral vaccine production," *Expert Rev. Vaccines*, vol. 14, no. 9, pp. 1181–1195, 2015.

[3] Z. Kis, C. Kontoravdi, R. Shattock, and N. Shah, "Resources, production scales and time required for producing RNA vaccines for the global pandemic demand," *Vaccines*, vol. 9, no. 1, p. 3, 2021.

[4] A. F. Pihl *et al.*, "High density Huh7.5 cell hollow fiber bioreactor culture for high-yield production of hepatitis C virus and studies of antivirals," (in eng), *Sci. Rep.*, vol. 8, no. 1, p. 17505, Nov. 2018.

[5] C. Gerke, P. N. Frantz, K. Ramsauer, and F. Tangy, "Measles-vectored vaccine approaches against viral infections: a focus on Chikungunya," (in eng), *Expert Rev. Vaccines*, vol. 18, no. 4, pp. 393–403, Apr. 2019.

[6] E. Suder, W. Furuyama, H. Feldmann, A. Marzi, and E. de Wit, "The vesicular stomatitis virus-based Ebola virus vaccine: from concept to clinical trials," (in eng), *Hum. Vaccines Immunother.*, vol. 14, no. 9, pp. 2107–2113, 2018.

[7] V. Lecouturier *et al.*, "Characterization of recombinant yellow fever-dengue vaccine viruses with human monoclonal antibodies targeting key conformational epitopes," (in eng), *Vaccine*, vol. 37, no. 32, pp. 4601–4609, Jul. 2019.

[8] D. Kuczera, J. P. Assolini, F. Tomiotto-Pellissier, W. R. Pavanelli, and G. F. Silveira, "Highlights for dengue immunopathogenesis: antibody-dependent enhancement, cytokine storm, and beyond," (in eng), *J. Interferon Cytokine Res.: Official J. Int.Soc. Interferon Cytokine Res.*, vol. 38, no. 2, pp. 69–80, Feb. 2018.

[9] WHO, "WHO Technical Report Series 978," in "*WHO Expert Commitee on Biological Standardization*," Geneva, Switzerland: World Health Organization, 2013.

[10] EMA, "ICH Topic Q 5 D, Quality of Biotechnological Products: Derivation and Characterisation of Cell Substrates Used for Production of Biotechnological/ Biological Products,"United Kingdom:European Medicines Agency London, 1998.

[11] FDA, "Guidance for Industry, Characterization and Qualification of Cell Substrates and Other Biological Materials Used in the Production of Viral Vaccines for Infectious Disease Indications," U.S. Department of Health and Human Services, Food and Drug Administration, Center for Biologics Evaluation and Research, 2010.

[12] Y. Genzel and U. Reichl, "Continuous cell lines as a production system for influenza vaccines," *Expert Rev. Vaccines*, vol. 8, no. 12, pp. 1681–1692, Dec. 2009.

[13] G. H. Markx and C. L. Davey, "The dielectric properties of biological cells at radiofrequencies: applications in biotechnology," *Enzyme Microbial. Technol.*, vol. 25, no. 3, pp. 161–171, 1999.

[14] D. A. M. Pais, P. R. S. Galrão, A. Kryzhanska, J. Barbau, I. A. Isidro, and P. M. Alves, "Holographic imaging of insect cell cultures: online non-invasive monitoring of adeno-associated virus production and cell concentration," *Processes*, vol. 8, no. 4, Article no. 487, 2020.

[15] D. Kuystermans, A. Mohd, and M. Al-Rubeai, "Automated flow cytometry for monitoring CHO cell cultures," *Methods*, vol. 56, no. 3, pp. 358–365, 2012.

[16] T. Bissinger *et al.*, "Semi-perfusion cultures of suspension MDCK cells enable high cell concentrations and efficient influenza A virus production," *Vaccine*, vol. 37, no. 47, pp. 7003–7010, 2019.

[17] A. Le Ru, D. Jacob, J. Transfiguracion, S. Ansorge, O. Henry, and A. A. Kamen, "Scalable production of influenza virus in HEK-293 cells for efficient vaccine manufacturing," *Vaccine*, vol. 28, no. 21, pp. 3661–3671, 2010.

[18] G. Gränicher, J. Coronel, F. Trampler, I. Jordan, Y. Genzel, and U. Reichl, "Performance of an acoustic settler versus a hollow fiber–based ATF technology for influenza virus production in perfusion," *Appl. Microbiol. Biotechnol.*, vol. 104, no. 11, pp. 4877–4888, 2020.

[19] A. Nikolay, A. Léon, K. Schwamborn, Y. Genzel, and U. Reichl, "Process intensification of EB66® cell cultivations leads to high-yield yellow fever and Zika virus production," *Appl. Microbiol. Biotechnol.*, vol. 102, no. 20, pp. 8725–8737, 2018.
[20] Y. Genzel *et al.*, "High cell density cultivations by alternating tangential flow (ATF) perfusion for influenza A virus production using suspension cells," *Vaccine*, vol. 32, no. 24, pp. 2770–2781, May 2014.
[21] K. Scharfenberg and R. Wagner, "A Reliable Strategy for The Achievement of Cell Lines Growing in Protein-Free Medium," in *Animal Cell Technology: Developments Towards the 21st Century*, E. C. Beuvery, J. B. Griffiths, and W. P. Zeijlemaker, Eds. Dordrecht: Springer Netherlands, 1995, pp. 619–623.
[22] V. Lohr, Y. Genzel, I. Behrendt, K. Scharfenberg, and U. Reichl, "A new MDCK suspension line cultivated in a fully defined medium in stirred-tank and wave bioreactor," *Vaccine*, vol. 28, no. 38, pp. 6256–6264, Aug. 2010.
[23] A. L. Caron, R. T. Biaggio, and K. Swiech, "Strategies to suspension serum-free adaptation of mammalian cell lines for recombinant glycoprotein production," (in eng), *Methods. Mol. Biol.*, vol. 1674, pp. 75–85, 2018.
[24] J. A. Howe *et al.*, "Matching complementing functions of transformed cells with stable expression of selected viral genes for production of E1-deleted adenovirus vectors," (in eng), *Virology*, vol. 345, no. 1, pp. 220–230, Feb. 2006.
[25] C. Chu, V. Lugovtsev, H. Golding, M. Betenbaugh, and J. Shiloach, "Conversion of MDCK cell line to suspension culture by transfecting with human siat7e gene and its application for influenza virus production," (in eng), *Proc. Nat. Acad. Sci. United States of America*, vol. 106, no. 35, pp. 14802–14807, Sep. 2009.
[26] P. B. Capstick, R. C. Telling, W. G. Chapman, and D. L. Stewart, "Growth of a cloned strain of Hamster kidney cells in suspended cultures and their susceptibility to the virus of foot-and-mouth disease," *Nature*, vol. 195, no. 4847, pp. 1163–1164, 1962.
[27] T. W. Pay, A. Boge, F. J. Menard, and P. J. Radlett, "Production of rabies vaccine by an industrial scale BHK 21 suspension cell culture process," (in eng), *Dev. Biol. Stand.*, vol. 60, pp. 171–174, 1985.
[28] A. Doroshenko and S. A. Halperin, "Trivalent MDCK cell culture-derived influenza vaccine Optaflu (Novartis Vaccines)," (in eng), *Expert Rev. Vaccines*, vol. 8, no. 6, pp. 679–688, Jun. 2009.
[29] EMA, "Flucelvax Tetra (influenza vaccine [surface antigen inactivated prepared in cell cultures])," vol. EMA/510023/2020, Amsterdam, NetherlandsEuropean Medicines Agency, 2020.
[30] FDA, "Sequirus, Flucelvax, Supplement approval," vol. BL 125408/366, U.S. Department of Health and Human Services, Food and Drug Administration, Center for Biologics Evaluation and Research, 2021.
[31] P. N. Barrett, W. Mundt, O. Kistner, and M. K. Howard, "Vero cell platform in vaccine production: moving towards cell culture-based viral vaccines," *Expert Rev. Vaccines*, vol. 8, no. 5, pp. 607–618, 2009.
[32] J. J. Ramsden, S.-Y. Li, J. E. Prenosil, and E. Heinzle, "Kinetics of adhesion and spreading of animal cells," *Biotechnol. Bioeng.*, vol. 43, no. 10, pp. 939–945, 1994.
[33] Cytiva. (2021). Microcarrier cell culture, principles and methods [Online]. Available: https://cdn.cytivalifesciences.com/dmm3bwsv3/AssetStream.aspx?mediaformatid=10061&destinationid=10016&assetid=11250
[34] F. Grinnell, "Cellular adhesiveness and extracellular substrata," (in eng), *Int. Rev. Cytol.*, vol. 53, pp. 65–144, 1978.
[35] D. E. Martens *et al.*, "Death rate in a small air-lift loop reactor of vero cells grown on solid microcarriers and in macroporous microcarriers," *Cytotechnology*, vol. 21, no. 1, pp. 45–59, 1996.

[36] STEMCELL. (2021). How We Define Our Culture Media and Supplements [Online]. Available: https://www.stemcell.com/how-do-we-define-our-media.html.

[37] A. F. Rodrigues, M. J. Carrondo, P. M. Alves, and A. S. Coroadinha, "Cellular targets for improved manufacturing of virus-based biopharmaceuticals in animal cells," *Trends Biotechnol.*, vol. 32, no. 12, pp. 602–607, Dec. 2014.

[38] J. Keenan, D. Pearson, and M. Clynes, "The role of recombinant proteins in the development of serum-free media," (in eng), *Cytotechnology*, vol. 50, no. 1-3, pp. 49–56, 2006.

[39] K. F. Wlaschin and W.-S. Hu, "Fedbatch Culture and Dynamic Nutrient Feeding," in *Cell Culture Engineering*, W.-S. Hu, Ed. Berlin, Heidelberg: Springer Berlin Heidelberg, 2006, pp. 43–74.

[40] S. P. DiBartola, "Chapter 3 – Disorders of Sodium and Water: Hypernatremia and Hyponatremia," in *Fluid, Electrolyte, and Acid-Base Disorders in Small Animal Practice (Third Edition)*, S. P. Dibartola, Ed. Saint Louis: W.B. Saunders, 2006, pp. 47–79.

[41] I. Nadeau, A. Garnier, J. Côté, B. Massie, C. Chavarie, and A. Kamen, "Improvement of recombinant protein production with the human adenovirus/293S expression system using fed-batch strategies," *Biotechnology and Bioengineering*, vol. 51, no. 6, pp. 613–623, 1996.

[42] F. Zagari, M. Jordan, M. Fau – Stettler, H. Stettler, M. Fau - Broly, F. M. Broly, H. Fau – Wurm, and F. M. Wurm, "Lactate metabolism shift in CHO cell culture: the role of mitochondrial oxidative activity," (in eng), no. 1876-4347 (Electronic), 2013.

[43] J. D. Young, "Metabolic flux rewiring in mammalian cell cultures," *Curr. Opin. Biotechnol.*, vol. 24, no. 6, pp. 1108–1115, 2013.

[44] C. Boero et al., "Design, development, and validation of an in-situ biosensor array for metabolite monitoring of cell cultures," *Biosens. Bioelectron.*, vol. 61, pp. 251–259, 2014/11/15/ 2014.

[45] S. S. Ozturk, J. C. Thrift, J. D. Blackie, and D. Naveh, "Real-time monitoring and control of glucose and lactate concentrations in a mammalian cell perfusion reactor," *Biotechnol. Bioeng.*, vol. 53, no. 4, pp. 372–378, 1997.

[46] A. Nikolay, T. Bissinger, G. Granicher, Y. Wu, Y. Genzel, and U. Reichl, "Perfusion Control for high cell density cultivation and viral vaccine production," *Methods Mol Biol*, vol. 2095, pp. 141–168, 2020.

[47] T. M. Larson, M. Gawlitzek, H. Evans, U. Albers, and J. Cacia, "Chemometric evaluation of online high-pressure liquid chromatography in mammalian cell cultures: Analysis of amino acids and glucose," *Biotechnol. Bioeng.*, vol. 77, no. 5, pp. 553–563, 2002.

[48] A. P. Teixeira, R. Oliveira, P. M. Alves, and M. J. T. Carrondo, "Advances in online monitoring and control of mammalian cell cultures: Supporting the PAT initiative," *Biotechnol. Adv.*, vol. 27, no. 6, pp. 726–732, 2009.

[49] L. Zhao, H.-Y. Fu, W. Zhou, and W.-S. Hu, "Advances in process monitoring tools for cell culture bioprocesses," *Engineering in Life Sciences*, vol. 15, no. 5, pp. 459–468, 2015.

[50] D. Eibl and R. Eibl, "Bioreactors for Mammalian Cells: General Overview," in *Cell and Tissue Reaction Engineering: With a Contribution by Martin Fussenegger and Wilfried Weber*. Berlin, Heidelberg: Springer Berlin Heidelberg, 2009, pp. 55–82.

[51] A. P. Sommer, M. K. Haddad, and H.-J. Fecht, "It is time for a change: Petri Dishes weaken cells," *J. Bionic Eng.*, vol. 9, no. 3, pp. 353–357, 2012.

[52] O.-W. Merten, "Advances in cell culture: anchorage dependence," (in eng), *Philos. Trans. R Soc. Lond. B Biol. Sci.*, vol. 370, no. 1661, p. 20140040, 2015.

[53] M. Butler, *Animal cell culture and technology*. London; New York: BIOS Scientific Publishers, 2004.

[54] J. Michl, K. C. Park, and P. Swietach, "Evidence-based guidelines for controlling pH in mammalian live-cell culture systems," *Commun. Biol.*, vol. 2, no. 1, p. 144, 2019.
[55] H. F. George and F. Qureshi, "Newton's Law of Viscosity, Newtonian and Non-Newtonian Fluids," in *Encyclopedia of Tribology*, Q. J. Wang and Y.-W. Chung, Eds. Boston, MA: Springer US, 2013, pp. 2416–2420.
[56] F. Garcia-Ochoa and E. Gomez, "Bioreactor scale-up and oxygen transfer rate in microbial processes: An overview," *Biotechnol. Adv.*, vol. 27, no. 2, pp. 153–176, 2009.
[57] M. Popović, H. Niebelschütz, and M. Reuß, "Oxygen solubilities in fermentation fluids," *Eur. J. Appl. Microbiol. Biotechnol.*, vol. 8, no. 1, pp. 1–15, 1979.
[58] E. Rischbieter, A. Schumpe, and V. Wunder, "Gas solubilities in aqueous solutions of organic substances," *J. Chem. Eng. Data*, vol. 41, no. 4, pp. 809–812, 1996.
[59] A. R. Lara, E. Galindo, O. T. Ramírez, and L. A. Palomares, "Living with heterogeneities in bioreactors," *Mol. Biotechnol.*, vol. 34, no. 3, pp. 355–381, 2006.
[60] A. R. Oller, C. W. Buser, M. A. Tyo, and W. G. Thilly, "Growth of mammalian cells at high oxygen concentrations," *J. Cell Sci.*, vol. 94, no. 1, pp. 43–49, 1989.
[61] A. R. Lara *et al.*, "Comparison of oxygen enriched air vs. pressure cultivations to increase oxygen transfer and to scale-up plasmid DNA production fermentations," *Eng. Life Sci.*, vol. 11, no. 4, pp. 382–386, 2011.
[62] A. Baez and J. Shiloach, "Effect of elevated oxygen concentration on bacteria, yeasts, and cells propagated for production of biological compounds," *Microbial Cell Factories*, vol. 13, no. 1, p. 181, 2014.
[63] A. N. Emery, D. C. H. Jan, and M. Al-Rueai, "Oxygenation of intensive cell-culture system," *Appl. Microbiol. Biotechnol.*, vol. 43, no. 6, pp. 1028–1033, 1995.
[64] J. C. Merchuk, A. Contreras, F. García, and E. Molina, "Studies of mixing in a concentric tube airlift bioreactor with different spargers," *Chem. Eng. Sci.*, vol. 53, no. 4, pp. 709–719, 1998.
[65] S. K. W. Oh, P. Vig, F. Chua, W. K. Teo, and M. G. S. Yap, "Substantial overproduction of antibodies by applying osmotic pressure and sodium butyrate," *Biotechnol. Bioeng.*, vol. 42, no. 5, pp. 601–610, 1993.
[66] L. Xie *et al.*, "Large-scale propagation of a replication-defective adenovirus vector in stirred-tank bioreactor PER.C6™ cell culture under sparging conditions," *Biotechnol. Bioeng.*, vol. 83, no. 1, pp. 45–52, 2003.
[67] D. Birch and N. Ahmed, "The Influence of sparger design and location on gas dispersion in stirred vessels," *Chem. Eng. Res. Des.*, vol. 75, no. 5, pp. 487–496, 1997.
[68] C. Sieblist, O. Hägeholz, M. Aehle, M. Jenzsch, M. Pohlscheidt, and A. Lübbert, "Insights into large-scale cell-culture reactors: II. Gas-phase mixing and CO2 stripping," *J. Biotechnol.*, vol. 6, no. 12, pp. 1547–1556, 2011.
[69] X. Zhang *et al.*, "Efficient oxygen transfer by surface aeration in shaken cylindrical containers for mammalian cell cultivation at volumetric scales up to 1000L," *Biochem. Eng. J.*, vol. 45, no. 1, pp. 41–47, 2009.
[70] T. Kumaresan and J. B. Joshi, "Effect of impeller design on the flow pattern and mixing in stirred tanks," *Chem. Eng. J.*, vol. 115, no. 3, pp. 173–193, 2006.
[71] P. Vrabel, R. van der Lans, K. Luyben, L. Boon, and A. W. Nienow, "Mixing in large-scale vessels stirred with multiple radial or radial and axial up-pumping impellers: modelling and measurements," (in English), *Chem. Eng. Sci.*, vol. 55, no. 23, pp. 5881–5896, 2000.
[72] A. W. Nienow and M. D. Lilly, "Power drawn by multiple impellers in sparged agitated vessels," *Biotechnol. Bioeng.*, vol. 21, no. 12, pp. 2341–2345, 1979.
[73] P. M. Armenante and G.-M. Chang, "Power consumption in agitated vessels provided with multiple-disk turbines," *Ind. Eng. Chem. Res.*, vol. 37, no. 1, pp. 284–291, 1998.

[74] T. Moucha, V. Linek, K. Erokhin, J. F. Rejl, and M. Fujasová, "Improved power and mass transfer correlations for design and scale-up of multi-impeller gas–liquid contactors," *Chem. Eng. Sci.*, vol. 64, no. 3, pp. 598–604, 2009.
[75] R. Gelves, A. Dietrich, and R. Takors, "Modeling of gas–liquid mass transfer in a stirred tank bioreactor agitated by a Rushton turbine or a new pitched blade impeller," *Bioprocess Biosyst. Eng.*, vol. 37, no. 3, pp. 365–375, 2014.
[76] T. Frensing, A. Pflugmacher, M. Bachmann, B. Peschel, and U. Reichl, "Impact of defective interfering particles on virus replication and antiviral host response in cell culture-based influenza vaccine production," *Appl. Microbiol. Biotechnol.*, vol. 98, no. 21, pp. 8999–9008, 2014.
[77] B. Hundt, C. Best, N. Schlawin, H. Kassner, Y. Genzel, and U. Reichl, "Establishment of a mink enteritis vaccine production process in stirred-tank reactor and Wave Bioreactor microcarrier culture in 1-10 L scale," (in eng), *Vaccine*, vol. 25, no. 20, pp. 3987–3995, May 16 2007.
[78] H. S. Lim, K. H. Chang, and J. H. Kim, "Effect of oxygen partial pressure on production of animal virus (VSV)," (in eng), *Cytotechnology*, vol. 31, no. 3, pp. 265–270, 1999.
[79] L. Pelz *et al.*, "Semi-continuous propagation of influenza A virus and its defective interfering particles: analyzing the dynamic competition to select candidates for antiviral therapy," (in eng), *J. Virol.*, vol. 95, no. 24, p. e0117421, 2021.
[80] C. F. Shen and A. Kamen, "Hyperosmotic pressure on HEK 293 cells during the growth phase, but not the production phase, improves adenovirus production," (in eng), *J Biotechnol.*, vol. 157, no. 1, pp. 228–236, Jan. 2012.
[81] Z. Sun, R. Zhou, S. Liang, K. M. McNeeley, and S. T. Sharfstein, "Hyperosmotic stress in murine hybridoma cells: effects on antibody transcription, translation, posttranslational processing, and the cell cycle," *Biotechnol. Prog.*, vol. 20, no. 2, pp. 576–589, 2004.
[82] T. R. Kiehl, D. Shen, S. F. Khattak, Z. Jian Li, and S. T. Sharfstein, "Observations of cell size dynamics under osmotic stress," (in eng), *Cytometry. Part A: J. Int. Soc. Anal. Cytol.*, vol. 79, no. 7, pp. 560–569, Jul. 2011.
[83] D. Vazquez-Ramirez, Y. Genzel, I. Jordan, V. Sandig, and U. Reichl, "High-cell-density cultivations to increase MVA virus production," *Vaccine*, vol. 36, no. 22, pp. 3124–3133, May 2018.
[84] T. Frensing, S. Y. Kupke, M. Bachmann, S. Fritzsche, L. E. Gallo-Ramirez, and U. Reichl, "Influenza virus intracellular replication dynamics, release kinetics, and particle morphology during propagation in MDCK cells," (in eng), *Appl. Microbiol. Biotechnol.*, vol. 100, no. 16, pp. 7181–7192, Aug. 2016.
[85] G. K. Atkin-Smith and I. K. H. Poon, "Disassembly of the dying: mechanisms and functions," *Trends Cell Biol.*, vol. 27, no. 2, pp. 151–162, 2017.
[86] N. W. Baylor, "The Regulatory Evaluation of Vaccines for Human Use," in *Vaccine Design*: Springer, 2016, pp. 773–787.
[87] D. Chiodin, E. M. Cox, A. V. Edmund, E. Kratz, and S. H. Lockwood, "Regulatory Affairs 101: introduction to Investigational New Drug Applications and Clinical Trial Applications," *Clin. Transl. Sci.*, vol. 12, no. 4, pp. 334–342, 2019.
[88] R. Wahid, R. Holt, R. Hjorth, and F. B. Scorza, "Chemistry, manufacturing and control (CMC) and clinical trial technical support for influenza vaccine manufacturers," *Vaccine*, vol. 34, no. 45, pp. 5430–5435, 2016.
[89] FDA, "Guidance for Industry: Content and Format of Chemistry," in *Manufacturing and Controls Information and Establishment Description Information for a Vaccine or related product*, U.S. Department of Health and Human Services, Food and Drug Administration, Center for Biologics Evaluation and Research, 1999.

[90] EMA, "Guideline on quality, non-clinical and clinical aspects of live recombinant viral vectored vaccines,"London, United Kingdom: European Medicines Agency, 2010.
[91] A. Offersgaard *et al.*, "SARS-CoV-2 production in a scalable high cell density bioreactor," *Vaccines*, vol. 9, no. 7, p. 706, 2021.
[92] L. Sanchez-Felipe *et al.*, "A single-dose live-attenuated YF17D-vectored SARS-CoV-2 vaccine candidate," *Nature*, vol. 590, no. 7845, pp. 320–325, 2021.

ABBREVIATIONS

AAV	Adeno-associated virus
ADME	Adsorption, distribution, metabolism, and excretion
AdV	Adenovirus
AFM	Animal component free medium
amn	Ammonia
BEI	Binary ethyleneimine
BHK	Baby hamster kidney
BSL	Biosafety level
cc	Cell cube
C1	Chromatography step 1
C2	Chromatography step 2
CDM	Chemically defined medium
CEF	Chicken embryo fibroblast
CFD	Computational fluid dynamics
CFF	Cross-flow filtration
CHO	Chinese hamster ovary
CMC	Chemistry, manufacturing and controls
CRISPR/Cas	Clustered regularly interspaced short palindromic repeats
CRO	Contract research organization
CTA	Clinical trial application
DEAE	Diethylethanolamine
DF	Depth filtration
DIP	Defective interfering particle
DMSO	Dimethyl sulfoxide
DNA	Deoxyribonucleic acid
DoE	Design of experiments
DSP	Downstream processing
ECGF	Endothelial cell growth factor
EGF	Epidermal growth factor
EMA	European Medicines Agency
EP	European Pharmacopeia
FBS	Fetal bovine serum
FCS	Fetal calf serum
FDA	Food and Drug Administration
Fn	Fibronectin
GCP	Good clinical practice

glc	Glucose
gln	Glutamine
GLP	Good laboratory practice
GMO	Genetically modified organism
GMP	Good manufacturing practice
HA	Hemagglutinin
HCD	High cell density
HCV	Hepatitis C virus
HFBR	Hollow fiber bioreactor
hpi	Hours post infection
HSV1	Herpes simplex virus 1
IAV	Influenza A virus
iBET	Instituto de Biologia Experimental e Tecnológica
ic	Icellis
IGF	Insulin-like growth factor
IND	Investigational new drug (application)
IL	Interleukin
lac	Lactate
MCB	Master cell bank
MEV	Mink enteritis virus
MHS	Multivalent heparan sulfate
MIR	Mid-infrared spectroscopy
MMR	Measles, mumps, rubella
mRNA	Messenger RNA
MSL	Master seed lot
µ-F	Microfiltration
MVA	Modified Vaccinia virus Ankara
N	Number of cell expansions for production
NIBSC	National Institute for Biological Standards and Control
NIR	Near-infrared spectroscopy
NMR	Nuclear magnetic resonance
NRC	National Research Council Canada
OSB	Orbital shaken bioreactor
PAT	Process analytical tools
PBBR	Packed bed bioreactor
PBS	Phosphate buffered saline
PCR	Polymerase chain reaction
PDGF	Platelet-derived growth factor
pDNA	Plasmid DNA
PFM	Protein free medium
pSTR	Perfusion STR
RB	Roller bottle
Rec	Recombinant
RNA	Ribonucleic acid
QMS	Quality management system
SARS	Severe acute respiratory syndrome

SARS-CoV-2	Severe acute respiratory syndrome corona virus type 2
SCM	Serum containing medium
SFM	Serum free medium
SOP	Standard operating procedure
STR	Stirred tank bioreactor
STV	Standard virus
USP	Upstream processing
VLP	Virus-like particles
VSV	Vesicular stomatitis virus
WCB	Working cell bank
WHO	World Health Organization
WSL	Working seed lot
XFM	Xeno free medium
YFV	Yellow fever virus
ZIKV	Zika virus

6 Upstream processing for viral vaccines-Process intensification

Sven Göbel and Lars Pelz
Bioprocess Engineering, Max Planck Institute for Dynamics of Complex Technical Systems, Magdeburg, Germany

Udo Reichl
Bioprocess Engineering, Max Planck Institute for Dynamics of Complex Technical Systems, Magdeburg, Germany and Chair for Bioprocess Engineering, Otto-von-Guericke-University Magdeburg, Magdeburg, Germany

Yvonne Genzel
Bioprocess Engineering, Max Planck Institute for Dynamics of Complex Technical Systems, Magdeburg, Germany

CONTENTS

DOI: 10.1201/9781003229797-6

Symbols

CSVY	Cell-specific virus yield	virions/cell
c_{Vir}	Virus particle concentration	virions/mL
c_S	Substrate concentration	mM
c_X	Cell concentration	cells/mL
D	Dilution rate	1/h
k_La	Oxygen transfer coefficient	1/h
MOI	Multiplicity of infection	infectious units/cell
PFU	Plaque forming units	PFU/mL
pO_2	Partial pressure of oxygen	%
RV	Reactor volume	mL
RT	Residence time	h
SRID	Single radial immunodiffusion	HAU/(µg/mL)
STY	Space-time yield	virions/RV/d
T	Temperature	°C
$TCID_{50}$	Tissue culture infectious dose 50	$TCID_{50}$/mL
TOH	Time of harvest	h
TOI	Time of infection	h
t_{tot}	Total process time	h
VVP	Volumetric virus productivity	virions/L/d
V_{tot}	Total volume of spent medium	L
wv	Working volume	L
µ	Specific cell growth rate	1/h
µmax	Maximum specific cell growth rate	1/h

6.1 INTRODUCTION

In the previous chapter, general aspects of cell culture-based virus production were presented. Increasing world population, pandemic threats, high costs, and new applications demanding large virus quantities, i.e. gene or cancer therapy, increase the pressure to identify innovative solutions to intensify the conventional virus production processes. In 2019, the yearly global vaccine production was estimated to be between 3.5 billion and 5.5 billion doses of all vaccines [1]. With the current 2019 outbreak of the coronavirus SARS-CoV-2, assuming a two-dose regiment just for COVID-19 vaccines, approximately 12 billion doses would be required to vaccinate the 6 billion vaccine-eligible humans worldwide. Moreover, emerging markets for gene therapies and gene-modified cell therapies further escalate the need for more efficient production processes. It becomes very clear that classical virus production capacities are not sufficient to meet this increasing demand. According to the EMA (European Medicines Agency), gene therapy medicines consist of a "vector or delivery formulation/system containing a genetic construct engineered to express a specific transgene (therapeutic sequence) for the regulation, repair, replacement, addition, or deletion of a genetic sequence" [2]. While CAR T-cell therapies (chimeric antigen receptor T-cell) already require a relatively high amount of vector genomes (1E12 viral genomes (vg) per injection), it has been found that direct injection for treatment of e.g., muscular dystrophy would require even higher dose inputs (up to 1E14 adeno-associated virus (AAV) vg per kg of patient bodyweight). With such high-dose requirements, the current state-of-the-art AAV manufacturing process could provide enough material to treat two patients with one production run [3]. What the final demand might be, will be evident in the short future, but still new ways to produce large amounts of viruses will be needed.

Over the last years, many promising bioreactor concepts have been established and process options are now available to significantly increase the yield in virus particle production. In addition, not only the production processes might change, but equally the formulation and the administration of products. In this chapter, process intensification towards higher volumetric virus productivity (VVP), the amount of virus particles produced considering the total volume of spent medium during cell growth and virus replication phase and the total process time, is discussed. Many lessons can be learned from improvements made in recombinant protein production, but likewise many specific requirements for virus production need to be considered additionally. Clearly, this is not a complete coverage of all relevant literature, only a selection of examples highlighting this topic is discussed, mainly from recent studies of our group.

6.2 MOTIVATION FOR PROCESS INTENSIFICATION

Following up on the proposed calculation exercise of chapter 5 to estimate how many vaccine doses per year can be obtained for a chosen production method and the respective virus-host cell system, it is obvious that even advanced production platforms are quickly limited for pandemic scenarios. While implementation of new technologies (e.g., mRNA vaccines) could mitigate problems arising from the limited production capacity of viral vectors, the current coronavirus disease 2019 (COVID-19) pandemic revealed that not only good manufacturing practice (GMP)-production capacities are limited, but also the supply of equipment and consumables, delivery, transport, and the availability of qualified and trained personnel. Streamlining, organization, resource management, and fair distribution are equally needed, whereas panic-buying and hoarding policies have to be avoided. For infectious diseases, where cell-specific virus yields (CSVY) are as low as those identified for SARS-CoV-2 [4] or Zika virus [5] (about 20 infectious virions/cell) or even lower as for hepatitis C virus (HCV) [6] and no alternatives such as mRNA vaccines are available yet, process intensification is a must. In the case of gene therapies, which require a highly specific delivery of genomic material into the target cell, viral vectors are often the preferred choice. Although several non-viral gene delivery systems, e.g., DNA or mRNA, have been developed in the last three decades, efficiency of gene transduction is usually less than in viral systems [7].

Process options and technologies for process intensification described in the following relate mainly to small-scale experiments performed by academic groups. While licensing new and innovative processes often involves high costs, in particular for the case additional clinical studies are required, we are convinced that the ever-increasing demand for virus particle-based products will force vaccine manufacturers to implement new technologies and consider advanced process options.

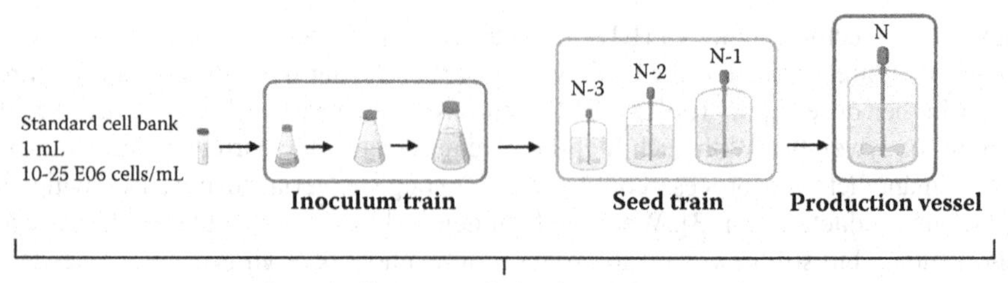

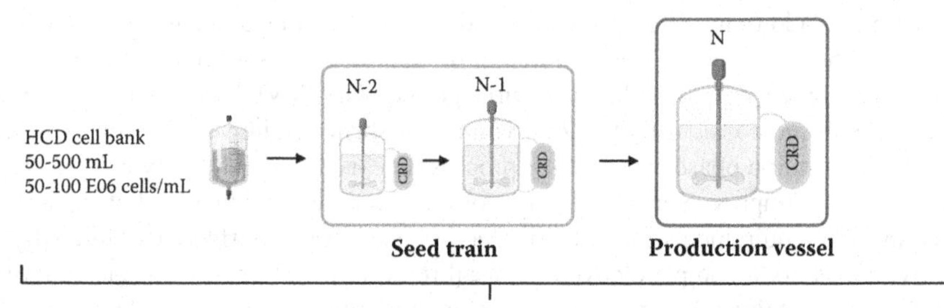

FIGURE 6.1 Comparison of a conventional seed train process with an intensified process using high-cell density cell banking and perfusion. HCD: high cell density; CRD: cell retention device. Production run (N: cell growth and virus production) and all scale-up runs (N-1, N-2, etc.) that involve cell growth only. Standard cell banks typically use 1–2 mL cryo flasks, whereas HCD cell banks use dedicated freezing bags with volumes up to 500 mL. Created with BioRender.com

In the case of virus production, process performance has to be improved at each stage of production,starting with upstream processing (USP) with increasing virus particle yields, CSVY and space-time yield (STY), continuing with purification strategies with a minimum number of unit operations and high recovery, debottlenecking and optimizing modular solutions for manufacturing [8]. For example, one general goal of process intensification in USP is to achieve high cell concentrations right from the start of thawing the cell bank, thereby reducing the time to reach the required cell concentrations at the final process scale with a minimum of scale-up steps (Figure 6.1). Based on high cell concentrations being reached in the final production vessel, overall virus titers and virus yields can be higher in case the CSVY can be kept at least constant. In this chapter, we will focus on USP and the related challenges with the aim to increase viral yields.

6.3 PARAMETERS FOR PROCESS COMPARISON AND EVALUATION

Before addressing advanced concepts for process optimization and intensification, tools or parameters for an unbiased comparison of virus production platforms are needed. Therefore, in Table 6.1, the most commonly used measured or calculated parameters are introduced using IAV as an example.

TABLE 6.1

Parameters considered for process optimization and comparison of productivities in virus production

Name	Typical standard deviation or equation	Units	Input for process optimization	Other comments	Ref.
Cell concentration	±6–10% (Vicell, Beckman Coulter)	cells/mL			Beckman Coulter
HA titer	±0.15 $\log_{10}$ (1:$2^{0.5}$ dilution series)	$\log_{10}$ HA units/100 µL	Allows estimation of IAV particle concentration and total number of virus particles. Quick and easy measurement	Only meaningful as time course. Depends on IAV strain. Different sources for the red blood cells are used (chicken, turkey, camels, seals), concentration not always clearly defined.	[9]
$TCID_{50}$	± 0.3 $\log_{10}$	$TCID_{50}$/mL	Infectious virus titer Time consuming assay	Only meaningful as time course, estimate for infectious virus particles, PFU and $TCID_{50}$ are not equivalent! Viral infectivity is a critical parameter to determine MOI.	[10,11]
PFU	± 0.2 $\log_{10}$	PFU/mL	Infectious virus titer Time consuming assay	Only meaningful as time course, estimate for infectious virus particles, PFU and TCID50 are not equivalent! Viral infectivity is a critical parameter of an infectious virus.	[12]
SRID	± 10%	HAU/(µg/mL)	Standard assay for vaccine release	SRID assays are used to determine potency of inactivated virus vaccines and are accepted by regulatory authorities	[13–15]
vRNA	± 25%	vg	Viral genome copy numbers	Only meaningful as time course, estimate for total virus particles, but not all RNA is packed into virus particles, more sensitive than HA assay	[16]

(*Continued*)

TABLE 6.1 (Continued)

Parameters considered for process optimization and comparison of productivities in virus production

Name	Typical standard deviation or equation	Units	Input for process optimization	Other comments	Ref.
Cell-specific virus yield (CSVY)	error propagation	virions/cell	Comparison of virus yields independent from cell concentration	Should be either total or infectious yield, high error from cell concentration and virus titer determination	
Space-time yield (STY)	error propagation	virions/RV/d (e.g., $TCID_{50}$/L/d)	Considers total process time and space (bioreactor working volume)	Does not consider perfusion medium	
Volumetric virus productivity (VVP)	error propagation	virions/L/d (e.g., $TCID_{50}$/L/d)	Considers total volume of medium needed and total process time	Considers all utilized perfusion medium	

Grey background: measured values, white background: derived parameters. IAV: Influenza A virus, $TCID_{50}$: Tissue culture infectious dose, SRID: single radial immunodiffusion, HA: hemagglutinin, PFU: plaque forming units. Values of standard deviation are typical values obtained by in-house assay validation or reported in literature. Log dilution assays are always limited by the dilution error. We are aware that also other definitions of CSVY, STY and VVP are possible, depending on the time point the maximum cell concentration is achieved, volume used and time range considered.

Besides the total number of cells at time of infection (TOI), the maximum total number of cells reached, the total number of virus particles, and/or the total number of infectious units, the CSVY is an important parameter to allow for a comparison of the performance of different cell lines at different cultivation conditions. As both cell concentration and titer measurements have a relatively high error, interpreting small differences might be challenging and not relevant. The same applies to derived parameters (like the CSVY) that involve error propagation. Nevertheless, if proper controls are used and assays are carried out carefully with reliable and precise techniques and in a systematic manner, the CSVY is a very good indicator for success in process optimization.

The space-time yield (STY) [virions/RV/d] is the amount of virus particles produced taking into account only the working volume of the bioreactor harvested (RV) and the total process time [d]. This performance indicator allows the comparison of different production platforms of various sizes. Whereas the volumetric virus productivity (VVP) [virions/L/d] is the amount of virus particles produced considering the total volume of spent medium (V_{tot}) [L] during cell growth and virus replication phase and the total process time (t_{tot}) [d]. As before, for meaningful comparisons, assay limitations and error propagation have to be taken into account.

6.4 FROM BATCH CULTURES TO INTENSIFIED PROCESSES

Generally, bioreactors are operated in discontinuous modes (batch, repeated batch, or fed-batch), continuous modes (chemostat), or perfusion modes [17]. In conventional cell culture-derived vaccine production, bioreactors are mainly operated in batch mode with STRs operated as a closed system with no addition of fresh media. Cells are usually grown to relatively low cell concentrations (1–2E06 cells/mL, maximum 5E06 cells/mL for suspension cells) and subsequently infected (Figure 6.2A). The ease of implementation, process robustness combined with low

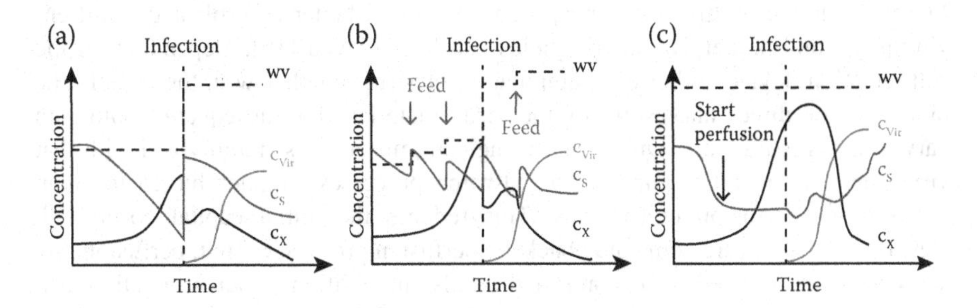

FIGURE 6.2 Schematic representation of virus production processes in batch (a), fed-batch (b), and perfusion (c) mode. Concentration profiles of various process parameters: wv, working volume; c_{Vir}, virus particle concentration; c_S, substrate concentration (glucose), c_X, cell concentration. Dashed vertical line shows TOI. For batch and fed-batch at TOI a medium exchange/dilution step by addition of medium results in a decrease of cell concentration and an increase in substrate level.

demands for sensors and few operative interventions are the main reasons for the wide use of batch processes in industry. Moreover, high VVP and high nutrient use can be achieved, thereby reducing the amount of required media. One major drawback is the large downtime between runs (cleaning, setup of next run, sterilization), which renders such processes often less economical. In addition, to plan, build, validate, and commission plants for conventional large-scale bioreactor operation in batch mode is costly and time consuming [18]. For this reason, there is a growing interest in realizing the benefits of fed-batch, perfusion, and continuous biomanufacturing. In addition, single-use production receives increasing attention.

A first step towards process intensification can be achieved by applying a fed-batch strategy to increase cell concentrations and virus yields. Here, an initial growth phase of the cells under batch conditions is followed by a (typically stepwise) addition of fresh medium or concentrated substrates. This leads to a corresponding increase in working volume and an extension of the cell growth phase as feeding provides additional nutrients to the cells (Figure 6.2B). This strategy has advantages over conventional batch production as higher cell concentrations before infection can be reached, which can result in higher virus titers as well as increased STY. However, as higher cell concentrations are achieved, accumulation of secondary metabolites like lactate and ammonia can inhibit cell growth and virus replication. Moreover, medium addition to bioreactors is limited due to restrictions in the total volume of vessels, aeration, and mixing.

One strategy to overcome most of the limitations of fed-batch processes is operation in perfusion mode. Cell retention devices are used to retain cells in the bioreactor throughout the run, a continuous renewal of medium without or with low cell losses can be achieved. After an initial batch phase, the perfusion mode is initiated where fresh medium is added to the bioreactor at a controlled rate, while spent medium is simultaneously removed at the same flow rate to keep the bioreactor volume constant (Figure 6.2C). As nutrient limitations and accumulation of unwanted toxic-by products are prevented, high cell concentrations can be reached. Moreover, longer cultivation times, higher viability, better control of the cell environment, and higher volumetric yields can be achieved [19]. Depending on the cell retention device, either co-accumulation of virus together with the cells in the bioreactor or direct harvesting of the virus material and subsequent cooling in harvest tanks is possible. Direct harvesting can impact virus stability and with that virus titer. All in all, compared to fed-batch processes, smaller bioreactors can be used for perfusion cultivations. This reduces the initial capital costs [19]. Nevertheless, there are some drawbacks to perfusion processes. First, perfusion rate and cell retention need to be controlled tightly. In addition, efficient aeration systems have to be used as surface aeration is not sufficient to supply oxygen in high cell density (HCD) cultures. While microspargers can ensure a high mass transfer coefficient (k_La) even for low gas flow rates, the generated microbubbles may lead to foaming problems. On the other hand, macrospargers resulting in larger bubble sizes might reduce foaming and sparging-related cell death, but require higher gas flow rates, to supply similar cell concentrations [20]. More on aeration and CO_2 control can be found in the previous chapter. Problems related to aeration but also mixing and pH control are particularly challenging at high cell concentrations with their high metabolic demand. Resulting heterogeneous cell populations (due to e.g., changes in cell size, cell aggregates, syncytia) can cause problems that require specific process control strategies to maintain optimal conditions for consistent virus production [20]. In combination with long operation times, technical failures are more likely to occur. As a consequence, operating staff needs to be specially trained. Another important point to consider is the use of media. Compared to recombinant protein production, there are only a handful of commercially available media for virus production. Even fewer media are optimized for cultivations performed at low perfusion rates to minimize media consumption. Furthermore, the very high virus titers that can be achieved in intensified processes might need additional biosafety measures.

Alternatively, viral yields can be increased utilizing true continuous production systems such as chemostats [21] or two-stage systems [22]. In a chemostat, new medium is supplied continuously, while consumed medium, virus, and cells are harvested. Here, the feed and harvest have the same flow rates to enable a constant working volume. In this system, the applied dilution rate (D) determines the specific cell growth rate (μ); hence, the cells are maintained in a steady-state condition (D is equal to μ). However, if the applied D is too high and approaches the maximum specific cell growth rate (μ_{max}), cell washout occurs, which has to be avoided. Chemostats can only be used for the production of non-lytic viruses as lytic viruses do not allow continuous cell growth. For lytic viruses, cell growth and virus propagation have to occur in separated vessels, which can be realized with a two-stage continuous stirred tank reactor (CSTR) cultivation system. Here, cells are grown in the first stirred tank bioreactor (STR) (operated as a chemostat) under steady-state conditions and continuously transferred into a second STR, where virus propagation takes place. The application of a two-stage CSTR was already reported for the production of baculovirus [23–25], poliovirus [26], influenza A virus (IAV) [22], and modified Vaccinia virus Ankara (MVA) [27]. For the genetically stable virus MVA, the production in a two-stage CSTR showed a higher concentration of virions produced compared to batch mode starting at 25 days of process time [28]. However, for IAV, the production led to periodic oscillations in virus titers, decreasing the VVP significantly compared to batch cultivations [22]. This was mainly due to the accumulation of defective interfering particles (DIPs, see previous chapter). This limits the usability of continuous processes for the production of IAV and other viruses spontaneously generating DIPs. Further limitations comprise the required high qualification level for technical staff and the increasing risk of contaminations due to the complexity of the setup and the prolonged process time.

A simplified overview of the most important parameters of all operation modes is given in Table 6.2. In summary, perfusion systems seem to be the most promising option to satisfy the ever-growing demand for viruses. As higher cell concentrations compared to batch and fed-batch processes are reached, higher VVP can be attained. As a result, smaller-sized equipment might be used, thereby reducing fixed costs and correlated peripheral costs e.g., facility, clean in-place operation, downtimes. Combined with today's trend to utilize single-use technology, the small footprint of perfusion systems is seen as an enabler for this development [18]. For viruses that only have very low CSVY such as HCV or flaviviruses, bioreactor

TABLE 6.2
Comparison of parameters for available bioreactor operation modes

	Batch	Fed-batch	Perfusion	Continuous
Volume	Constant	Increasing	Constant/Increasing	Constant
Cell removal	No	No	No	Yes
Specific cell growth rate	Decreases	Decreases	Decreases	Constant
Maximum cell conc. (x1E06 cells/mL)	1–10	2–20	10–160	2–5
Cells	Adherent/ suspension	Adherent/ suspension	Adherent/ suspension	Suspension
Qualification staff	Low	Modest	High	Highest
Contamination risk	Low	Modest	High	Highest
Foot-print	Large	Smaller	Smallest	Small
Medium requirement	Medium	Concentrated feed	Concentrated feed & medium	Medium
Equipment	Larger volume, reduced auxiliary equipment	Larger volume, reduced auxiliary equipment	Lower volume, increased auxiliary equipment	Lower volume, increased auxiliary equipment
Sensors	°C, pH, pO_2	°C, pH, pO_2	°C, pH, pO_2,	°C, pH, pO_2
Options for single-use*	Yes	Yes	No, some yes	No, maybe

Note
* Options of single-use bioreactors is rather limited by size than by operation mode (single-use bags up to 3,000 L).
pO_2: partial pressure of oxygen.

operation in perfusion mode might be the only option for economical production of vaccines. However, for perfusion systems to be competitive in real-life industrial applications, robust and efficient retention methods are required to maintain high densities of viable cells inside the bioreactor. An overview on today's available cell retention devices (CRDs) is given below (Figure 6.3).

6.5 CRITICAL FACTORS FOR VIRUS PRODUCTION AT HIGH CELL DENSITY

Virus production processes differ significantly from those of most cell culture-based biologicals. Although these processes commonly use animal cells, most biological products are accumulated in the cell culture broth during the cell-proliferation phase in batch or fed-batch mode and harvested once peak concentrations are reached. In contrast, virus production typically requires a biphasic process, which comprises a

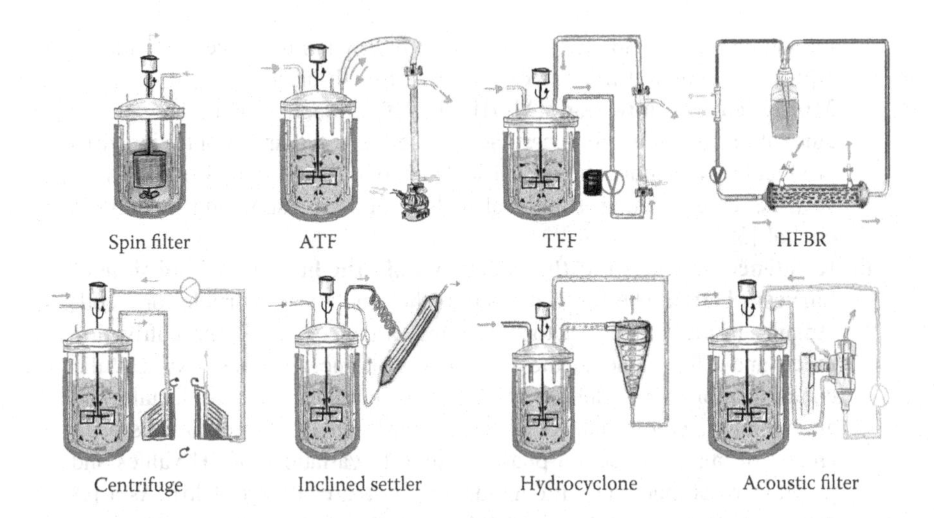

FIGURE 6.3 Available cell retention devices/options and their connections to a bioreactor vessel. Gray arrows indicate the direction of flow, blue arrows indicate cooling systems or other flows (e.g., air for ATF diaphragm pump) with respective flow directions. The symbol (V) enclosed by a circle represents a pump. Orange dots represent cells. Abbreviations: ATF: alternating tangential flow filtration; TFF: tangential flow filtration; HFBR: hollow fiber bioreactor, here the bioreactor itself is responsible for the cell retention, therefore not explicitly further discussed as cell retention device.

cell growth phase followed by a virus replication phase that is initiated with addition of a seed virus (see Figure 6.2A). Following intracellular virus replication, progeny virions are assembled and released. In the case of lytic viruses, this process is coupled with cell lysis resulting in release of contaminants including host cell proteins and cellular DNA [29]. For the establishment of HCD cultures with the respective high cell concentration (>1E07 cells/mL) and subsequent virus production phase, the cell growth phase will be extended depending on cell inoculation concentration and the respective maximum cell concentration envisaged. The most important parameter is the CSVY, which should not decrease by increasing cell concentration at TOI. However, the so-called "cell-density effect," a reduction of the CSVY at high cell concentrations, has been reported repeatedly for previous virus production processes [29–34]. While the exact cause for this decrease still often remained unclear, it can often be related to the accumulation of inhibitory factors such as ammonia and lactate or a limitation of nutrients [33]. In addition, it was often speculated that still unknown inhibitors of virus replication would accumulate [31,33]. Chapter 5 outlined three key factors that are critical to intensify the production process and maximize the virus yields [29]:

i. **Metabolic state of the cells and cell concentration at TOI:** Well-nourished cells with an appropriate growth and metabolic state as well as a sufficient supply of extracellular substrates is necessary to achieve maximum virus yields [35]. To achieve high virus titers, cells should be typically infected at the late exponential growth phase.

ii. **Multiplicity of infection (MOI) at TOI:** Infection with an optimal amount of infectious virus particles per cell is necessary to optimize virus production. This concerns in particular maximum titers and time of harvest. As before, ample supply of cells with a corresponding medium is crucial [35].

iii. **Residence time (RT) of the virions within the bioreactor and time of harvest (TOH):** The RT is defined as the time a virus particle (or a cell) remains inside the bioreactor/CRD and is dependent on the cultivation mode and CRD. Due to the lytic nature of most viruses, extracellular contaminations (cell debris, inhibiting metabolites such as lactate and ammonia, enzymes such as proteases, host cell DNA) continuously increase during the infection phase, leading to variations of pH values and possible virus inactivation, degradation, or aggregation [29,35]. As a result, an early TOH and a short RT is beneficial for viral vaccines (e.g., live attenuated vaccines, viral vectors), where potency is defined by infectivity. Moreover, the higher the contamination level the more difficult purification will become.

Perfusion processes involving a complete medium exchange prior to infection [11,36–38] or a continuous medium renewal during cell proliferation and virus propagation constitute the foundation for an optimal metabolic state of the cells. In addition, medium renewal can prevent the accumulation of unwanted inhibitors of cell growth and virus replication and with that reduce the so-called "cell-density effect".

Many recent studies investigating HCD processes for viral vaccine production have successfully demonstrated that the CSVY can be maintained or even increased. Moreover, intensification strategies such as adaptation of various feeding schemes after infection (e.g., "hybrid fed-batch/perfusion [37]) can further increase both CSVY and VVP compared to perfusion-only strategies. To shorten seed train timelines and maximize plant flexibility, inoculation procedures such as FASTEC (frozen accelerated seed train for execution of a campaign) consisting of HCD inoculation with disposable cryopreserved bags are increasingly finding application [39]. The use of online capacitance probes for monitoring the cell concentration/volume during cell growth and virus production allows for increased process robustness through improved control of substrate concentrations for example [11,40,41]. In some cases, signals obtained can be correlated with the time of virus particle release or maximum virus titers, which could be used to improve virus harvesting strategies [41,42].

Besides the selection of appropriate process strategies, culture medium is the most important factor in cell culture technology. The medium selected needs to ensure all basic cell functions, proliferation, cell survival, and has a direct influence on the product yield and quality. Compared to approaches requiring genetic engineering of cell lines, media optimization is simple, delivers relatively fast results, and is reasonably inexpensive at laboratory scale. Nevertheless, for host cell lines used in virus production, there are currently only a few media commercially available that support all process needs. In particular, medium development should consider specific requirements of cells to achieve high cell concentrations versus those relevant to accomplish high virus yields. Furthermore, there is certainly a need for medium formulations, which can support high cell concentrations at low volumetric perfusion rates [18].

Regarding DSP, adverse consequences cannot be excluded for intensified viral HCD processes. This concerns in particular, potential problems with subsequent unit operations such as filtering or chromatography, i.e., filter blocking or losses in purification steps, and is directly related to the incremental increase in contaminating host cell DNA and proteins with increasing cell concentrations. However, for processes operated in perfusion mode, high medium exchange rates after virus infection can mitigate the accumulation of such contaminants in the virus harvest. For example, Gränicher et al. showed that process intensification in MVA production had no negative impact regarding cell clarification, host cell DNA removal, and purification compared to a batch process [41]. Furthermore, the ratio of viral genome copy numbers to infectious virions and antigen glycosylation were not affected by HCD cultivation [37]. Nevertheless, small-scale studies addressing the impact of process changes on the performance of unit operations in DSP are required to avoid virus losses and guarantee that an intensified process will consistently result in a product that meets its predetermined specifications and quality attributes are to ensure safety and efficacy of vaccines.

6.6 CELL RETENTION DEVICES

The retention of cells inside of the bioreactor is critical to reach high cell concentrations. Therefore, the selection of the appropriate retention device including the corresponding parameters for its operation is of utmost importance. A detailed overview of various CRDs is given in Table 6.3. For industrial processes, CRDs need to comply to GMP requirements, should be commercially available at various sizes (preferably in single-use), combine a high perfusion capacity (at least 1,000 L d^{-1}) with a high retention efficiency while not damaging the cells, allow high-yield production, and operate over a complete run without maintenance [43]. Nowadays, a large variety of CRDs are available (Figure 6.3). Membrane-based systems such as spin-filters [32], tangential flow filtration (TFF) [5,44], or alternating tangential flow filtration (ATF) [37,38,45] are currently the most commonly considered systems for virus vaccine production. Hollow-fiber bioreactors (HFBRs) [46] follow the same idea; however, here the bioreactor itself is responsible for the cell retention within the extracapillary space. Due to the lytic nature of most viruses, their size (up to 350 nm for MVA), and their surface properties, the usage of membrane-based retention devices has been shown to be challenging [37,38]. Particularly, membrane clogging and unwanted virus accumulation inside the bioreactor or the modules are well known drawbacks of most membrane-based retention systems. Alternatively, retention technologies that make use of density differences for separation can be used. Such systems do not use a physical barrier and can allow sustainable long-term operation [43]. Examples for such devices are acoustic settlers [47–50], centrifuges, hydrocyclones [51], and inclined settlers [52,53] (Figure 6.3). Moreover, those devices

TABLE 6.3
Cell retention devices/options and some important parameters for their use

System	Working volume range (L)	Max. cell conc. (x 1E06 cells/mL)	Max. perfusion rate (L/d)	Cell debris removal	Residence time in equipment	Single use	Additional equipment	Cells used	Virus production demonstrated	Reference
Spin filter	0.5–500	80	0.5–500	No	n.a.	No	Peristaltic pump	CHO, Hybridoma, BHK21	RABV	[32,54]
HFBR	0.004–0.1	72	288	No	Process time	Yes	Duet pump, incubator, oxygenator	MDCK, AGE1.CR, HEK293, Huh7.5	IAV, MVA, HIV, HCV	[6,28,46,55]
PBBR	0.5–100	150	200	No	Constant	Yes	Peristaltic pumps, controller, incubator	Vero, HEK293, MRC-5, Huh7.5, MDCK, CHO	IAV, SARS-CoV-2, HCV, VSV, Ad5, AAV, IPV, RABV, HAV, CHIKV	[4,6,56–59]
TFF	0.01–2000	200	100	No	10 s	Yes	Levitronix pump, peristaltic pumps	AGE1.CR, BHK21, HEK293, CHO	IAV, YFV, Ad5, ZIKV	[5,44,60]
ATF	0.5–5000	130–360	1200	No	1–2 min	Yes	ATF controller, peristaltic pumps	MDCK, AGE1.CR, BHK21, EB66, CHO, PBG.PK2.1	IAV, MVA, ZIKV, YFV, Ad5	[5,36–38, 44,45,61]

Inclined settler	0.5–3000	20, 50	3000	Yes	>30 min	No	Cooling loop, peristaltic pump	CHO, HEK293, AGE1.CR, BHK21, Hybridoma	IAV	[62]
Acoustic filter	0.5–200	50	1000	Yes	>10 min	No	Cooling, controller, peristaltic pumps	HEK293, AGE1.CR	IAV, MVA, Ad5, LV	[41,48–50]
Hydrocyclone	5–500	50	2200	Yes	0.03–0.1 s	No	Peristaltic pump	CHO	No	
Centrifuge	1–1000	30	3700	Yes	2–9 min	Yes	Peristaltic pump	Hybridoma, CHO	No	

White fields: Membrane-based/separation by size (PBBR use of macrocarriers for cell attachment); grey fields: separation by sedimentation. HFBR: Hollow-fiber bioreactor, PBBR: packed-bed bioreactor, TFF: tangential flow filtration, ATF: alternating tangential flow filtration, CHO: Chinese hamster ovary cells, MDCK: Madin Darby canine kidney cells, HEK293: human embryonic kidney cells, Huh7.5: human hepatocellular carcinoma cells, AGE1.CR: avian duck cells, IAV: influenza A virus, CHIKV: chikungunya virus, HIV: human immunodeficiency virus, HAV: hepatitis A virus, HCV: hepatitis C virus, MVA: modified vaccinia virus Ankara, ZIKV: Zika virus, YFV: yellow fever virus, IPV: inactivated Polio virus, AD5: adenovirus-5, LV: lentivirus; RABV: rabies virus, n.a.: not applicable. Adapted from [17,19,43,52,53,63]. Homepages of Fibercell systems, Sartorius, Repligen, and other manufacturers.

would allow for continuous virus harvest, which has been shown to have potential advantages. For example, Gränicher et al. (2020) demonstrated an increase of CSVY of at least 1.5 through continuous removal of IAVs using an acoustic settler compared to an ATF system [50]. Whether continuous harvest of virus particles with short RTs or retention within the bioreactor during the whole production phase is beneficial depends on the virus and has to be evaluated beforehand [29].

6.6.1 Spin Filters

Spin filters were first introduced as a cell retention device for HCD cultivations of mammalian suspension cells by Himmelfarb et al. in 1969 [64]. Here, a cylindrical membrane placed inside or outside the bioreactor is rotated around the same axis as the impeller (see Figure 6.4). Rotation is initiated either by mounting the spin filter on the impeller shaft or an independent motor [53]. Cell-free permeate is pumped into the cylinder and removed through the harvest line while cells are retained in the vessel, outside of the spin filter by the membrane. There are two major differences between spin filters and other cross-flow filters. First, the rotation of the cylinder is used to create a fluid flow relative to the filter surface which is placed inside the bioreactor. Therefore, the cells remain in a controlled environment, the retentate cell concentration is

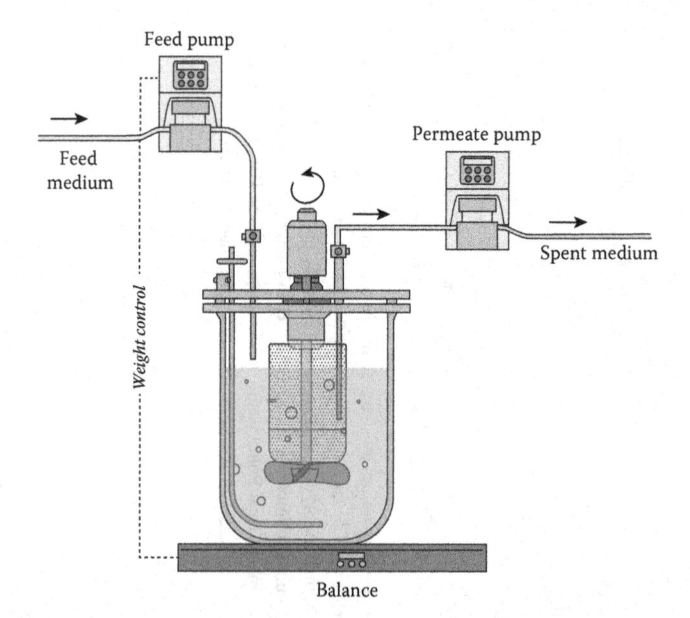

FIGURE 6.4 Schematic illustration of a spin filter setup. Cell-free supernatant is constantly removed from the interior of the cylindrical spin filter, allowing fresh medium to be added to the bioreactor. Cells are retained on the outside of the cylinder. In this setup, impeller and filter rotate independently. Figure adapted from [65].

homogenously distributed over the complete filter surface. Second, the transmembrane pressure drop can be controlled independently from the cross-flow velocity [53].

Both the cells and particles/cell debris in the broth are exposed to a multitude of various forces: gravity force, axial force caused by impeller rotation, centrifugal force created by the filter rotation, and radial force due to the perfusion flux [52]. In case external filters are used, additional secondary flows (so called Taylor vortices) are formed, which are supposed to produce abrasion effects preventing fouling, although this has been shown to play a minor role in the filtration performance [63]. Implementation of spin filters resulted in high cell concentrations in a variety of experiments. However, short perfusion times and premature termination of many runs highlight the major concerns regarding the use this technology: fouling and clogging [43,53]. Since replacement of internal units is impossible, spin-filters should be designed and operated to reduce both risks [52]. Fouling is mainly caused by deposition of dead cells and nucleic acid on the filter [19], while clogging can occur either if cells accumulate in the pores when the filtration flux exceeds the retention capacity of the filter or when pores are increasingly narrowing down due to cell growth on the filter surface [43]. Several points have to be considered for efficient operation of spin filters: Hydrophobic plastic instead of stainless steel should be used as a filter screen material due to lower binding of proteins, nucleic acid, and cells [52,66]. By using larger pore sizes (up to 50 μm) the filter longevity can be increased while allowing a selective retention of viable cells [52]. Moreover, this could allow virus particles to pass through the membrane. As for cross-flow filters, the risk of filter fouling increases with cell concentration and perfusion rate [53,67]. Transmembrane pressure differences in the range of 0.5–1 bar were found for maximum perfusion fluxes. At smaller (reduction of the driving force) or at larger differential pressures (improved cake layer formation), less favorable filtration conditions are present. In particular, it is critical not to use too high perfusion rates. Henzler et al. calculated maximum perfusion rates in relation to specific membrane areas based on literature data compiled by Castilho and Medronho and Voisard on cultivations with spin filters [43,52,63]. They showed that high perfusion rates can only be achieved with very large filter areas. This is highly unfavorable for intensified virus production processes and scale-up, as high cell concentrations require high perfusion rates. Fouling caused by high perfusion rates could be partially compensated by increasing the tangential rotation speed. However, too high rotation speeds were shown to decrease cell retention efficiencies [52]. Nevertheless, successful scale-up was demonstrated up to 500 L at a rate of 1 RV/d [67].

While spin filters are mainly used for the production of monoclonal antibodies, some studies investigated their use for the production of an experimental rabies vaccine using suspension BHK21 cells [32,54]. Despite much progress regarding the design, the scale-up and the operation of spin filters over the last decade, fouling or retention problems still persist which could be particularly problematic for the production of viruses.

6.6.2 Tangential Flow Filtration

Originally developed as a rapid and efficient DSP method for separation and purification of biomolecules, tangential flow filtration (TFF) can be applied in a wide range of biological processes including cell retention for perfusion cultures. In the TFF system, cell broth is pumped through a dip tube in the lumen of a hollow fiber module via a peristaltic pump. A simplified overview and description of the setup is shown in Figure 6.5. This leads to a perpendicular flow of the cell broth to the permeation direction [19,52]. The hollow fiber module consists of multiple fibers that are mounted in parallel in the cartridge. Each cartridge has a specific cut-off (typically 0.1–10 μm) retaining the cells within the hollow fibers, while spent medium is withdrawn as permeate that is passing through the membrane. Due to the relatively high shear stress at the filter surface caused by the tangential flow of the cell broth, filter fouling is reduced, allowing continuous operation and high filtration flux. Indeed, TFF systems have shown reduced fouling compared to spin filters. However, shear rates and thus feed flow rates are limited by the shear stress sensitivity of animal cells (Castilho 2002). Although various mammalian and insect cell lines were shown to survive hydrodynamic stress up to $\gamma = 3{,}000\ s^{-1}$ [68] or $\gamma = 4{,}000\ s^{-1}$ [69], values as low as $\gamma = 620\ s^{-1}$ were shown to be harmful for suspension HEK293 cells after adenovirus infection [60]. Moreover, since recirculation is based on peristaltic pumps, high local shear stress areas are generated by the pumps further limiting the use for the cultivation of shear stress sensitive cell lines. This issue can be overcome by implementation of low shear stress centrifugal

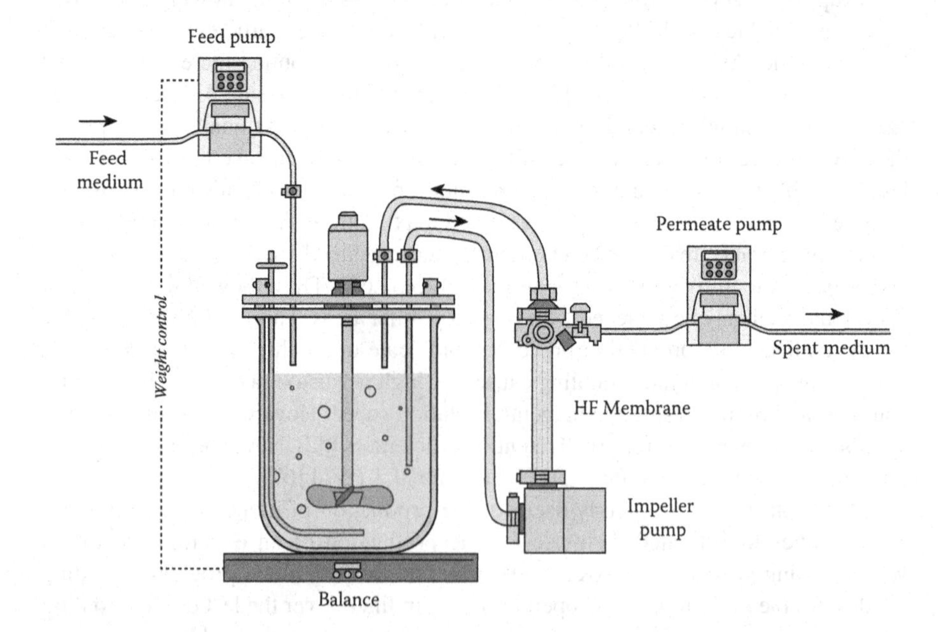

FIGURE 6.5 Schematic illustration of a setup of a TFF as CRD. Peristaltic pumps or centrifugal pumps are typically operated in constant or pulsed unidirectional flow. Cell-free supernatant is constantly removed in a direction orthogonal and radial to the hollow fiber filters, allowing fresh medium to be added to the bioreactor. Compared to ATF setups (see Figure 6.6) the working volume in the bioreactor remains constant. Figure adapted from [65].

pumps with magnetic levitation (e.g., Puralev®, Levitronix). By use of such a centrifugal pump, Coronel et al. demonstrated similar production yields of IAV in orbital shaken bioreactors for both ATF and TFF systems, respectively [44]. Nevertheless, and despite of the partial cleaning of membranes by the tangential flow, membrane fowling is common in TFF systems at HCD and during virus production due to the accumulation of cell debris caused by cell lysis. During the virus propagation phase, almost complete retention of virus particles in the membrane was commonly observed [70]. Therefore, in particular for virus production, material and structural properties of the hollow fiber membranes selected are critical for efficient production. A systematic overview on different membrane materials, and structural and physicochemical properties with respect to filter fouling and virus harvesting is given by Nikolay et al. (2020). With new membranes/materials available, it will certainly be worthwhile to re-evaluate this method.

Although many problems including fouling, shear stress and product retention are not completely solved yet, simple scale-up, good reproducibility, and suitability for various cell lines and viruses make TFF an attractive option for cell retention in virus production.

6.6.3 Alternating Tangential Flow

First introduced in 2000 by Shevitz, the ATF systems have been since then applied in a variety of research and industry projects [71]. The external, pressure-based ATF system consists of a hollow fiber unit, which is positioned between the bioreactor and a diaphragm pump. A simplified overview and description of the ATF setup is shown in Figure 6.6. The diaphragm pump pulls and pushes the cell suspension in the hollow-fiber unit in an alternating way. This process can be divided into two phases: exhaust cycle and pressure cycle [19]. Active filtration occurs during the pressure cycle. A vacuum causes the convex diaphragm to be pulled downwards into the chamber, thereby increasing the volume of the liquid chamber and drawing cell suspension into the filter unit. In the filter unit, tangential filtration of the cell broth occurs. The following exhaust cycle creates the alternating nature of the ATF system by pushing the diaphragm with pressurized air back into the liquid chamber. This backflush transfers the filtrate back into the bioreactor without the need for an additional peristaltic pump thereby reducing the shear stress for cells. Moreover, the entire filter length is backflushed reducing the risk of blocking of the membrane pores [72].

A peristaltic pump removes constantly the separated permeate and fresh medium can be added directly into the bioreactor, while the cells are retained in the hollow fiber unit. For cross-flow filtration, cake-formation (increasing thickness with increasing axial distance) and tubular liquid flow profiles can occur that lead to oscillating fluctuations in pressure and sporadically declining fluxes [72]. The cross-flow velocity minimizes the tendency of membrane fouling, but even in ATF systems high cell concentrations and long growth cycles might ultimately lead to filter clogging.

As mentioned before, the ATF system is very well-established and commercially available. Hollow fiber units can be obtained from several suppliers. In animal cell culture, excellent separation performance was demonstrated in HCD cultivations up to 2E08 cells/mL, and currently most perfusion-based processes for recombinant

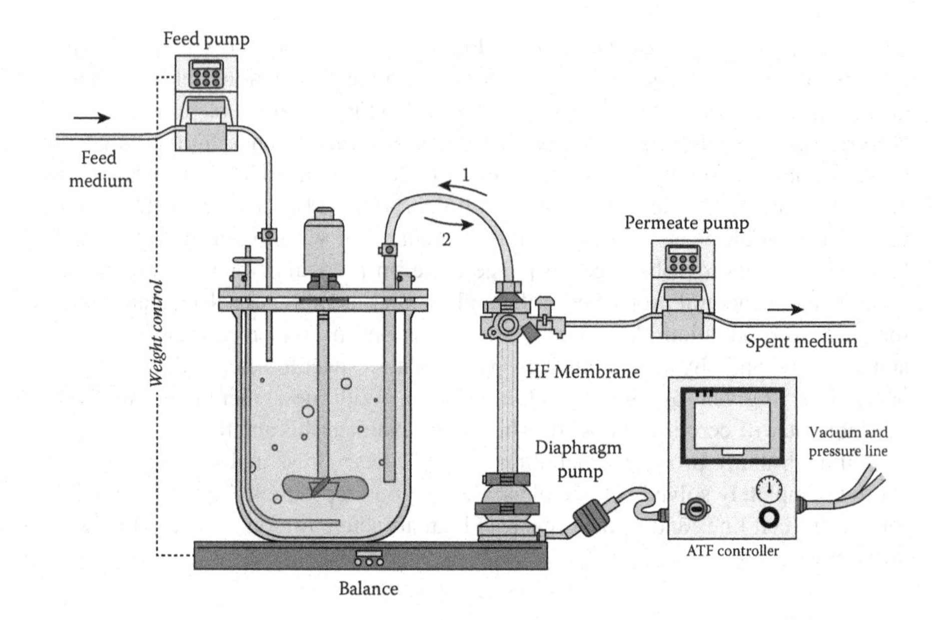

FIGURE 6.6 Schematic illustration of an ATF setup and a hollow-fiber unit. The diaphragm pump pushes the liquid in a bidirectional flow direction: 1) Exhaust cycle: Pressurized air is pressing the diaphragm into the liquid chamber, resulting in a backflush of the cells and medium into the bioreactor. 2) Pressure cycle: A vacuum pulls the diaphragm into the liquid chamber, resulting in an inflow of cell broth into the filter unit. Cell-free supernatant is removed through the membrane pores during the pressure cycle, allowing fresh medium to be added to the bioreactor. Figure adapted from [65].

protein production use ATF systems [28,73,74]. Vázquez-Ramírez et al. used an ATF-based perfusion process for the production of MVA using suspension AGE1.CR.pIX cells, reaching up to 50E06 cells/mL with high infectious MVA titers around 1E10 $TCID_{50}$/mL, thus demonstrating the potential of ATF systems also in virus production [37]. However, several drawbacks limit the usability of ATF systems, especially for the production of viruses. Due to the lytic nature of the virus production process and larger sizes of some viruses (100–400 nm), membranes tend to clog. Larger pore sizes in the membrane (currently available pore sizes: 45 kDa-650 nm) could allow continuous harvesting of virus particles, facilitating higher CSVY and virus stability due to shortened RT in the bioreactor. This was further investigated by Genzel et al. by comparing three different ATF membrane cut-offs in the range of 45 kDa-650 nm for IAV production at 25E06 cells/mL. Pore sizes of 0.65 µm still not allowed to harvest IAV virus particles of about 0.2 µm size through the membrane. This clearly illustrates the need for larger cut-offs or other membrane materials [38]. A recent study performed by Hein et al. described a novel tubular membrane (Artemis Biosystems), which allowed 100% of produced virus particles (IAV) to pass through [40]. Finally, compared to TFF systems, the shear stress generated by ATF systems is lower by a factor of 0.64 which can be advantageous for shear stress sensitive cell lines such as suspension HEK293 cells [75]. Furthermore, often the fact that only one tube is connected instead of two ports for a TFF system is considered as an advantage.

The biggest challenge for the use of ATF systems are cultivations at very high cell concentrations. With increasing concentrations, the viscosity of the cell suspension increases and the pull or compression capacity of the diaphragm pump decreases [76]. Moreover, long-term cultivation can increase the risk for potential adsorption of media components to the membranes, leading to changes in the medium composition. However, cell densities used in virus production processes are not too high, yet. As most of the studies in virus production so far were done in small-scale systems, it is still to be seen, how both ATF and TFF systems will perform at larger scales.

6.6.4 Disc Centrifuge

Centrifuges use centripetal acceleration to separate particles of greater and lesser density. Therefore, they can be very efficient for separating cells from culture media. Centrifuges are commonly applied for harvesting animal cells or as a CRD in long-term perfusion cultivations [52]. As separation is based on the sedimentation of cells in a centrifugal field, the separation efficiency is greatly improved compared to gravitational settling. In the early 1990s, Westfalia Separator AG [77] and Alfa Laval [63] developed special centrifuges (disk-stack centrifuges and Centritech®) for continuous cultivations with animal cells. Regardless of the centrifuge type, the main streams of the separator can be divided into three parts, as shown in Figure 6.7: The feed stream containing cells is pumped from the bioreactor into the centrifuge. Centrifugal forces push the cells outward and separate them as the underflow, which can be pumped back into the bioreactor. The cell-free supernatant can be harvested via the overflow. The main advantages of using centrifuges as a CRD for perfusion cultivations are the lack of fouling or clogging, a tightly controllable separation rate by g-force and feed flow rate, and a separation of viable and dead cells by adjusting the g-force [17]. As separation occurs under a centrifugal field, cells are subjected to relatively high shear stress. Another drawback, particularly for stainless steel centrifuges, is a certain sterility risk due to the necessity of an additional waste line for permeate withdrawal. The combination of fast moving parts with the risk of sealing breakages or gaps further complicate sterilization procedures.

The disk-stack centrifuges by Westfalia (now GEA Westfalia Separator Group GmbH) were investigated in two studies for the perfusion cultivation of hybridoma, HeLa, and CHO cells [77,78]. No negative effects on cell viability and cell growth were detected despite high angular velocities and flow rates for either study. However, decreased retention efficiencies were observed by Björling et al. due to clogging in the concentrate channel [78]. Centritech® centrifuges were originally developed by Alfa Laval, but are available today from CARR/Pneumatic Scale Corporation (USA) after several rounds of acquisitions and purchases [19]. They were developed to minimize shear stress and consist of a disposable insert bag, which is mechanically fixed on a conical rotor in continuous rotation [19,52]. The cell broth enters the insert bag at one top end and the overflow is harvested at the other top end. Concentrated cells are

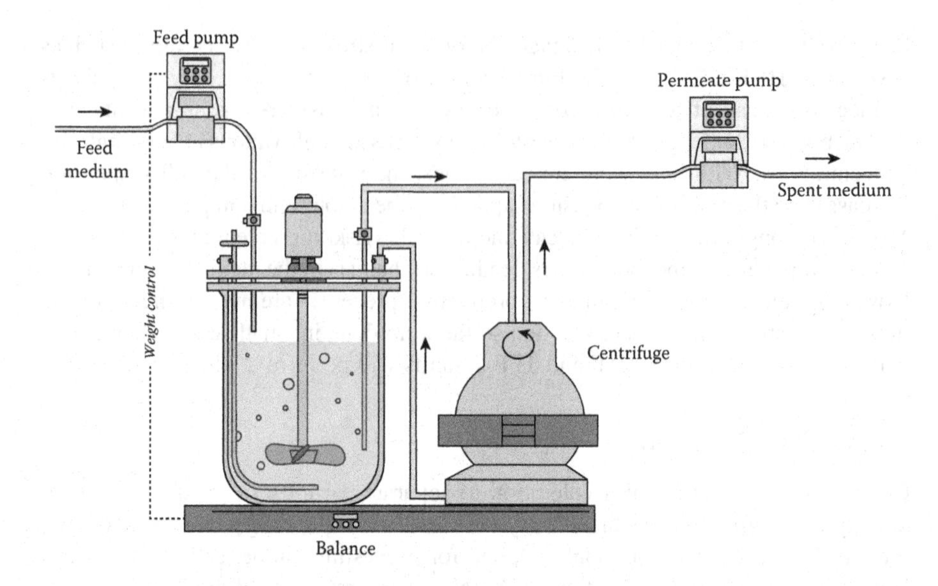

FIGURE 6.7 Schematic illustration of a disc centrifuge setup for perfusion processes. The main streams of the centrifuge can be divided into three parts: The feed stream containing cells is pumped from the bioreactor into the centrifuge. Centrifugal forces push the cells outward and separate them as the underflow, which can be pumped back into the bioreactor. Cell-free supernatant is constantly removed from the overflow, allowing fresh medium to be added to the bioreactor. Figure adapted from [65].

withdrawn through the underflow pipe at the bottom of the insert bag [52]. As no seals are required for connection of the single use insert bag and the conical rotor, and the inserts have been approved by the FDA, these systems are attractive for large-scale virus production [19,43]. To our knowledge no published study reported details regarding the use of centrifuges for the cultivation of viruses in perfusion mode, so far. However, they have been routinely used in commercial perfusion processes for production of labile proteins [19]. Drawbacks of the Centritech® system as well as the other centrifuges are the high investment costs. Furthermore, long residence times of the cells inside the centrifuge (up to 9 min) and high wall shear stress in the insert bags are disadvantageous. Moreover, reliability and robustness of the perfusion process is solely dependent on the quality of the insert bag, which has to be replaced at every 20 million revolutions or 31 days of continuous operation [19,53]. Compared to a filtration-based perfusion system, usage of the Centritech® system has been shown to reduce the productivity in MAb production by 30% [79].

Centrifuges can also be used as separate external "pseudo" retention devices for small-scale screening systems. Particularly for research purposes including initial process development and high-throughput screening, small-scale semi-perfusion shake flask cultivations are a convenient alternative to full perfusion processes [11,19]. Using small-scale systems (shake flasks or spin tubes), total medium usage is significantly reduced, while even for HCD cultures, cell viabilities similar to bioreactor perfusion systems can be achieved [11]. In semi-perfusion, cells are retained using a manual centrifugation step, followed by partial or total discarding of supernatant and resuspension in fresh media [19]. Moreover, parallel experiments can easily be performed. The lack of online probes and the limited working volume usually lead to rigid exchange regimes considering predefined growth or uptake rates for media exchanges. This rigid batch-wise regime often results in profiles of (over-)feeding with abrupt metabolite gradients [11], although current progress around shake flask cultures online monitoring instrumentation might overcome these hurdles [80]. Nevertheless, semi-perfusion cultures have been successfully used for perfusion process development for a variety of cell lines and viruses [36,45,61,65,81]. Foremost, it enabled to verify in a simple manner if medium and cells can reach higher cell concentrations and if then the CSVY can be kept as compared to batch mode.

6.6.5 Inclined Settler

Inclined settlers exploit the difference in density between the culture media and cells by providing a quiescent liquid area in which the cells settle from suspension. As no physical barrier is used for retention, clogging or mechanical shear stress does not occur. This allows longer operation times and selective retention of viable cells (non-viable cells have an approximately twofold lower sedimentation velocity) [53]. However, the small size and low density of mammalian cells (with 1.03–1.08 g/cm^3 around 5% higher than that of the medium) result in low gravitational settling velocities (1–15 cm/h) [43,53]. This limitation is circumvented by inclined settlers, which include several closely packed inclined flat plates (lamella), positioned at an angle to the vertical. The effective sedimentation path and upward fluid velocity is reduced (Boycott effect), thereby resulting in a more rapid sedimentation [52]. A simplified overview and description of the inclined settler setup is shown in Figure 6.8. The cells enter the inclined settler through the lower part of the device and gravitational forces drive the cells toward the lower surface of each lamella. Once settled, the cells slide down in a layer towards the plate periphery and are transferred back into the bioreactor [52]. Supernatant containing small particles (e.g., non-viable cells or virions) is removed from the top part of the settler. The main disadvantages are the tendency of cells to adhere to the lamella, low perfusion rates, and long RTs in un-oxygenated, unmixed environments [19,53]. Applying special coatings to the lamella and vibrations are commonly used to reduce cell adhesion [19]. Moreover, by cooling down the cells that exit the bioreactor in a heat exchanger, sedimentation velocity is increased and oxygen limitations can be overcome by reducing cell metabolism [82]. Large-scale GMP-compliant inclined settlers were first developed by Bayer AG [63] in the 1980s and are successfully used for the production of recombinant blood factors with operation times up to 3–5 months [83]. They also find application in the seed train of fed-batch processes at scales up to 3,000 L/d [84].

Only few studies investigated the application of inclined settlers for the production of viruses. Coronel et al. used an inclined settler-based perfusion process for the production of IAV in suspension AGE1.CR.pIX cells up to 50E06 cells/mL with cell retention efficiencies over 96%. No virus retention was observed and

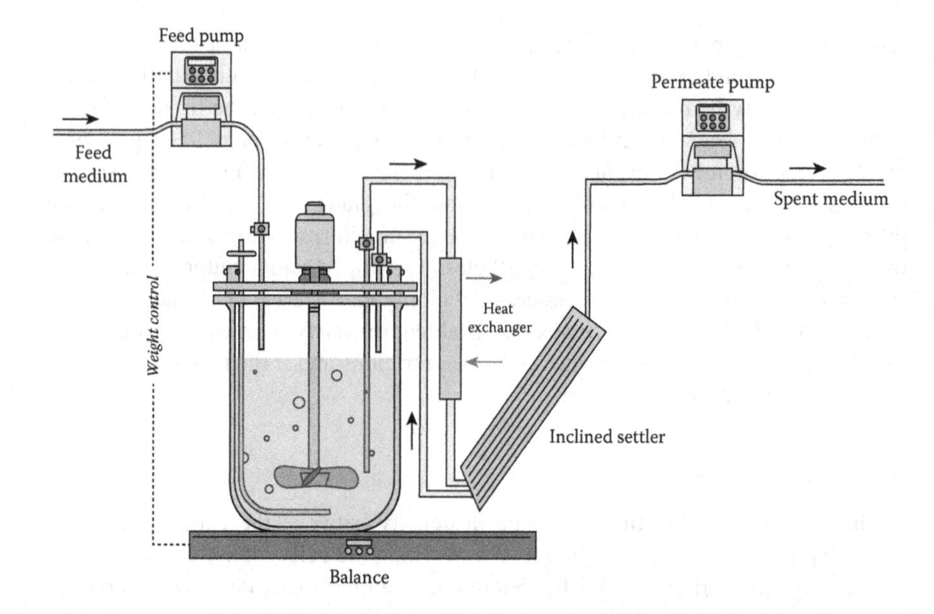

FIGURE 6.8 Schematic illustration of an inclined settler setup for perfusion processes. The cell broth is pumped through in the lower part of the settler and gravitational forces drive the cells towards the lower surface of each lamella. Once settled, the cells slide down in a layer towards the plate periphery and are transferred back into the bioreactor. Cell-free supernatant is constantly removed from the top part of the settler, allowing fresh medium to be added to the bioreactor. Figure adapted from [65].

similar yields compared to an ATF-based process were achieved. The usage of a heat exchanger was identified as a crucial parameter to obtain high virus titers [62]. However, at the used scale, the dead volume within this CRD was very large. How this translates at large-scale processes still has to be investigated. Another study by Alvim et al. applied an inclined settler to the production of yellow fever virus like particles (VLPs) using suspension HEK293 cells. Cell concentrations of 42E06 cells/mL were obtained further highlighting the potential of inclined settlers for the production of viruses and VLPs [85].

6.6.6 Hydrocyclone

The hydrocyclone is the simplest type of centrifugal separators, in which the rotating motion is induced by the liquid flowing through one or more tangentially arranged inlets at the head of the system. The apparatus consists of a conical section connected to a cylindrical portion at the upper part through which clarified supernatant can be harvested. A simplified overview and description of the hydrocyclone setup is shown in Figure 6.9.

By injecting the cell broth tangentially at a high flow rate, a strong swirling downward vortex is generated. Resulting centrifugal forces disperse cells radially

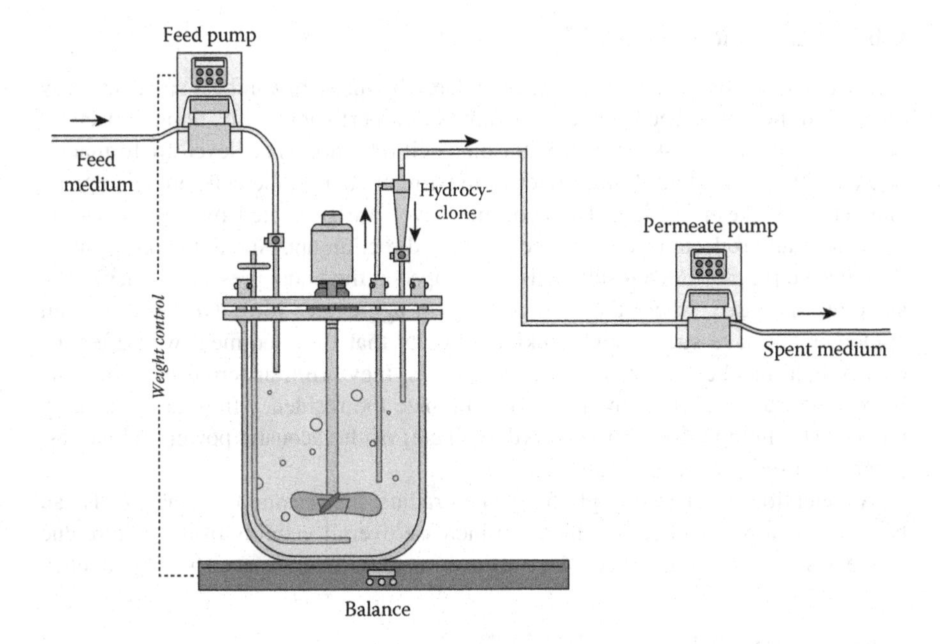

FIGURE 6.9 Schematic illustration of a hydrocyclone setup for perfusion processes. Tangential injection of the feed results in the creation of a flow vortex in the conical section of the hydrocyclone. Resulting centrifugal forces separate cells from the supernatant. Cell-free supernatant is constantly removed through the upper part, cells settle down, and are then transferred back into the bioreactor, fresh medium is added accordingly. Figure adapted from [65].

towards the wall into the boundary layer. As the lower opening of the hydrocyclone is extremely narrow, a complete discharge of the liquid is not possible and only the boundary layer carrying heavier particles (e.g., cells) can escape. The rest of the liquid reverses its vertical direction and flows in a strong vortex motion upwards and out through the overflow pipe carrying the smaller particles (e.g., virions) [52]. As centrifugal and drag forces are proportional to particle-size and particle volume, the separation performance could allow separation of viable and dead cells. Due to their simple structure, hydrocyclones are easy to install, have a small footprint, are in situ sterilizable, and easy to clean and re-use. Moreover, they would allow a continuous harvest of virions and high perfusion rates while maintaining a good retention efficiency at high flow rates. Major drawbacks are the high-pressure drops and high shear stresses to which cells are submitted. However, several studies demonstrated resistance of various mammalian cells to those pressure drops mainly due to the extremely short RTs inside the hydrocyclone (0.03–0.1 s) [52]. As good retention efficiency is only reached at high flow rates (typically exceeding 50–100 L/h), larger bioreactors are necessary limiting their use in small-scale research applications. To our knowledge, no research investigating the use of hydrocyclones for the production of viruses has been published so far.

6.6.7 Acoustic Settler

The use of acoustic settlers has been considered by many researchers since the early 1990s and they have been successfully applied to perfusion cultivations of bacteria and mammalian cells [43,86,87]. Ultrasonic cell retention is achieved by formation of a standing wave field. In the zones of maximum energy, the cells are aggregated and retained like in a filter. The acoustic waves are generated by a piezoelectric transducer and reflected from a reflector in the direction opposite to the propagation of the wave [63,86]. As a result, cells are trapped at the nodal pressure planes of the standing 3D wave, where they merge to form aggregates [86]. The resulting cell agglomerates have such a high sinking velocity that they sediment well after the energy field has been switched off. Afterwards, they are transferred back into the bioreactor via the return flow [63]. To minimize the residence time of the cells in the acoustic field periodic turn-offs (duty cycle) of the acoustic power and harvest pump are used [86] (see Figure 6.10).

As retention depends strongly on the cell radius, the retention of viable cells can be better than for dead cells, which can increase overall viability of the culture due to the washout of dead cells. Most acoustic settlers operate at a fixed frequency

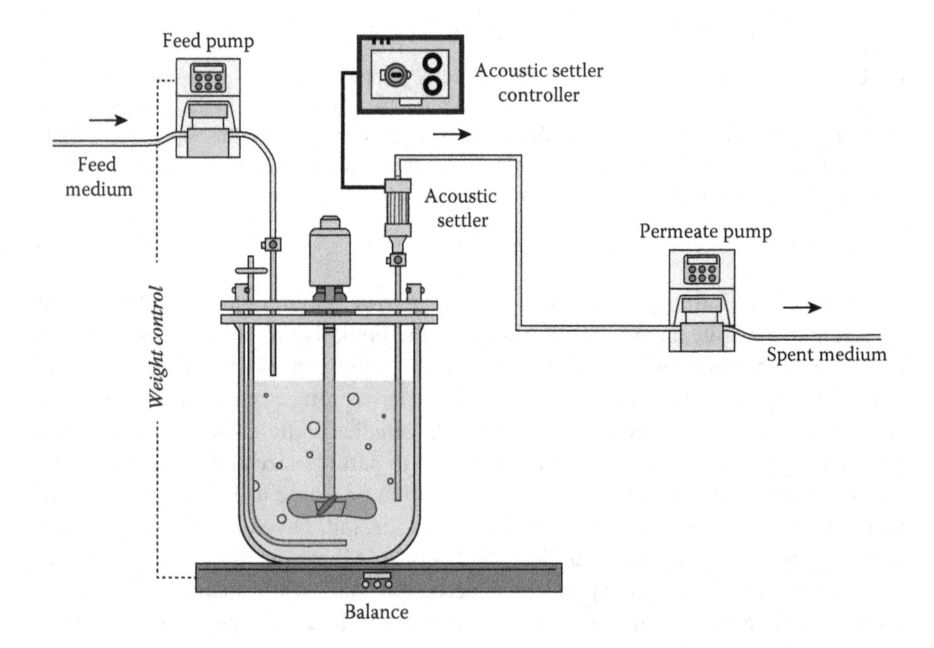

FIGURE 6.10 Schematic illustration of an acoustic settler setup and acoustic filter chamber for perfusion processes. The acoustic waves are generated by a piezoelectric transducer and reflected from a reflector in the direction opposite to the propagation of the wave. As a result, cells are trapped at the nodal pressure planes of the standing 3D wave where they merge to form aggregates. Cell-free supernatant is removed in an intermittent way, allowing fresh medium to be added to the bioreactor. Figure adapted from [65].

(between 2–3 MHz), and the acoustic power is the only internal parameter that can influence the retention efficiency. Frequencies below this range can lead to cavitation resulting in cell damage and additional heating of the medium. External parameters that can affect the separation efficiency are the duty cycle, flow rate, backflush frequency, and recirculation rate [86]. As described before, duty cycles are the stop times of the acoustic filter and harvest pump, allowing the transfer of aggregated cells back to the bioreactor. Even though these duty cycles have little influence on the acoustic settler performance, it was shown to be crucial to increase IAV yields [50]. To avoid nutrient depletion at high cell concentrations during the cell growth phase and virus production phase, high flow medium withdrawal rates through the acoustic filter need to be achieved. However, this results in a lower separation efficiency as more cells will be washed out. This decrease in separation efficiency could be mitigated by increasing the acoustic power input. However, this leads to increased heat production and increased temperature in the acoustic chamber. Alternatively, the backflush frequency, which describes the number of times the acoustic chamber is cleared of sedimented cells per hour, could be enhanced. But this would also lower the separation efficiency. The last external parameter, the recirculation rate, describes the flow rate with which sedimented cells in the acoustic chamber are recycled back into the bioreactor [86].

Acoustic settlers have no physical barrier and no moving parts are required for cell retention and are, therefore, less susceptible to mechanical failure and fouling [88]. Moreover, they are easily cleaned and sterilized-in-place further facilitating their integration in perfusion processes. Maximum perfusion rates of 10, 50, 200, and 1,000 L d^{-1} can be achieved over long periods at a high cell retention efficiency and with high cell viabilities [63]. A fully integrated virus production process was recently established by Gränicher et al., who used an acoustic settler to reach HCD and to preclarify the virus containing harvest for subsequent in-line purification steps [41].

However, several issues limit the usability of this CRD: The power required to generate the standing wave field is associated with a heating effect, which can damage cells and virus particles. Solutions such as air cooling and water circulation have been implemented for lab and pilot scales (200 L), but remain an issue for larger scales [19]. Same as for other CRDs, acoustic filters are external devices, where cells in the external loop are exposed to uncontrolled conditions and nutrient gradients for short periods. Dalm et al. described significant oxygen concentration gradients for recirculation pump rates smaller than 6 RV d^{-1} [88]. If the acoustic settler is used for HCD cultivations, more cells will be in the settling chamber. Due to the compactness and aggregation of the cells at the nodal pressure planes, high cell concentrations could lead to an attenuation of the acoustic wave, which could weaken the standing 3D wave, resulting in a lower retention efficiency [86]. Compared to other CRDs, operation of acoustic filters seems slightly more complex and less "plug-and-play," requiring specially trained staff and constant maintenance.

6.7 USE OF DISPOSABLES

Single-use bioreactors (SUBs) have been adopted into several cell-culture–based production systems including commercial vaccine production (also described in the

TABLE 6.4
Pros and cons of disposable bioreactors

Reported advantages		Reported disadvantages
Comparable process yields	Elimination of costs and downtime related to cleaning operations	Scalability limitations: 6000 L is currently largest working volume
Elimination of CIP/SIP water and steam costs	Reduction of capital equipment design, installation, and validation cost	Unproven process robustness
Shorter plant commission and start-up times	Reduction of initial capital investment and cost of goods	Materials inventory, storage, and repetitive purchases required
Reduction of plant footprint	Improvement of compliance values (reduces error potential)	Vendor dependency, possible supply chain shortages
Flexible footprint	Ease of material transfer between diverse biosafety levels	Novel validation demands and process layout and flow designs
Reduction of overall carbon footprint	Ease of GMP "cold chain logistics" and "good storage practice"	Observed variability in vendor "maturity" and capabilities
Reduction of production facility service requirements	Simplified and accelerated product changeover and turnaround	Development of new industry and regulatory standards and definitions
Enables flexible approaches in process flow and layout	Reduction of process flow and equipment modification costs	Product containment failure concerns
Rapid site-to-site facility transfer	Support of existing process sensing, monitoring, modeling, and control	Connectivity issues: Standardization and continued solutions required
Plug-and-play approach	Increased plant capacity, flexibility, and campaigning	Pharmacopoeia standards: Leachables and extractables
Good scalability within available system sizes	Ease in reconfiguration and extension of production facilities	Open questions about carbon footprint and plastic waste
Reduction of contamination risk	Reduction of lot and product cross contamination risk	Standards for bag/tubing manifold systems
Elimination of hazardous cleaning materials	Increased sterility assurance (irradiation over steam)	Lack of robust ON/OFF sterile connectors

Reprinted and adapted from [90]. CIP: clean in place, SIP: steam in place.

previous chapter). The list of reported advantages of implementing SUBs and other disposables is long (see Table 6.4) and still growing as more and more adoptions take place. Several studies demonstrated competitive virus and recombinant protein yields plus comparable cell growth characteristics for SUBs compared to stainless steel bioreactors [89]. The use of SUBs entails unique physical and chemical criteria that must be fulfilled: welds and polymeric multilayer films in the plastic bags must be compliant with pharmacopoeia standards (e.g., free of any leachables and extractables). Special attention must be paid to safety precautions to prevent bag bursting, especially for the production of infectious materials [89]. In principal, SUBs are simple to install and universally applicable with low capital costs, reduced validation efforts, low downtimes, and flexible footprints. By eliminating cleaning in-place operations, faster turnarounds between batches or switching between products (for vaccines between viruses) with reduced cross-contamination risks can be achieved. The use of disposables including SUBs is therefore often seen as an enabler for continuous manufacturing of biologicals seamlessly linking upstream and downstream operations [18]. Moreover, commission of new production facilities that fully rely on disposables is typically fast as the complexity involved in planning, installation, and validation is drastically reduced. Particularly for vaccine production, this has significant advantages in case of pandemics (as recently demonstrated by the COVID-19 pandemic). Also, the transfer between areas of different biosafety levels is facilitated [90]. It is envisioned that such systems could even be placed at the site of need in containers readily transported on trucks. Manufacturers can order custom-made solutions with different disposable sensors, sparging systems, and mechanical agitation systems (top- or bottom-mounted impellers, rockers, orbital shakers, etc.) from several vendors [89]. One example for SUBs is the SB-X single-use OSB (orbital shaken bioreactor) series from Kuhner shaker that allows disposable-based vaccine productions up to 2,500 L scale (see Figure 6.11).

Despite the many advantages, there are still some challenges with the use of SUBs that are listed (Table 6.4). Direct transfer of large-volume vaccine production processes into disposable systems is often not possible due to volume restrictions and can only be circumvented by using multiple parallel systems. For optimization of the latter, additional laboratory-scale investigations might be required. The parallel use of many smaller SUBs also enhances the complexity of process operations, and increases the risk of leaky tubings and welds and of plastic bag breakage [89,90]. Finally, key issues such as GMP regulatory requirements, standards in bag and tubing manifold systems, need for standardization in extractables and leachables (E&L) testing procedures, lack of ON/OFF sterile connectors, relevant validation studies, etc. must be addressed. Furthermore, in particular for HCD virus vaccine

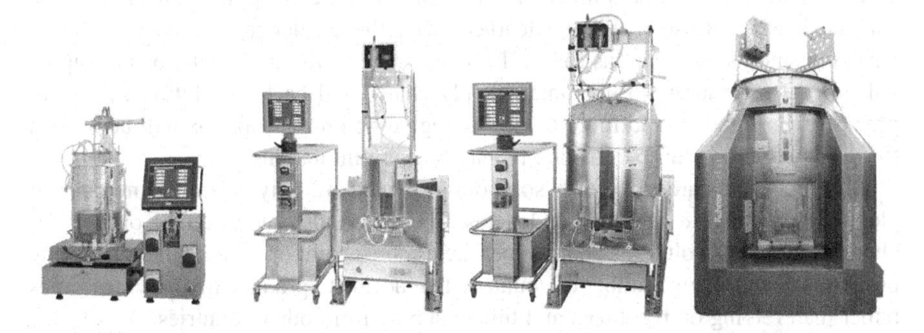

FIGURE 6.11 Possible scale-up line for SUBs. Here orbital shaken bioreactors 4–12 L, 15–50 L, 50–200 L, and 500–2500 L scale from Kuhner shaker. Alternatively, STRs or wave or other bioreactor types as SUBs are available. Pictures were kindly provided by Kuhner shaker.

manufacturing, biosafety measures during cultivation and for inactivation will need to be considered carefully. From the manufacturing and regulatory perspective, guarantee of supply and the request for a back-up system is often limiting large-scale production with highly innovative SUB systems. In addition, the environmental impact of single-use technologies has to be considered.

6.8 CONCLUSION

Over the previous decades, a deeper understanding for animal cell culture was gained, allowing the development of robust basal growth media. Cell concentrations exceeding 1E07 cells/mL are reached in batch operations with suspension cells growing in serum-free or chemically defined media. This improved cell handling laid the foundation for intensified process operations, which were already successfully implemented for several approved biotherapeutics, i.e., recombinant proteins. For viral vaccine and/or vector production, initial goals for process intensification comprised increased cell concentrations, increased viral titers, and increased VVPs and STYs. Although most vaccine production processes still rely on adherent cell cultures or chicken eggs, the establishment of intensified virus production platforms offers many advantages. Facilities with significantly reduced footprints and lower volumes could be used, allowing a rapid production of large numbers of virions. Achieving higher cell concentrations could reduce bioreactor sizes, and therefore the use of SUB for commercial productions. This could give manufactures the opportunity to flexibly adapt to product changeovers and remain in the same facility throughout the entire clinical development. Higher VVPs could increase production capacities while simultaneously reducing costs. However, several challenges remain. In order to go for process intensification, new upstream and downstream tools that are capable of culturing and processing high cell concentrations and high virus titers are needed. Choice of the right cell retention device will depend on the cell and virus type used. Critical process parameters need to be identified for each cell retention device and individually optimized. Impact of high cell concentrations on productivity, quality and efficacy, and process-related impurities, also on downstream applications, have to be evaluated. Many lessons will be learned from the challenges of the COVID-19 pandemic and its implications for viral vaccine manufacturing. Limitations in supply will impact regulations to guarantee supply chains and backup solutions. A better process understanding and more flexibility regarding process options will be needed to avoid production limitations due to supply chain limitations.

A tendency towards platform solutions provided by only a limited number of global players will increase dependencies. This could ultimately reduce options for alternative process solutions and hamper the establishment of vaccine manufacturing processes in a growing number of countries that decide to produce their own vaccines rather than relying on resources and timely supply from other countries.

The next years will be exiting for cell culture-based virus production for vaccination and therapy. We are convinced that the establishment of HCD cultures and process intensification is mandatory to provide safe and potent vaccines to supply an increasing world population and to support future developments in gene and cancer therapy.

REFERENCES

[1] WHO, *Global Vaccine Market Report.*Geneva, Switzerland: World Health Organization, 2019, [Online]. Available: https://apps.who.int/iris/handle/10665/311278

[2] EMA, *Guideline on Quality, Non-Clinical and Clinical Requirements for Investigational Advanced Therapy Medicinal Products in Clinical Trials.* London, United Kingdom: European Medicines Agency, 2019.

[3] W. F. Kaemmerer, "How will the field of gene therapy survive its success?," (in eng), *Bioeng. Transl. Med.*, vol. 3, no. 2, pp. 166–177, 2018.

[4] A. Offersgaard *et al.*, "SARS-CoV-2 production in a scalable high cell density bioreactor," *Vaccines*, vol. 9, no. 7, p. 706, 2021.

[5] A. Nikolay, A. Léon, K. Schwamborn, Y. Genzel, and U. Reichl, "Process intensification of EB66® cell cultivations leads to high-yield yellow fever and Zika virus production," *Appl. Microbiol. Biotechnol.*, vol. 102, no. 20, pp. 8725–8737, 2018.

[6] A. F. Pihl *et al.*, "High density Huh7.5 cell hollow fiber bioreactor culture for high-yield production of hepatitis C virus and studies of antivirals," *Sci. Rep.*, vol. 8, no. 1, p. 17505, 2018.

[7] N. Nayerossadat, T. Maedeh, and P. A. Ali, "Viral and nonviral delivery systems for gene delivery," (in eng), *Adv. Biomed. Res.*, vol. 1, p. 27, 2012.

[8] D. Radhakrishnan, E. A. Wells, and A. S. Robinson, "Strategies to enhance productivity and modify product quality in therapeutic proteins," *Curr. Opin. Chem. Eng.*, vol. 22, pp. 81–88, 2018.

[9] B. Kalbfuss, A. Knöchlein, T. Kröber, and U. Reichl, "Monitoring influenza virus content in vaccine production: Precise assays for the quantitation of hemagglutination and neuraminidase activity," *Biologicals*, vol. 36, no. 3, pp. 145–161, 2008.

[10] Y. Genzel, J. Rödig, E. Rapp, and U. Reichl, "Vaccine Production: Upstream Processing with Adherent or Suspension Cell Lines," in *Animal Cell Biotechnology: Methods and Protocols*, R. Pörtner, Ed. Totowa, NJ: Humana Press, 2014, pp. 371–393.

[11] A. Nikolay, T. Bissinger, G. Granicher, Y. Wu, Y. Genzel, and U. Reichl, "Perfusion control for high cell density cultivation and viral vaccine production," *Methods Mol. Bio.l*, vol. 2095, pp. 141–168, 2020.

[12] L. Pelz *et al.*, "Semi-continuous propagation of influenza A virus and its defective interfering particles: analyzing the dynamic competition to select candidates for antiviral therapy," (in eng), *J. Virol.*, vol. 95, p. e01174-21, Sep. 2021.

[13] R. J. Hill, "An evaluation of a method of quantitative radial immunodiffusion," (in eng), *Immunochemistry*, vol. 5, no. 2, pp. 185–202, Mar. 1968.

[14] J. M. Wood, G. C. Schild, R. W. Newman, and V. Seagroatt, "An improved single-radial-immunodiffusion technique for the assay of influenza haemagglutinin antigen: application for potency determinations of inactivated whole virus and subunit vaccines," (in eng), *J. Biol. Stand.*, vol. 5, no. 3, pp. 237–247, 1977.

[15] J. M. Wood and J. P. Weir, "Standardisation of inactivated influenza vaccines-Learning from history," (in eng), *Influenza Other Respir. Viruses*, vol. 12, no. 2, pp. 195–201, Mar. 2018.

[16] F. Tapia, T. Laske, M. A. Wasik, M. Rammhold, Y. Genzel, and U. Reichl, "Production of defective interfering particles of influenza A virus in parallel continuous cultures at two residence times—Insights from qPCR measurements and viral dynamics modeling," (in English), *Front. Bioeng. Biotechnol.* vol. 7, Article no. 275, 2019.

[17] R. Pörtner, *Bioreactors for Mammalian cells*, vol. 9, 2015. ISBN 978-3-319-10319-8.

[18] K. B. Konstantinov and C. L. Cooney, "White paper on continuous bioprocessing. May 20-21, 2014 continuous manufacturing symposium," *J. Pharm. Sci.*, vol. 104, no. 3, pp. 813–820, Mar. 2015.
[19] V. Chotteau, "Perfusion Processes," in *Animal Cell Culture*, M. Al-Rubeai, Ed. Cham: Springer International Publishing, 2015, pp. 407–443.
[20] S. S. Ozturk, "Engineering challenges in high density cell culture systems," *Cytotechnology*, vol. 22, no. 1–3, pp. 3–16, Jan. 1996.
[21] D. G. Kilburn and A. L. van Wezel, "The effect of growth rate in continuous-flow cultures on the replication of rubella virus in BHK cells," *J. Gen. Virol.*, vol. 9, no. 1, pp. 1–7, Oct. 1970.
[22] T. Frensing *et al.*, "Continuous influenza virus production in cell culture shows a periodic accumulation of defective interfering particles," *PLoS One*, vol. 8, no. 9, p. e72288, 2013.
[23] G. P. Pijlman, J. de Vrij, F. J. van den End, J. M. Vlak, and D. E. Martens, "Evaluation of baculovirus expression vectors with enhanced stability in continuous cascaded insect-cell bioreactors," *Biotechnol. Bioeng.*, vol. 87, no. 6, pp. 743–753, Sep. 2004.
[24] R. Kompier, J. Tramper, and J. M. Vlak, "A continuous process for the production of baculovirus using insect-cell cultures," *Biotechnol. Lett.*, vol. 10, no. 12, pp. 849–854, 1988.
[25] F. L. van Lier, E. J. van den End, C. D. de Gooijer, J. M. Vlak, and J. Tramper, "Continuous production of baculovirus in a cascade of insect-cell reactors," *Appl. Microbiol Biotechnol.*, vol. 33, no. 1, pp. 43–47, Apr. 1990.
[26] G. B. Gori, "Continuous cultivation of virus in cell suspensions by use of the lysostat," (in eng), *Appl. Microbiol.*, vol. 13, no. 6, pp. 909–917, Nov. 1965.
[27] F. Tapia, I. Jordan, Y. Genzel, and U. Reichl, "Efficient and stable production of modified vaccinia Ankara virus in two-stage semi-continuous and in continuous stirred tank cultivation systems," *PLoS One*, vol. 12, no. 8, p. e0182553, 2017.
[28] G. Gränicher, F. Tapia, I. Behrendt, I. Jordan, Y. Genzel, and U. Reichl, "Production of modified vaccinia Ankara virus by intensified cell cultures: A comparison of platform technologies for viral vector production," (in eng), *J. Biotechnol.* , vol. 16, no. 1, p. e2000024, Special Issue: Biomanufacturing of Gene Therapy Vectors, 2020.
[29] F. Tapia, D. Vazquez-Ramirez, Y. Genzel, and U. Reichl, "Bioreactors for high cell density and continuous multi-stage cultivations: Options for process intensification in cell culture-based viral vaccine production," *Appl. Microbiol. Biotechnol.*, vol. 100, no. 5, pp. 2121–2132, Mar. 2016.
[30] T. B. Ferreira, M. J. Carrondo, and P. M. Alves, "Effect of ammonia production on intracellular pH: Consequent effect on adenovirus vector production," *J. Biotechnol.*, vol. 129, no. 3, pp. 433–438, May 2007.
[31] I. Nadeau and A. Kamen, "Production of adenovirus vector for gene therapy," *Biotechnol. Adv.*, vol. 20, no. 7–8, pp. 475–489, Jan. 2003.
[32] P. Perrin, S. Madhusudana, C. Gontier-Jallet, S. Petres, N. Tordo, and O.-W. Merten, "An experimental rabies vaccine produced with a new BHK-21 suspension cell culture process: Use of serum-free medium and perfusion-reactor system," *Vaccine*, vol. 13, no. 13, pp. 1244–1250, 1995.
[33] A. Bock, J. Schulze-Horsel, J. Schwarzer, E. Rapp, Y. Genzel, and U. Reichl, "High-density microcarrier cell cultures for influenza virus production," *Biotechnol. Prog.*, vol. 27, no. 1, pp. 241–250, 2011.
[34] H. A. Wood, L. B. Johnston, and J. P. Burand, "Inhibition of Autographa californica nuclear polyhedrosis virus replication in high-density Trichoplusia ni cell cultures," (in eng), *Virology*, vol. 119, no. 2, pp. 245–254, Jun. 1982.

[35] J. G. Aunins, "Viral vaccine production in cell culture," *Encyclopedia Cell Technol.*, 2003, 10.1002/0471250570.spi105

[36] D. Vazquez-Ramirez, Y. Genzel, I. Jordan, V. Sandig, and U. Reichl, "High-cell-density cultivations to increase MVA virus production," *Vaccine*, vol. 36, no. 22, pp. 3124–3133, May 2018.

[37] D. Vazquez-Ramirez, I. Jordan, V. Sandig, Y. Genzel, and U. Reichl, "High titer MVA and influenza A virus production using a hybrid fed-batch/perfusion strategy with an ATF system," *Appl. Microbiol. Biotechnol.*, vol. 103, no. 7, pp. 3025–3035, Apr. 2019.

[38] Y. Genzel *et al.*, "High cell density cultivations by alternating tangential flow (ATF) perfusion for influenza A virus production using suspension cells," *Vaccine*, vol. 32, no. 24, pp. 2770–2781, May 2014.

[39] G. Seth *et al.*, "Development of a new bioprocess scheme using frozen seed train intermediates to initiate CHO cell culture manufacturing campaigns," *Biotechnol. Bioeng.*, vol. 110, no. 5, pp. 1376–1385, May 2013.

[40] M. D. Hein, A. Chawla, M. Cattaneo, S. Y. Kupke, Y. Genzel, and U. Reichl, "Cell culture-based production of defective interfering influenza A virus particles in perfusion mode using an alternating tangential flow filtration system," *Appl. Microbiol. Biotechnol.*, vol. 105, no. 19, pp. 7251–7264, 2021.

[41] G. Gränicher *et al.*, "A high cell density perfusion process for Modified Vaccinia virus Ankara production: Process integration with inline DNA digestion and cost analysis," (in eng), *Biotechnol. Bioeng.*, vol. 118, no. 12, pp. 4720–4731, Sep. 2021.

[42] E. Petiot, S. Ansorge, M. Rosa-Calatrava, and A. Kamen, "Critical phases of viral production processes monitored by capacitance," *J. Biotechnol.*, vol. 242, pp. 19–29, Jan. 2017.

[43] D. Voisard, F. Meuwly, P. A. Ruffieux, G. Baer, and A. Kadouri, "Potential of cell retention techniques for large-scale high-density perfusion culture of suspended mammalian cells," (in eng), *Biotechnol. Bioeng.*, vol. 82, no. 7, pp. 751–765, Jun. 2003.

[44] J. Coronel *et al.*, "Influenza A virus production in a single-use orbital shaken bioreactor with ATF or TFF perfusion systems," *Vaccine*, vol. 37, no. 47, pp. 7011–7018, 2019.

[45] G. Gränicher *et al.*, "Efficient influenza A virus production in high cell density using the novel porcine suspension cell line PBG.PK2.1," (in eng), *Vaccine*, vol. 37, no. 47, pp. 7019–7028, Nov. 2019.

[46] F. Tapia *et al.*, "Production of high-titer human influenza A virus with adherent and suspension MDCK cells cultured in a single-use hollow fiber bioreactor," *Vaccine*, vol. 32, no. 8, pp. 1003–1011, Feb. 2014.

[47] E. Petiot and A. Kamen, "Real-time monitoring of influenza virus production kinetics in HEK293 cell cultures," *Biotechnol. Prog.*, vol. 29, no. 1, pp. 275–284, 2013.

[48] A. P. Manceur *et al.*, "Scalable Lentiviral Vector Production Using Stable HEK293SF Producer Cell Lines," (in eng), *Hum. Gene Ther. Methods*, vol. 28, no. 6, pp. 330–339, 2017.

[49] O. Henry, E. Dormond, M. Perrier, and A. Kamen, "Insights into adenoviral vector production kinetics in acoustic filter-based perfusion cultures," (in eng), *Biotechnol. Bioeng.*, vol. 86, no. 7, pp. 765–774, Jun. 2004.

[50] G. Gränicher, J. Coronel, F. Trampler, I. Jordan, Y. Genzel, and U. Reichl, "Performance of an acoustic settler versus a hollow fiber–based ATF technology for influenza virus production in perfusion," *Appl. Microbiol. Biotechnol.*, vol. 104, no. 11, pp. 4877–4888, 2020.

[51] E. A. Elsayed, R. D. A. Medronho, R. Wagner, and W. D. Deckwer, "Use of hydrocyclones for Mammalian cell retention: Separation efficiency and cell viability (Part 1)," *Eng. Life Sci.*, vol. 6, pp. 347–354, 2006.

[52] L. R. Castilho and R. A. Medronho, "Cell retention devices for suspended-cell perfusion cultures," (in eng), *Adv. Biochem. Eng./Biotechnol.*, vol. 74, pp. 129–169, 2002.

[53] S. M. Woodside, B. D. Bowen, and J. M. Piret, "Mammalian cell retention devices for stirred perfusion bioreactors," (in eng), *Cytotechnology*, vol. 28, no. 1–3, pp. 163–175, Nov. 1998.

[54] O. W. Merten, J. V. Kierulff, N. Castignolles, and P. Perrin, "Evaluation of the new serum-free medium (MDSS2) for the production of different biologicals: use of various cell lines," (in eng), *Cytotechnology*, vol. 14, no. 1, pp. 47–59, 1994.

[55] M. Leong, W. Babbitt, and G. Vyas, "A hollow-fiber bioreactor for expanding HIV-1 in human lymphocytes used in preparing an inactivated vaccine candidate," (in eng), *Biologicals*, vol. 35, no. 4, pp. 227–233, Oct. 2007.

[56] B. Sun *et al.*, "Production of influenza H1N1 vaccine from MDCK cells using a novel disposable packed-bed bioreactor," (in eng), *Appl Microbiol. Biotechnol.*, vol. 97, no. 3, pp. 1063–1070, Feb. 2013.

[57] S. Kiesslich, J. P. Vila-Chã Losa, J. F. Gélinas, and A. A. Kamen, "Serum-free production of rVSV-ZEBOV in Vero cells: Microcarrier bioreactor versus scale-X™ hydro fixed-bed," (in eng), *J. Biotechnol.*, vol. 310, pp. 32–39, Feb. 2020.

[58] H. P. Lesch *et al.*, "Process Development of Adenoviral Vector Production in Fixed Bed Bioreactor: From Bench to Commercial Scale," (in eng), *Hum. Gene Ther.*, vol. 26, no. 8, pp. 560–571, Aug. 2015.

[59] R. Rajendran *et al.*, "Assessment of packed bed bioreactor systems in the production of viral vaccines," (in eng), *AMB Express*, vol. 4, p. 25, 2014.

[60] V. Cortin, J. Thibault, D. Jacob, and A. Garnier, "High-Titer Adenovirus Vector Production in 293S Cell Perfusion Culture," *Biotechnology Progress*, vol. 20, no. 3, pp. 858–863, 2004. 10.1021/bp034237l

[61] Y. Wu, T. Bissinger, Y. Genzel, X. Liu, U. Reichl, and W.-S. Tan, "High cell density perfusion process for high yield of influenza A virus production using MDCK suspension cells," *Appl. Microbiol. Biotechnol.*, vol. 105, no. 4, pp. 1421–1434, 2021.

[62] J. Coronel, G. Gränicher, V. Sandig, T. Noll, Y. Genzel, and U. Reichl, "Application of an inclined settler for cell culture-based influenza A virus production in perfusion mode," (in eng), *Front. Bioeng. Biotechnol.*, vol. 8, p. 672, 2020.

[63] H.-J. Henzler, "Kontinuierliche Fermentation mit tierischen Zellen. Teil 2. Techniken und Methoden der Zellrückhaltung," *Chemie Ingenieur Technik*, vol. 84, no. 9, pp. 1482–1496, 2012.

[64] P. Himmelfarb, P. S. Thayer, and H. E. Martin, "Spin filter culture: the propagation of mammalian cells in suspension," (in eng), *Science (New York, N.Y.)*, vol. 164, no. 3879, pp. 555–557, May 1969.

[65] A. Nikolay, *Intensified Yellow Fever and Zika Virus Production in Animal Cell Culture,* 2020.

[66] L. R. Esclade, S. Carrel, and P. Péringer, "Influence of the screen material on the fouling of spin filters," (in eng), *Biotechnol. Bioeng.*, vol. 38, no. 2, pp. 159–168, Jun. 1991.

[67] Y. M. Deo, M. D. Mahadevan, and R. Fuchs, "Practical considerations in operation and scale-up of spin-filter based bioreactors for monoclonal antibody production," (in eng), *Biotechnol. Prog.*, vol. 12, no. 1, pp. 57–64, Jan-Feb. 1996.

[68] B. Maiorella, G. Dorin, A. Carion, and D. Harano, "Crossflow microfiltration of animal cells," *Biotechnol. Bioeng.*, vol. 37, no. 2, pp. 121–126, 1991.

[69] R. van Reis, L. C. Leonard, C. C. Hsu, and S. E. Builder, "Industrial scale harvest of proteins from mammalian cell culture by tangential flow filtration," *Biotechnol. Bioeng.*, vol. 38, no. 4, pp. 413–422, 1991.

[70] A. Nikolay, J. de Grooth, Y. Genzel, J. A. Wood, and U. Reichl, "Virus harvesting in perfusion culture: Choosing the right type of hollow fiber membrane," *Biotechnol. Bioeng.*, vol. 117, no. 10, pp. 3040–3052, 2020.
[71] J. Shevitz, "Fluid filtration system patent 6544424," Grant, April 8, 2003, Refined Technology Company, Jerry Shevitz, 2000.
[72] S. R. Hadpe, A. K. Sharma, V. V. Mohite, and A. S. Rathore, "ATF for cell culture harvest clarification: mechanistic modelling and comparison with TFF," *J. Chem. Technol. Biotechnol.*, vol. 92, no. 4, pp. 732–740, 2017.
[73] J. Warnock and M. Al-Rubeai, "Production of Biologics from Animal Cell Cultures," in *Applications of Cell Immobilisation Biotechnology*, V. Nedović and R. Willaert, Eds. Dordrecht: Springer Netherlands, 2005, pp. 423–438.
[74] J. M. Bielser, M. Wolf, J. Souquet, H. Broly, and M. Morbidelli, "Perfusion mammalian cell culture for recombinant protein manufacturing - A critical review," (in eng), *Biotechnol. Adv.*, vol. 36, no. 4, pp. 1328–1340, Jul–Aug. 2018.
[75] C. Zhan *et al.*, "Low shear stress increases recombinant protein production and high shear stress increases apoptosis in human cells," (in eng), *iScience*, vol. 23, no. 11, p. 101653, Nov. 2020.
[76] M. F. Clincke, C. Mölleryd, Y. Zhang, E. Lindskog, K. Walsh, and V. Chotteau, "Very high density of CHO cells in perfusion by ATF or TFF in WAVE bioreactor™. Part I. Effect of the cell density on the process," (in eng), *Biotechnol. Prog*, vol. 29, no. 3, pp. 754–767, May-Jun. 2013.
[77] V. Jäger, "High Density Perfusion Culture of Animal Cells Using a Novel Continuous Flow Centrifuge," in *Animal Cell Technology: Basic & Applied Aspects: Proceedings of the Fourth Annual Meeting of the Japanese Association for Animal Cell Technology, Fukuoka, Japan, 13–15 November 1991*, H. Murakami, S. Shirahata, and H. Tachibana, Eds. Dordrecht: Springer Netherlands, 1992, pp. 209–216.
[78] T. Björling, U. Dudel, and C. Fenge, "Evaluation of a cell separator in large scale perfusion culture," in *Animal Cell Technology:Developments Towards the 21st Century*, E. C. Beuvery, J. B. Griffiths, W. P. Zeijlemaker, Eds. Dordrecht: Springer, 1995, pp. 671–675.
[79] M. Johnson, S. Lanthier, B. Massie, G. Lefebvre, and A. A. Kamen, "Use of the centritech lab centrifuge for perfusion culture of hybridoma cells in protein-free medium," *Biotechnol. Prog.*, vol. 12, no. 6, pp. 855–864, 1996.
[80] G. Jänicke, C. Sauter, R. Bux, and J. Haas, "Characterisation of shake flasks for cultivation of animal cell cultures," *Dordrecht*, 2007. Netherlands: Springer, pp. 727–731.
[81] T. Bissinger *et al.*, "Semi-perfusion cultures of suspension MDCK cells enable high cell concentrations and efficient influenza A virus production," *Vaccine*, vol. 37, no. 47, pp. 7003–7010, 2019
[82] Y. Shen and K. Yanagimachi, "CFD-aided cell settler design optimization and scale-up: effect of geometric design and operational variables on separation performance," (in eng), *Biotechnol. Prog.*, vol. 27, no. 5, pp. 1282–1296, Sep-Oct. 2011.
[83] J. H. Vogel *et al.*, "A new large-scale manufacturing platform for complex biopharmaceuticals," (in eng), *Biotechnol. Bioeng.*, vol. 109, no. 12, pp. 3049–3058, Dec. 2012.
[84] M. Pohlscheidt *et al.*, "Optimizing capacity utilization by large scale 3000 L perfusion in seed train bioreactors," (in eng), *Biotechnol. Prog.*, vol. 29, no. 1, pp. 222–229, Jan-Feb. 2013.
[85] R. G. F. Alvim, T. M. Lima, J. L. Silva, G. A. P. de Oliveira, and L. R. Castilho, "Process intensification for the production of yellow fever virus-like particles as potential recombinant vaccine antigen," (in eng), *Biotechnol. Bioeng.*, vol. 118, no. 9, pp. 3581–3592, Sep. 2021.

[86] I. Z. Shirgaonkar, S. Lanthier, and A. Kamen, "Acoustic cell filter: a proven cell retention technology for perfusion of animal cell cultures," *Biotechnol. Adv.*, vol. 22, no. 6, pp. 433–444, 2004.

[87] Z. Wang, "Two approaches for cell retention in perfusion culture systems," Cleveland State University, 2009.

[88] M. C. Dalm, S. M. Cuijten, W. M. van Grunsven, J. Tramper, and D. E. Martens, "Effect of feed and bleed rate on hybridoma cells in an acoustic perfusion bioreactor: part I. Cell density, viability, and cell-cycle distribution," *Biotechnol. Bioeng.*, vol. 88, no. 5, pp. 547–557, Dec. 2004.

[89] L. E. Gallo-Ramirez, A. Nikolay, Y. Genzel, and U. Reichl, "Bioreactor concepts for cell culture-based viral vaccine production," *Expert Rev. Vaccines*, vol. 14, no. 9, pp. 1181–1195, 2015.

[90] W. Whitford, "Using disposables in cell-culture–based vaccine production," *BioProcess Int.*, vol. 8, 2010.

ABBREVIATIONS

AAV	Adeno-associated virus
AD5	Adenovirus 5
AGE1.CR	Avian duck cells
ATF	Alternating tangential flow filtration
CAR T-cell	Chimeric antigen receptor T cells
CHIKV	Chikungunya virus
CHO	Chinese hamster ovary
CIP	Clean in place
COVID-19	Coronavirus disease 2019
CRD	Cell retention device
CSTR	Continuous stirred tank reactor
DIP	Defective interfering particle
DNA	Deoxyribonucleic acid
DSP	Downstream processing
E&L	Extractables and leachables
EMA	European Medicines Agency
FASTEC	Frozen accelerated seed train for execution of a campaign
FDA	Food and Drug Administration
GMP	Good manufacturing practice
HA	Hemagglutinin
HAU	Hemagglutinin units
HAV	Hepatitis A virus
HCD	High cell density
HCV	Hepatitis C virus
HFBR	Hollow fiber bioreactor
HEK293	Human embryonic kidney
HIV	Human immunodeficiency virus
Huh7.5	Human hepatocellular carcinoma cells
IAV	Influenza A virus
IPV	Inactivated polio virus

LV	Lentivirus
MAb	Monoclonal antibody
MDCK	Madin Darby canine kidney
MRC-5	Human diploid lung cells (Medical Research Council cell strain 5)
mRNA	Messenger RNA
MVA	Modified Vaccinia virus Ankara
N	Number of cell expansions for production
OSB	Orbital shaken bioreactor
PBBR	Packed bed bioreactor
RABV	Rabies virus
RNA	Ribonucleic acid
SARS-CoV-2	Severe acute respiratory syndrome corona virus type 2
SIP	Steam in place
SRID	Single radial immunodiffusion
STR	Stirred tank bioreactor
SUB	Single use bioreactor
TFF	Tangential flow filtration
USP	Upstream processing
vg	Viral genomes
VLP	Virus-like particles
VSV	Vesicular stomatitis virus
YFV	Yellow fever virus
ZIKV	Zika virus

7 Downstream processing of viral-based vaccines

Rita P. Fernandes and Cristina Peixoto
iBET, Instituto de Biologia Experimental e Tecnológica, Apartado 12, 2781-901 Oeiras, Portugal

Instituto de Tecnologia Química e Biológica António Xavier, Universidade Nova de Lisboa, Av. da República, 2780-157 Oeiras, Portugal

Piergiuseppe Nestola
Sartorius Stedim Switzerland AG, Ringstrasse, Tagelswangen, Switzerland

CONTENTS

7.1 INTRODUCTION

This chapter will explore the available downstream processing (DSP) strategies for the purification of viral vectors, used for vaccine production, in both industry and academia. The drivers for process design are the maximization of product recovery, scalability, and robustness while operating in ideal conditions for maintaining

DOI: 10.1201/9781003229797-7

product stability. The ultimate goal of the development of DSP is to reach a final manufacturing process that is in accordance with the guidelines of the regulatory authorities for full-scale production of the biotherapeutic target into the market. The following chapter will cover the different operation units, from the harvest of the product from cell culture until the sterile filtration to formulate the final product. Besides a well designed purification process, there is the need for good analytical tools and quality control in parallel, so this topic will be also briefly discussed in this chapter, focusing on the importance of high throughput process development. In the end, the future of purification technology will be also highlighted regarding topics as new facility design and process intensification.

7.1.1 Viral-Based Vaccines Manufacturing—Current Challenges in Downstream Processing

For viral-based vaccines (VBVs), the downstream processing is a challenging task. It involves the purification of high amounts of complex particles assembled from proteins, lipids, and nucleic acids thanks to the combination of several validated purification procedures. Moreover, the biological complexity and diversity of the viral-based products can directly impact the selected purification process schemes. For example, lower amounts of host cell DNA and host cell protein (HCP) are present in enveloped viral vector's production systems than for non-enveloped viruses. However, such enveloped viruses are more sensitive to shear stress; thus, achieving high yields of purified viral products is more challenging (Figure 7.1). All these factors have to be taken into account during process development, while decreasing the cost per dose.

As DSP can account for up to 60–70% of the overall production cost, its optimization will extensively contribute to more cost-effective processes There have been great achievements in the implementation of robust vaccine manufacturing processes using animal cells to increasing viral product titer and quality. However, much less effort has been put to the essential downstream processing. Thus, this constitutes currently a major bottleneck together with the development of dedicated analytical tools. Moreover, the growing pressure to develop cost-efficient processes has brought the need to improve purification strategies, when compared with the traditional purification methods, that will be discussed later in this chapter.

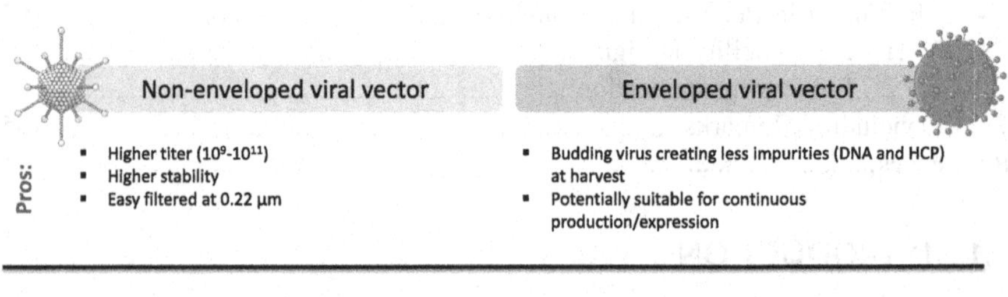

FIGURE 7.1 Comparison (pros and cons) between enveloped and non-enveloped viral vectors.

The main aim of DSP is to eliminate the impurities, which can be divided into two subgroups: process-related and product-related impurities [1]. Product-related contaminants are related to variants that differ from the desired product, such as dimers, aggregates, empty capsids, product with heterogeneity of post-translational modifications (glycosylation, phosphorylation and acylation). Process-related contaminants include all the potential ones added or derived from the manufacturing process during up- or downstream, such as host cell protein (HCP), host cell DNA (HCDNA), cell culture media components, endotoxins, anti-foam reagents, stabilizers, excipients, proteases, and nucleases, which do not have properties comparable to those of the desired product with respect to activity, efficacy, and safety. Residual host cell DNA, HCP, and endotoxins are the impurities more strictly controlled by the authorities. The maximum residual amounts of host-cell DNA depend on the cellular production platform. Nevertheless, it should be between 100 pg/dose to 10 ng/dose of sizes not above 200 base pairs [2]. Regarding HCP, no strict limitation have been established by regulatory authorities. However, most therapeutic proteins reviewed by FDA have been reported to contain ELISA-based host cell protein level between 1–100 ppm [3]. Also, a recommendation of endotoxin level below 20 EU/mL for recombinant sub-unit vaccines and below 10 EU/mL for genetic vectors is suggested; although higher endotoxin levels may be acceptable in certain cases [4].

Finally, the goal is to obtain a product with high purity, potency, and quality that can meet the stringent guidelines of the regulatory authorities, the U.S. Food and Drug Administration (FDA) and the European medicines Agency (EMA) [5,6].

7.1.2 Traditional Methods for Virus Purification

Complexity of the purification processes holds in the fact that viruses are quite large when compared with other biomolecules present in the harvested cell lysate such as proteins, peptides, sugars, and nucleic acids. For example, proteins typically range from 0.005 to 0.006 × 10^6 Da, whereas viruses are around 5 × 10^6 Da. They present sizes up to 1,000 nm, although the most common viruses produced have sizes between 20–400 nm [7]. This is why the first purification strategies implemented in vaccine manufacturing were based on virus size and density properties. Ultracentrifugation and density gradient centrifugation, later combined with filtration techniques, were the first methods to be used for the purification of viral particles.

Density gradient centrifugation, either used with sucrose, caesium chloride (CsCl), or iodixanol gradients, is a well-known and established traditional purification technique. It is generally used to purify limited amounts of bulk for preclinical applications. These gradients were used to process many viral particle preparations, including adenovirus [8], AAVs [9], and influenza [10], but also several other viral particles [11]. Figure 7.2 shows a typical result of CsCl purification of adeno- (AdVs) and adeno-associated virus (AAVs) with their respective analyses of the final purified viruses by transmission electron microscopy (TEM) images. A purification strategy

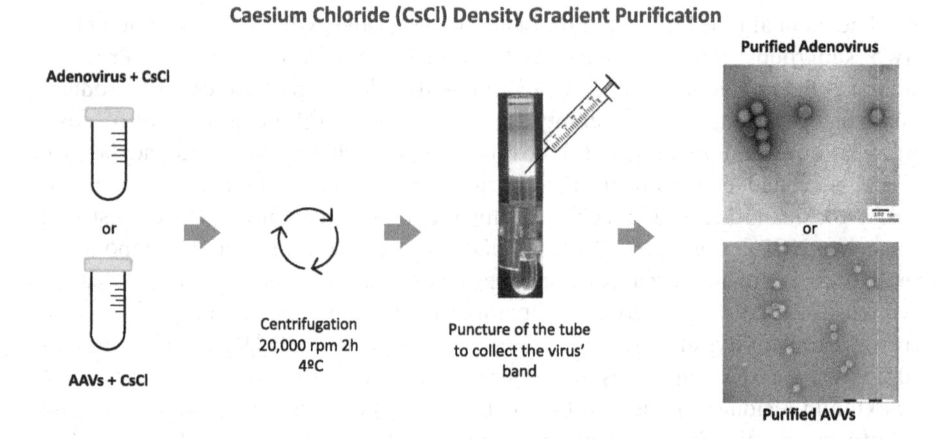

FIGURE 7.2 Density gradient ultracentrifugation using caesium chloride to purify adenovirus and AAVs.

using iodixanol (OptiPrep™) density gradient was also established for the retrovirus purification. This method requires fewer manipulations, when compared with other density gradient mediums, and leads to higher titers and lower processing times due to the lower viscosity of iodixanol. Given the lack of toxicity of this compound, there is no need for its removal [12].

Although high-speed preparative centrifugation has been used for the past decades for virus purification, its use presents several drawbacks. Indeed, it is related to high initial investment costs, limited scalability, and expensive maintenance, being suitable for laboratory scale to obtain high concentrated virus preparations [13,14]. Additionally, viruses may suffer a loss of infectivity during the long period required to fractionate the gradients, commonly between 16 and 90 hours.

Although ultracentrifugation was progressively replaced by new operations for manufacturing applications in the last two decades, it is still the method of choice for new vaccine candidates. Indeed, for such cases, partial knowledge of their physicochemical properties, especially for early clinical trials, ultracentrifugation is still the safer purification approach. For that reason, single-use ultracentrifuges were developed while single-use manufacturing processes were deployed in the industries, like for example Ksep® (Sartorius Stedim) and CARR UniFuge® (Pneumatic Scale Angelus). Both equipments have the advantage to apply low shear on cells, enabling the harvest of intact cells as product or discard them as a by-product. This is especially important for budding enveloped viruses, where a cell lysis step is usually not required.

However, the trend in the industry is trying to avoid the centrifugation processes, and increasingly exploring the chromatographic and membrane-based purification techniques.

7.2 PURIFICATION STRATEGIES FOR VIRUSES

The purification of viruses is not a simple task given their biological complexity compared with other biotherapeutics, like antibodies (mAbs). Indeed, they differ in size, shape, and surface structure. Additionally, physicochemical features like isoelectric point (pI), surface hydrophobicity, and the presence of viral envelope can play important roles in the design of the DSP train. Given this biodiversity, there are no universal purification processes for viral vaccines. However, there are common steps and unit operations providing a good starting point, as depicted in Figure 7.3. Furthermore, achieving the high levels of purity required for viral-product clinical application requires a complex cascade of unit operations, that go further than the traditional purification methods discussed before. The purification scheme usually includes a unit operation for clarification, intermediate purification (capture and concentration), and polishing of the product of interest. However, the number and order of the unit operations might vary based on the virus of interest as well as final therapeutic dose and usage, given the fact that the impurities specifications are based on the dose and final use.

In the following sections, each unit operation will be described and the technology trends in vaccine manufacturing will be mentioned.

7.2.1 Harvest and Clarification

When the virus or viral vector is harvested from the cell culture, at the end of the upstream processing (USP), the resulting harvest material contains, apart from the viral product of interest, processing reagents (media components), cell debris, and host cell-related molecules (host cell proteins, DNA). All these contaminants present a health risk if present in a final formulation of a vaccine candidate. So, there is a need for their removal to a certain extent before that virus preparation is used as a biotherapeutic, specifications being set by regulatory authorities. The time of harvest (TOH), as defined in Chapter 6 as the upstream process development–process intensification, should be established, taking into consideration different factors like product quality, process reproducibility, virus stability, and not only the process productivity [15]. Harvesting later in the upstream cell culture process will lead to reduced cell viabilities and consequently to an increase in host-cell contaminants i.e., host cell protein (HCP), host cell DNA (HCDNA), and cell debris. This might strongly affect to following efficiency of the purification process.

Viruses used for vaccination are generally produced by cell infection, and their release is dependent upon the virus cycle since they can be found intra- or extracellularly. In the case of lytic viruses assembled intra-cellularly, a cell lysis step is necessary to release the neosynthetized particles as for adenovirus or adeno-associated virus. Such cell lysis or cell disruption is thus performed prior or after clarification step. It can be performed by mechanical (homogenization, sonication) or chemical (freeze-thaw cycles, detergent addition) methods [16]. Considering adenovirus manufacturing, at a laboratory scale, the purification process consists of a cell lysis step using freeze and thaw methodologies, followed by density gradient ultracentrifugation and a final desalting step [17,18]. This process is successful at a small scale, achieving high purity level while maintaining a low total to infectious viral particle ratio (below 30). However, this methodology has strong limitations associated with the processing time and scalability. This is why cell lysis at a larger scale is commonly performed by the addition of detergents [19]. One detergent widely used is a mild non-ionic detergent

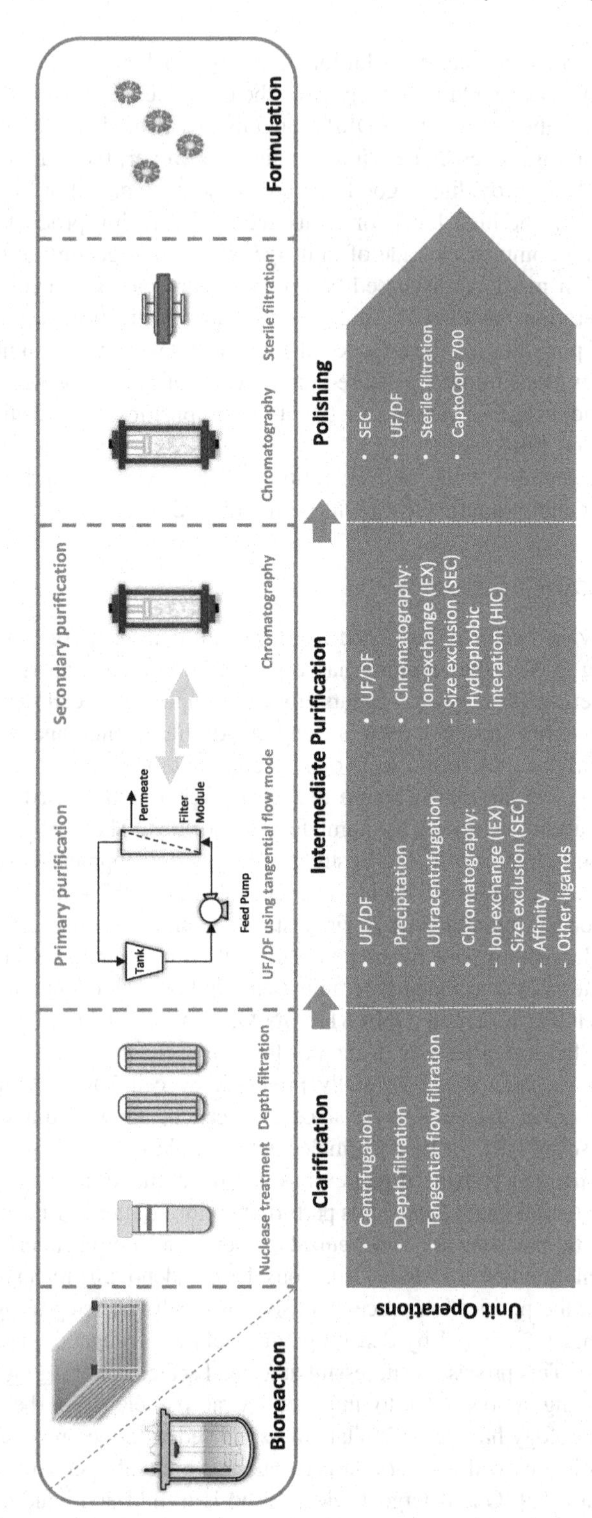

FIGURE 7.3 Schematic representation of the DSP train in viral vaccine manufacturing. The process scheme shows the different steps of a downstream process ranging from the initial product harvested to concentration, intermediate purification, and polishing unit operations with respective possible methods in each unit operation.

known as Triton X-100. It solubilizes the cell membranes allowing the viruses release. Several other applications of Triton X-100, commonly formulated at 0.1% exist, as the protein solubilization, the viral sub-unit preparation, and the enveloped virus inactivation [20,21]. However, there is evidence that Triton X-100 can have undesirable effects on the environment due to endocrine disruption properties during its degradation. Thus, it was added to the REACH (Registration, Evaluation, Authorisation and Restrictions of Chemicals) list, forbidding its use from 4 January 2021, forcing the companies to work on eco-friendly substitutes. One alternative recently reported was Polysorbate 20, which is a stable, non-toxic, and non-ionic surfactant widely used in domestic and pharmaceutical applications [22]. In this work, the efficacy of Polysorbate 20 was evaluated and compared with Triton X-100. Results showed no negative effects on the adenovirus's purification train and an increased virus recovery and impurities' removal. Other alternatives, such as sodium deoxycholate for the AAVs' purification [23] and CHAPS for adenovirus's purification [24] have already been applied. Nevertheless, detergents added to the culture should be removed, as they can have an impact on the next downstream operation. The increase in virus yield thus obtained should compensate for the extra efforts to achieve the required purification level for the viral vaccine.

After cell lysis, there is an increase in the impurities level (host cell proteins, host cell DNA, and cell debris) needing to be removed by centrifugation or filtration technologies such as depth filtration or microfiltration. Although clarification is not always categorized as a downstream operation and is sometimes neglected in vaccine manufacturing, this unit operation is of high importance in virus purification. In fact, an optimized and efficient clarification step can strongly impact the overall purification process' costs and also reduce the manufacturing footprint, by reducing the subsequent size of chromatography columns and buffers' consumption.

Given the biological diversity of vaccine types, several series of operations are required to achieve an efficient clarification [25]. The first operation aims to remove larger particles, like remaining cells and cell debris usually applying centrifugation. Afterwards, a second step is used to remove low molecular weight particles thanks to filtration techniques. Centrifugation is frequently used as the method of choice for clarification of cell culture production-based products at large scales as it can be operated both in batch or continuous mode. However, several drawbacks of this technique makes of filtration techniques new methods of choice.

First, the cost of investment for large-scale centrifugation for equipment is high compared to filtration techniques and their sanitization procedures are critical and laborious. Then, even though centrifugation is possible to operate at large scales, the limited sample capacity of the centrifuge is still one of its main disadvantages. Thus, given the improvement in upstream processing technology, enabling higher titers, filtration techniques have gained interest in vaccine clarification.

Filtration techniques used in clarification through membrane microfiltration can be performed either by normal flow filtration (NFF, also known as dead-end filtration) or tangential flow filtration (TFF, also known as cross-flow filtration). For viral-based products, filtration membranes have pore sizes in the range of 0.1–10 μm. Depth filtration is normally used in dead-end mode, and is usually preferred for crude cell supernatant rich in high biomass content, due to their higher capacity to retain impurities within the filter. Their mode of operation is not only based on size exclusion, but also hydrophobic and electrostatic interactions. Some filter aids can be added to depth filters, such as diatomaceous earth (DE), that revealed to improve the filter efficiency in retaining particles [26]. Depth filters play mainly with two factors for particle retention: size exclusion and adsorption. Some filters, such as Sartopure® PP3, stand out given their very low unspecific binding. This allows retaining most of cells and cells debris [27].

On the other hand, microfiltration membranes using TFF retain only impurities larger than the pore size and have a lower impurity holding capacity. They are, therefore, suitable for a secondary clarification step. Both kinds of clarification filters (depth filters and membrane devices) are suitable for scale-up in vaccine manufacturing and both have already been incorporated in the manufacturing processes of viral vaccines [28–30]. Generally, clarification is a complex step with a number of available technologies. As presented in Table 7.1, there is no universal solution since the selection of the filters and operation parameters requires critical handling.

As discussed previously, it is of importance to wisely select the TOH, as, it can strongly impact clarification step. Indeed, harvesting cell cultures with low cell viability increase the presence of cell debris in the extracellular medium thus reducing filter capacity and further blockage. To assess the clarification process efficiency, solution turbidy monitoring is an important parameter. It also enables the detection of the filter capacity which is related to the fouling of the membrane [38]. Membrane fouling is commonly the consequence of the formation of a polarized layer on the filter surface due to impurities' accumulation present in the cell supernatant. Operating membrane microfiltration through tangential flow filtration (TFF) allows avoiding such formation of the polarized layer since the cross-flow decreases the formation of a "cake" on the membrane surface.

Membrane devices such as hollow fibers or cassettes can be used for the clarification step using TFF. However, the latest upstream technologies advances (Chapter 5), especially the high cell densities processes, are challenging such filtration operations. In the last decade, several virus's production systems have been described to produce at cell densities above 10^7 cell/ml, namely PER.C6 cell line grown up to 10^8 cell/mL for HIV vaccine candidate [39]; MDCK cells infected with the influenza virus at 5×10^7 cell/ml [40] and, more recently, CR.pIX cells reaching 2.5×10^7 cells/mL for modified vaccinia Ankara production [41]. All of these productions employed an advanced alternating tangential flow (ATF) perfusion system, using a hollow-fiber device. Another filtration technology widely applied to high-cell densities is body feed filtration (BFF). Here few filter aids are added to the crude bulk, such as diatomaceous earth (DE), to enhance the filter capacity [42]. Another novel technology of Repligen for fed-batch clarification is TFDF™, which combines benefits of tangential flow (TF) and depth filtration (DF) in a single-use system with pore size ranging from 2 to 5 µm. This system was successfully applied to separate lentiviral vectors from cell debris in batch and perfusion production modes [43].

During vaccine manufacturing, there is an extra concern with host-cell DNA removal depending on a risk assessment to evaluate possible side effects with the

TABLE 7.1
Combination of technologies used for clarification of viral-based vaccines at pilot (1–20 L) or production scale (more than 20 L)

Vaccine	Production system	Vaccine type	Product location	Primary clarification	Secondary clarification	References
Hepatitis C	Insect cell culture	Virus-like particle	Extracellular	0.2 μm hollow fiber cartridges (TFF)	300 kDa cassettes (TFF)	[31]
Yellow fever	Vero cell culture	Inactivated virus	Extracellular	Sartopure® PP2 (8.0 μm), NFF	Sartoclean® CA (3.0 μm + 0.8 μm), NFF	[28]
Rotavirus	Insect cell culture	Virus-like particle	Intracellular	Centrifugation at 1000 g for 10 min at 4°C	Ultracentrifugation 100,000 g for 1 h at 4°C	[16]
Influenza virus	MDCK cell culture	Inactivated virus	Extracellular	0.65 μm polypropylene depth filter (NFF)	NA	[32]
Enterovirus 71	Vero cell culture	Killed virus	Extracellular	0.65 μm filter (NFF)	NA	[33]
Rotavirus	Vero cell culture	Live virus	Extracellular	Centrifugation at 2831 g for 30 min and 4424 g for 10 min at 4°C	0.45 μm hollow fiber (TFF)	[34]
Poliovirus	Vero cell culture	Inactivated virus	Extracellular	Diatomaceous earth filter with 75 μm pore size, or Millistak + ® C0HC (Depth filter)	0.45 and 0.22 μm filtration (NFF)	[35]
Rabies	Vero cell culture	Inactivated virus	Extracellular	8 μm MF-Millipore™ membrane filter	0.45 μm Durapore® membrane filter	[36]
Hepatitis C	Insect cell culture	Virus-like particle	Extracellular	Optiscale depth filtration devices with Polygard® CN membrane material	300 kDa cassettes (TFF)	[37]

(Adapted from Ref. [25]). NA – Not applied.

amount of DNA present in the final product. Also, DNA removal early in the process will prevent the formation of complexes with the virus [44]. Nuclease treatment, such as Benzonase®, is one option frequently used to reduce the DNA content. However, there have been emerging alternative options, such as Denarase® and, more recently salt activated nuclease (SAN), that showed improved results in AAV production when compared with the other two commercially available endonucleases [45]. Oxford Biomedica developed a novel strategy that overcomes the nuclease addition step using secreted nucleases (SecNuc®) in co-production with viral vector manufacturing. Although it is a challenging step, these alternatives can overcome the challenges by reducing the overall manufacturing costs. The nuclease addition step can be performed before or after the clarification step as it is an additional impurity that should be removed during the next purification steps. Notwithstanding, the industry witnessing the end of Benzonase's patent and the appearance of several competitors will certainly cause a substantial decrease in nuclease price in the future. Other more conventional and cheaper alternatives for host-cell DNA removal have been proposed. As examples, selective precipitation and flocculation can replace nuclease addition and compensate for the higher costs through the process [2]. Kröber et al. developed a method for DNA precipitation with polyethylamine (PEI) together with a specific unit operation which allowed to reach a 500-fold decrease of host-cell DNA at a pilot scale [46]. This technique for DNA clearance could also work with single-use centrifugation, allowing low-speed and consequent high-titer viruses.

7.2.2 Intermediate Purification—Virus Concentration and Chromatography

Viral vaccine manufacturing requires a combination of several downstream purification strategies to achieve the desired purity and virus concentration. After clarification, the intermediate purification strategy is highly dependent on the final vaccine specifications. To move away from the ultracentrifugation techniques, the knowledge acquired with chromatography in protein purification has been adapted to viral particles. However, classical chromatography devices and matrices were initially developed for smaller biomolecules, such as proteins. Thus, there has been a need for evolution of these processes, to be applied to more complex products like viruses [47].

Furthermore, to reach the required virus dose, concentration methods are necessary. The intermediate purification of viral vaccines usually comprises a concentration using membrane filtration processes and/or a chromatography method.

Currently, ultrafiltration (UF) is the key technique for the concentration and diafiltration of viral particles, either for laboratory or large-scale bioprocesses. It allows the removal of low molecular weight impurities, the reduction of the volumetric volume, and enables buffer exchange of viral solution. This operation is a pressure-driven separation method, with transmembrane pressure (TMP), feed, retentate, and permeate flux as well as the molecular weight cut-off (MWCO) as key parameters. The MWCO is the mean pore size of a normal distribution, which depends on the membrane material and the manufacturer. For virus purification, 100–750 kDa pore size range is the typical range that gives reasonable recoveries (70–85%). The

UF membranes can be manufactured using various polymers, including polysulfone (PS), regenerated cellulose (RC), polyethersulfone (PES), and polyvinylidene fluoride (PVDF).

A good solution for concentration and diafiltration of viruses is the use of low adsorptive membranes especially in the case of enveloped viruses. As an example, the biotechnology company Sartorius proposes Hydrosart® cassettes that exhibit a novel technology based on stabilized cellulose membranes. Such technology enables high flow rates and high product recovery by reducing the fouling effect. It already showed promising results for adenovirus purification with a recovery of 87% [48]. More recently, this technology has also become available in Sartocon® ECO format for low shear operations, which is characterized by being extremely hydrophilic and prone to non-protein binding.

Despite the efforts in developing robust purification processes for viral biotherapeutics, most of the process development has been focused on chromatographic steps. However, recently Carvalho et al. developed a fully optimized filtration cascade process for an influenza vaccine candidate [49]. Several companies are investing efforts to develop devices suitable for vaccine product candidates, such as Cytiva's 750 C hollow-fiber membranes. These membranes were optimized to propose a more open structure, enabling higher removal of host-cell DNA [50].

Another operation mode of ultrafiltration is the single-pass TFF (SPTFF), a method that enables the concentration of the product of interest in a single pump pass as an inlet concentration step (Figure 7.4). This technology enables to achieve high concentration (up to 30 times) and recovery, control, and reduce the in-process volumes, operating in continuous mode [51]. This technology can be operated either with cassettes or hollow fibers. In the absence of recirculation, thus applying less shear stress to the circulating solution, it makes a promising strategy for the fragile enveloped viruses.

Chromatography is probably the most popular technique of the intermediate purification step. Separation methods based on chromatography have been used in the biological field for decades for the purification of recombinant proteins and

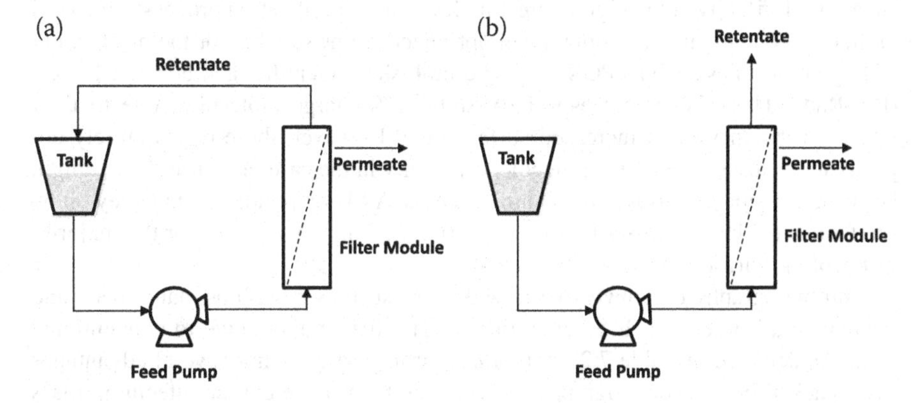

FIGURE 7.4 Different designs of tangential flow filtration: (a) traditional batch tangential flow filtration and (b) single-pass tangential flow filtration.

monoclonal antibodies (mAbs). Chromatography is based on different interactions between the target viral particle and components placed on the stationary phase. These interactions rely on the surface properties of the viral particle (adsorptive chromatography) or their size (size exclusion chromatography). In the case of adsorptive chromatography, the viral particles' solution is passed through a solid phase (stationary phase) coated with functional groups that capture the viral particle (also known as bind-and-elude mode or positive chromatography) or capture the impurities (also known as flow-through mode or negative chromatography). This depends on the global charge of the virus and the operation mode. In this case, cationic or anionic exchangers can be used. Ion exchange chromatography (IEX) is the most used chromatographic technique in vaccines and viral vectors manufacturing, exploring the reversible interaction between a charged particle surface and an oppositely charged matrix. Besides IEX, there are other interactions which have been exploited such as affinity (AC), hydrophobic interaction (HIC), mixed-mode (MMC), and size exclusion (SEC). Affinity chromatography, used for long time in protein purification, has recently gained special prominence in the complex biotherapeutics field due to their specificity separation, high degree of purity and recovery, and consequent contribution to the reduction of necessary unit operations [52,53].

The flow-through mode chromatography avoids some of the issues addressed with other chromatography separation methods, being transversal for different biological systems. Harsh elution conditions may be omitted and, consequently, the risk of immunogenicity/infectivity loss and aggregation is also reduced [54]. This strategy has been recently evaluated for the purification of hepatitis C virus-like particles. In this process, the product is recovered in the flow-through, avoiding harsh elution conditions and the risk of virus's disassembly [37].

The stationary phase can be physically structured as packed beds, membrane adsorbers, and monoliths, as shown in Figure 7.5. The most commonly used are packed beds, which consist of beads physically packed into a chromatographic column. They were extensively used for protein separation, but viruses are generally larger than proteins, between 20 nm to 400 nm, resulting in low binding capacities [55]. Given the high interest in developing purification processes for viral particles, there are already some novel optimized resins suitable for the purification of large molecules such as POROS® (ThermoFisher Scientific) and NUVIA HP-Q® (Bio-Rad). These beads are especially suitable for large molecules, due to their rigidity, robustness, and increased surface area. However, these resins largely depend on the size of the target molecule, especially in the case of a virus. They might be ideal for small viruses (20–30 nm) such as AAV or poliovirus but they show limitations for larger viruses (above 80–100 nm), which is the case for the majority of oncolytic viruses.

Chromatography on convective flow devices such as monoliths, nanofibers, and membrane adsorbers has been emerging as an efficient alternative to conventional ones. As depicted in Table 7.2, membrane chromatography has several advantages over packed-bed chromatography. Owing to their different architecture, mass transport through the pores/channels takes place mainly by convection overcoming virus particle diffusion issues faced by the traditional packed-bed chromatography.

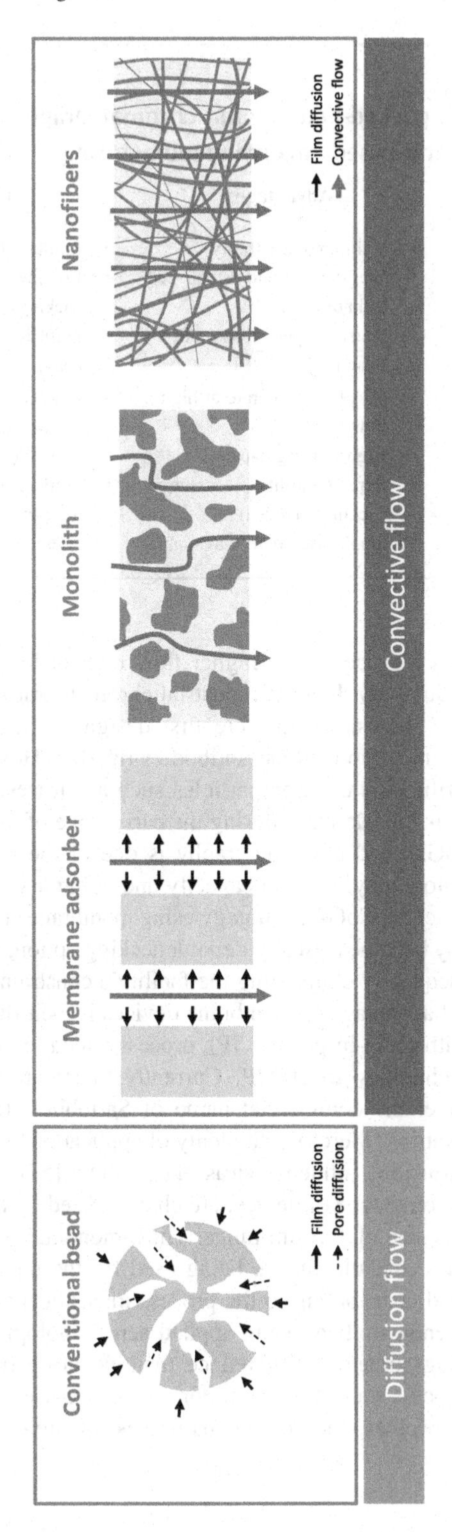

FIGURE 7.5 Chromatography technologies for biotherapeutics manufacturing.

TABLE 7.2
Comparison between packed-bed column chromatography and membrane chromatography in viral-based vaccine manufacturing

	Advantages	Limitations
Packed-bed column chromatography	• Unlimited scalability • Well-known and established technology • Several types of possible ligands	• Limited flow rates • Complex operation (column packing) • Diffusion limited (conventional bead)
Membrane chromatography	• Possible to operate at high flow rates • High binding capacity • Reproducibility (no column packing required) • Single-use devices available	• Limited choice of membrane chemistry commercially available • Limited scalability (restricted to the commercially available sizes)

This allows performing a process with higher flow rates at lower pressure drops, resulting in higher productivity. Briefly, industrial chromatography is dominated by porous bead stationary phases, which were first designed for the purification of therapeutic proteins such as monoclonal antibodies (mAbs). However, alternatives are needed for the purification of large particles such as viruses.

Something to have in consideration during the early stage of bioprocess design is the cost of goods (CoGs), and chromatography is one of the unit operations that increases it. In chromatography, column capacity and column sizing are two of the crucial factors that impact the CoGs. A strategy using membrane chromatography in a flow-through mode may become a greatly debottlenecking option, by reducing buffer consumption and consequently eliminating the facility's constraints caused by plant capacity. Another great advantage of membrane devices is their disposability, which minimizes the costs with clean-in-place (CIP), process validation efforts, facilitating the transfer of the technology to cGMP. Currently, there are three well-known membrane devices under the commercial name of Sartobind® (Sartorius Stedim), Mustang® (Pall), and Natrix® (Merck) with plenty of applications, for example for the purification of adenovirus [56], influenza virus-like particles [57], and lentivirus [58]. Monoliths, also a convective type of device, are characterized by their highly porous structure, with channels size of 1–5 μm range. Thus, monoliths are gaining a lot of interest in viral particles' manufacturing, being used for the purification of several viral-based products and contributing to the process intensification [59,60]. Process intensification is the trendy strategy being applied across biological manufacturing, that involves combining strategies that reduce manufacturing time, improve efficiency, and increase process standardization. For this, single-use devices, shorter and efficient purification processes, and continuous processing are concepts extensively into exploitation.

7.2.3 Polishing

A GMP compliance purification process often includes one or more final steps for minor impurities removal, named polishing steps. Besides of removal of trace impurities, the aim of this part of the process is also to ensure the final product formulation, stabilization, and sterilization. This is challenging since the impurities still present at this stage are very closely related to the product of interest, such as aggregates and wrongly assembled particles. Consequently, they are more difficult to remove. Thus, the resolution is especially important to guarantee a successful separation, so chromatography techniques are the workhorses at this step. Size exclusion chromatography (SEC) is the most used technique to remove low molecular weight impurities, where the virus elutes in the void volume, while the smaller impurities elute afterward. UF/DF (TFF) can be also used for the same purpose since both methods rely on size differences. However, SEC has limitations in scale-up, since the load should not exceed the 10% of column bed volume. Both techniques can be used for buffer exchange since the final formulation is a crucial factor to guarantee not only the particles' stability but also a proper immune response. The formulation and stabilization field is gaining special attention as a research field of interest since the use of adjuvants contributes to an improved immune response [61–63].

Adsorptive chromatography can be also used for the polishing step, but is usually preferably operated in negative mode. Ion-exchange chromatography (IEC) is also a widely used technique for resolving host cell protein-virus particle, host cell DNA-virus, or even damaged from intact viral particles. There are several described processes in the literature, using SEC followed by IEC as a polishing step [32,64]. Multi-modal or mixed-mode (MM) resins are a novel technology, combining simultaneously various types of interaction in the same chromatographic media, such as ionic interaction, hydrogen bonding, and hydrophobic interaction. One well-known example of mixed-mode media is ceramic hydroxyapatite, which has already proven its worth for virus purification [65]. Another example of MM is Capto Core 700 resin, which plays with both binding property and size exclusion. This resin has been used in the negative mode for the polishing of biotherapeutics [13,54,66].

Sterilizing filtration is usually the final process, where the product is filtered through a 0.2 μm filter for the final removal of bioburden. This step is exclusively dependent on the particle size, so particles larger than 200 nm cannot be sterile-filtered. This is the case for Vaccinia virus and Poxvirus, which impose to work in a closed-system process to guarantee sterility or bioburden control since the beginning of the process. The membrane material, similar to the other filtration techniques already described in this chapter, is one of the essential parameters which should be optimized to avoid non-specific interaction between the viral particles and the membrane matrix. Another critical parameter is the transmembrane pressure (TMP), which can increase uncontrollably at constant flux, due to filter clogging. There is still some limitation in sterile filtration in viral-based bioprocesses, since either by virus adsorption to the filter or presence of aggregates that decrease the total product recovery [67].

7.3 ANALYTICS TOOLS ON PROCESS UNDERSTANDING AND HIGH THROUGHPUT PROCESS-DEVELOPMENT

Biotechnology and pharmaceutical companies are increasingly facing external pressure to develop vaccines not only in shorter process development timelines but also for them to be more affordable and accessible worldwide. Thus, it is fundamental to give solid steps in process development from the early beginning, which can be potentiated through in-parallel tools and strategies contributing to process understanding and characterization. The process understanding can be enhanced through novel strategies such as process analytical technology (PAT), which consists of designing, analyzing, and controlling manufacturing through timely measurements of critical quality and performance attributes. The goal is to achieve a better control strategy of the process and the effect of scale-up. (This subject is extensively described in Chapter 8).

Nowadays in the pharmaceutical industry, the development is performed by the concept of quality by design (QbD), where it begins with predefined objectives, emphasizes product and process understanding and control based on risk management [2,68]. Even for downstream processing optimization, which essentially relies on physicochemical properties not entirely known, are strategies such as design of experiments (DoE) that contribute to the increase of knowledge. For experimental evaluation, high throughput screening (HTS) and high throughput experimentation (HTE) are promising tools for process development [69].

The use of HTS and HTE for process development (PD) is the strategy currently being applied in big biopharma companies. This strategy is a much smarter way to generate process knowledge, settle in mechanistic modeling-based approaches that will allow a safer tech transfer due to process automation, reduction of human intervention, and consequently minimize the risk of process failure. For this mechanistic and predictable view of process design, it is important to have good analytical tools that enable process automation and optimization.

Analytical tools are fundamental for both process development and process understanding, being also essential to reach a characterized final product regarding its quality and safety (Figure 7.6). Besides the traditional quantification methods, size exclusion HPLC, capillary zone electrophoresis, dynamic light scattering (DLS), surface plasmon resonance (SPR), asymmetry flow fractionation, and electron microscopy are methods used to complement the full characterization of the viral particles. These assays allow to evaluate product fragmentation, aggregation, and variants that contribute to microheterogeneity [70]. Mass spectrometry-based methods are one of the powerful methods that can provide valuable insights about virus particle composition, structure, conformational stability, assembly, maturation, interactions with other viral and cellular biomolecules, and changes induced in viruses by external factors as bioprocess operation conditions [71,72]. Analytical ultracentrifugation (AUC) is a highly accurate technique to quantify empty and genome-containing capsids, as this method has been the method of choice for adenovirus and adeno-associated virus [73,74].

The tendency in analytical tools is improving the PAT tools for monitoring and process developing (up- and downstream) in a faster manner, and also the translation of tools currently used in the analytical setting, towards process monitoring and PAT.

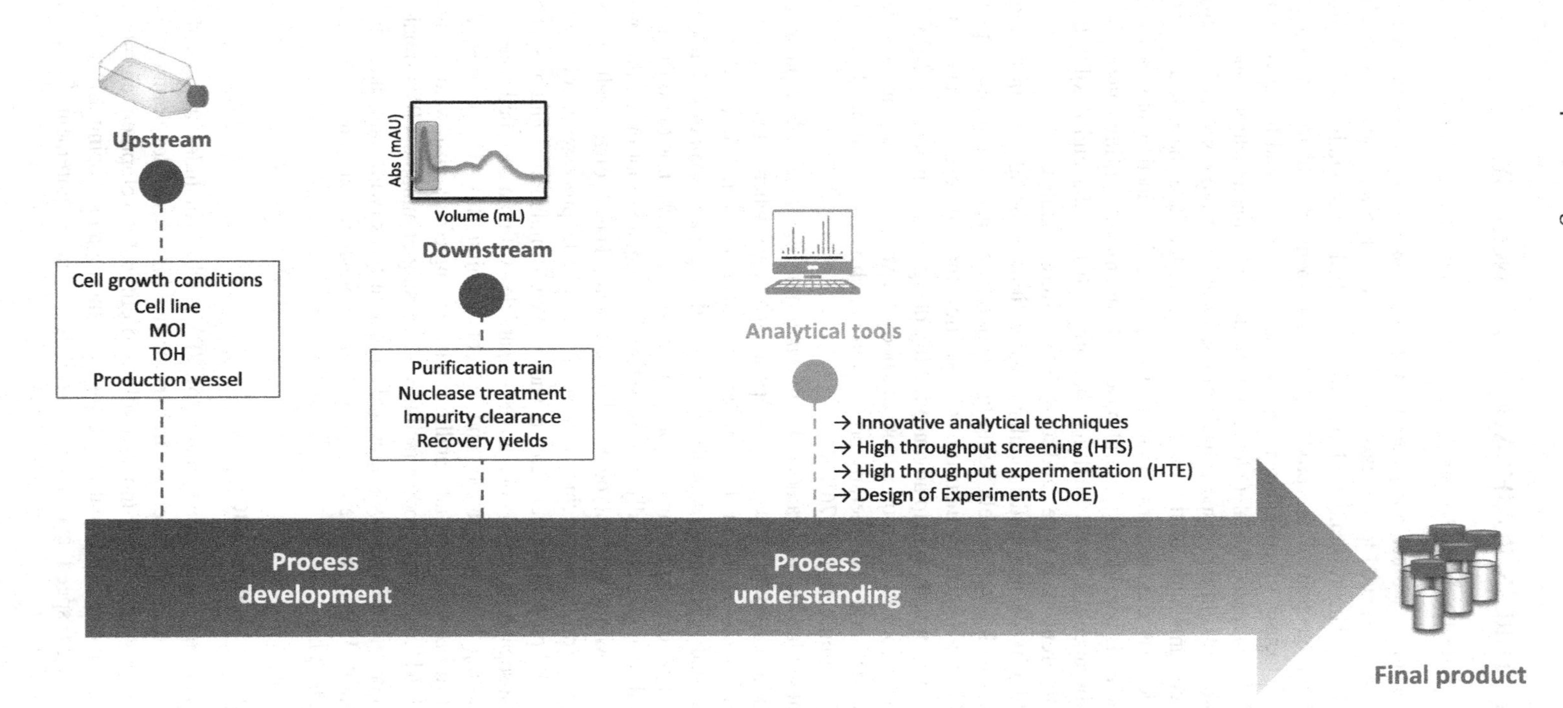

FIGURE 7.6 Key parameters (process development allied to process understanding) with robust analytical tools to achieve a final biotherapeutic product.

7.4 THE FUTURE IN PURIFICATION TECHNOLOGY

7.4.1 New Facility Design

Viral-based vaccine manufacturing under GMP conditions requires an additional biosafety level when compared to the classic mAbs industry. Biosafety level 2 (BSL2) is often the standard for many viral-based manufacturing. BSL 1 (generally used of mAbs) focuses on the protection of the product from the operators and the environment, whereas BSL 2 focuses on containment, protecting the operators and the environment from the product. Increased biosafety level classification adds manufacturing and facility complexity and costs. In this respect, having closed processes are essential to mitigate any possible contamination of the product and also to facilitate the risk assessment evaluation which has to be performed for any new product introduction into the facility.

Besides the BSL consideration, viral-based vaccine manufacturing and in general viral vector therapeutics needs facilities that are both modular and flexible to accommodate different virus types and as well as different scales.

Modular facility designs and modular construction are being used in biologics manufacturing, with a key benefit being faster time from inception to start-up. The BioPhorum Group hosted an industry collaboration that proposed a standardized, modular design approach to help advance facility design in the industry [75]. While the initial project looked at a monoclonal antibody (mAb) facility, it plans to extend the concept to cell and gene therapy production which is similar in needs to viral-based vaccine manufacturing [76].

With the modular facility framework, the ballroom concept is a key pillar. The ballroom approach features a large open operational space where closed processing equipment can be co-located in the same space. The industry has accepted that the functionally closed upstream production trains for therapeutic proteins, but not viral operations, can be deployed using an open ballroom approach. For the operation of VBVs in a ballroom concept, fully closed systems are required (not only functionally closed), together with a very thorough risk assessment. This is often challenging to achieve in many viral-based vaccine processes [77]. To date, processes involving host cell infection, viral production, purification, and product formulation should be spatially segregated in a separate room to contain vector particles within a specified zone in the facility. Regarding heat, ventilation, and air conditioning (HVAC) systems, these spaces should utilize dedicated air handling units or single-pass airflow to minimize contamination risks. For multiproduct facilities, processing of multiple VBVs should be performed either on a temporally segregated campaign basis (with sanitization in between) or in parallel but completely segregated viral production spaces for each product campaign produced.

7.4.2 Process Intensification

Given the increasing worldwide demand for VBVs, which includes attenuated and inactivated viral vaccines, as well as viral vector vaccines, faster development times are required to progress VBVs more rapidly into clinical development and then to market. This has meant VBV manufacturing is changing and is being driven by the need for increased speed and greater flexibility. The requirement for greater flexibility has led to new types of manufacturing facilities that can accommodate a growing range of different types of VBVs and can be reconfigured easily to take high clinical trial attrition rates into account. To meet these challenges, vaccine manufacturers, for example, those developing SARS-CoV-2 vaccines in cell cultures, are adopting strategies including process intensification as they believe this will help achieve higher product titers while reducing manufacturing footprint thus making larger numbers of doses of VBVs more readily available.

An advantage of using cell cultures to produce VBVs is that they could potentially be manufactured using process intensification. This is an approach to process development originally pioneered in the chemical industry by the Process Technology Group at Imperial Chemical Industries (ICI) in the United Kingdom. The aim was to reduce plant size while increasing productivity, thus decreasing the cost of goods (CoGs) by lessening capital investment and overhead costs [2]. Today, multiple definitions of process intensification have been developed but all have the ultimate goal of increasing productivity.

Process intensification could be used for vaccine manufacturing, to utilize facilities with a smaller plant footprint, and less scale-up volumes to rapidly produce a large number of doses required for mass vaccination campaigns. This is because there are more 2,000 L scale bioreactors than 20,000 L good manufacturing practice (GMP) compliant bioreactors which can be accessed quickly. Therefore, utilizing an intensified cell culture process could improve overall manufacturing yield to produce 10–20 doses of vaccine/mL, making it possible to perform 2,000 L runs and produce sufficient vaccine for small-scale trials. For example, by increasing the final titer of VBVs by 1 log a potential scale-up could be reduced from 20,000 L to 2,000 L, which would have a significant effect on overall production timelines [78].

Process intensification also offers the opportunity for more localized production of VBVs in low to middle income (LMI) countries because it can provide more doses/mL, potentially making the cost per dose lower. This will help the transfer towards LMI countries, where having a fully closed, high titer process is essential to facilitate GMP requirements, prevent unwanted contamination, and avoid the need for scale-up. In this case, a scale-out would be feasible and more easily manageable than a scale-up especially in a setting where bioprocessing skills and knowledge may be lacking.

Process intensification for VBV production is built on three pillars: equipment, mode of operation, and technology (Figure 7.7). Many manufacturers are assessing and adopting these pillars around process intensification, generally in a step-wise approach.

For downstream clarification of VBVs produced in intensified cell cultures by either enveloped, shear sensitive viruses, for example the SARS-CoV-2 or measles virus, or lytic non-enveloped viruses including adenoviruses, cell harvest and clarification can become a bottleneck. The development of high cell densities processes ($\geq 20 \times 10^6$ cells/mL) present several challenges because increased cell numbers produce an increase in host cell proteins (HCPs), DNA and protein aggregates, which can foul and clog filters, requiring sizeable filter areas and large buffer volumes for VBV purification.

The most common option to help remove the excess DNA before filtration is to use nuclease, but this is an expensive enzyme, and if a large amount of DNA is present then higher amounts of nuclease could be required making the

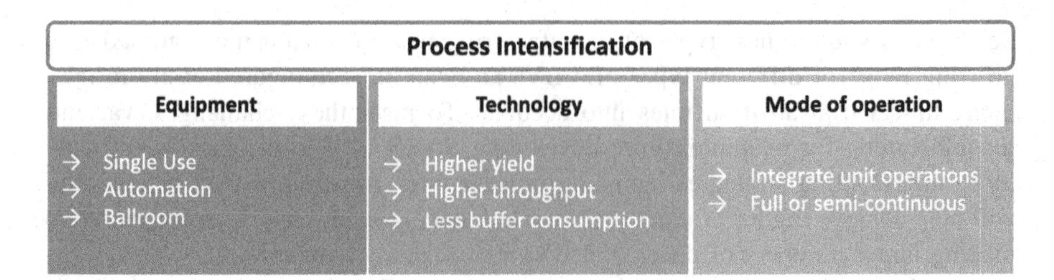

FIGURE 7.7 The three pillars of process intensification with viral-based vaccines (VBVs) (adapted from "Integrated Technologies to Accelerate Process Intensification for Viral Vaccine Manufacturing," Sartorius Stedim).

manufacturing process prohibitively expensive. One alternative to using nucleases with lytic viruses, such as adenoviruses being used to produce SARS-CoV-2 vaccines candidate is to use flocculants and centrifugation. Centrifuges that operate by balancing out centrifugal force and medium flow to keep cells contained in an expanded bed are recommended, as these exert lower shear forces on the viral product to minimize loss of infectivity. If classic filter trains do not work or a huge surface area is required, centrifugation is advised as the first step in cell harvest for high cell density as it reduces the filter area, eliminating pre-rinse steps to efficiently separate cells, enabling the use of smaller footprint facilities where less buffer volume and buffer storage are required. A suitable centrifuge for this application is the kSep® single-use (Sartorius Stedim), sterile centrifuge, or similar system such as Unifuge® (Pneumatic Scale Angelus).

For example, in the DSP purification different approaches can be taken to intensify the process [79]. With high binding capacity in mind, membrane adsorbers can be used in place of conventional resin-based affinity chromatography for the purification of VBVs [80]. Membrane adsorbers with beads that have large pore sizes (3–5 μm) are recommended as these provide maximum ligand accessibility for viruses and eliminate the need for diffusion through resin pores. Membrane adsorbers with large pore sizes such as Sartobind® Q or Mustang® Q can operate at volumetric flow rates that are typically 20-fold higher than classical resins. Studies with adenovirus have shown that compared to resins, they can achieve a 10-fold higher viral binding capacity using 58% less buffer, without compromising virus purity [79,81]. Additionally, a purification process using a NatriFlo® HD-Q membrane enabled the reduction of process time from 9 hours to 30 minutes, still improving the virus's recovery from 65–70% to 90% [82]. Besides the small features of the different membrane adsorbers available, the huge and remarkable benefit is the high productivity with the smallest footprint.

In addition to continuous centrifuges and membrane absorbers or other convective media (Monolith, fiber technologies, etc.), a key role is played by the TFF step. Usually, tangential flow filtration is a unit operation that can add a significant portion to the cost of goods. With this respect, the use of SPTFF can greatly reduce the buffer consumption as well as the operation time for concentration and/diafiltration. Overall intensified processes are the results of smooth and straightforward unit operation connection, reducing the overall footprint and increasing the process productivity.

7.5 CONCLUDING REMARKS

The progressive higher demand for prophylactic and therapeutic vaccines in the last decades is pressuring the biotechnological industry to accelerate the process development phase. There is a real need to develop purification processes with higher productivity and robustness, to address the challenges related to the biological diversity of the viral-based vaccine candidates under development. The current commercial-scale manufacture of biological products is performed in batch processes. Although the batch operation promotes an easier design, optimization of the unit operations and off-line quality control of the target product, the DSP at a research level is focusing to develop continuous processes. The adoption of continuous bioprocessing is expected to relieve the economic and regulatory challenges that are being faced in the biopharmaceutical industry [83]. Allied to this, high throughput process development (HTPD) has evolved to include both upstream and downstream development, and big biopharma companies are using such an approach in their process development. Regarding downstream processing, efforts are being made to allow better purification processes with the development of new and improved materials as well as new operating modes. In the case of manufacturing facilities, the tendency is for the adoption of a multimodal type that allows a flexible approach, capable of accommodating several processes. This means less risk to the companies, either if the final product is still waiting for approval or the emergence of competitors. However, a multimodal facility brings other challenges, like the requirement of a flexible operator team, having to learn numerous different processes, a higher probability of cross-contamination, and more complex supply chains.

Overall, the current trends in the biopharmaceutical industry, especially in the vaccines field, are driving the implementation of high-capacity processes that enable the reducing costs and equipment footprint but always prioritizing the process efficiency and product quality.

REFERENCES

[1] D. G. Bracewell and M. Smales, "The challenges of product- and process-related impurities to an evolving biopharmaceutical industry," *Bioanalysis*, vol. 5, pp. 123–126, 2013.

[2] P. Nestola, C. Peixoto, R. R. J. S. Silva, P. M. Alves, J. P. B. Mota, and M. J. T. Carrondo, "Improved virus purification processes for vaccines and gene therapy," *Biotechnol. Bioeng.*, vol. 112, no. 5, pp. 843–857, 2015.

[3] A. A. Shukla, C. Jiang, J. Ma, M. Rubacha, L. Flansburg, and S. S. Lee, "Demonstration of robust host cell protein clearance in biopharmaceutical downstream processes," *Biotechnol Prog.*, vol. 24, no. 3, pp. 615–622, 2008.

[4] L. A. Britoand M. Singh "Acceptable levels of endotoxin in vaccine formulations during preclinical research," *J. Pharm. Sci.* vol. 100, no. 1, pp. 34–37, 2011. doi: 10.1002/jps.22267

[5] FDA, "Guidance for industry cell characterization and qualification of cell substrates and other biological materials used in the production of viral vaccines for infectious disease indications," FDA-2006-D-0223, Issued by: Center for Biologics Evaluation and Research, no. February, pp. 1–50, 2010.

[6] EMA, "Guideline on Quality, non-clinical and clinical aspects of live recombinant viral vectored vaccines," *Reproduction*, vol. 44, no. November, pp. 1–14, 2009.
[7] R. Morenweiser, "Downstream processing of viral vectors and vaccines," *Gene Ther.*, vol. 12, pp. S103–S110, 2005.
[8] H. Ugai *et al.*, "Purification of infectious adenovirus in two hours by ultracentrifugation and tangential flow filtration," *Biochem. Biophys. Res. Commun.*, vol. 331, no. 4, pp. 1053–1060, 2005.
[9] S. Zolotukhin *et al.*, "Recombinant adeno-associated virus purification using novel methods improves infectious titer and yield," *Gene Ther.*, vol. 6, no. 6, pp. 973–985, 1999.
[10] C. B. Reimer, R. S. Baker, T. E. Newlin, and M. L. Havens, "Influenza virus purification with the zonal ultracentrifuge," *Science*, vol. 152, no. 3727, pp. 1379 LP–1381, Jun. 1966.
[11] J. Hilfenhaus, R. Köhler, and F. Behrens, "Large-scale purification of animal viruses in the RK-model zonal ultracentrifuge: II. Influenza, mumps and Newcastle disease viruses," *J. Biol. Stand.*, vol. 4, no. 4, pp. 273–283, 1976.
[12] M. M. Segura, M. Mangion, B. Gaillet, and A. Garnier, "New developments in lentiviral vector design, production and purification," *Expert Opin. Biol. Ther.*, vol. 13, no. 7, pp. 987–1011, Aug. 2013.
[13] G. Iyer *et al.*, "Reduced surface area chromatography for flow-through purification of viruses and virus like particles," *J. Chromatogr. A*, vol. 1218, no. 26, pp. 3973–3981, 2011.
[14] R. M. Anderson, P. J. Scannon, and J. T. Matthews, "Planning for pandemics of infectious diseases 30 Years of commercial experience," 2006.
[15] A. C. Silva *et al.*, "Adenovirus vector production and purification," *Curr. Gene Ther.*, vol. 10, no. 6, pp. 437–455, Dec. 2010.
[16] C. Peixoto, M. F. Q. Sousa, A. C. Silva, M. J. T. Carrondo, and P. M. Alves, "Downstream processing of triple layered rotavirus like particles," *J. Biotechnol.*, vol. 127, no. 3, pp. 452–461, 2007.
[17] O.-W. Merten, M. Schweizer, P. Chahal, and A. Kamen, "Manufacturing of viral vectors: part II. Downstream processing and safety aspects," *Pharm. Bioprocess.*, vol. 2, no. 3, pp. 237–251, 2014.
[18] A. Kamen and O. Henry, "Development and optimization of an adenovirus production process," *J. Gene Med.*, vol. 6, no. SUPPL. 1, pp. 184–192, 2004.
[19] N. E. Altaras, J. G. Aunins, R. K. Evans, A. Kamen, J. O. Konz, and J. J. Wolf, "Production and formulation of adenovirus vectors," *Adv. Biochem. Eng. Biotechnol.*, vol. 99, no. November, pp. 193–260, 2005.
[20] E. J. D'Hondt and H. B. Engelmann, "Process for producting influeza vaccine," 2010.
[21] F. Colavita *et al.*, "Evaluation of the inactivation effect of Triton X-100 on Ebola virus infectivity," *J. Clin. Virol.*, vol. 86, pp. 27–30, 2017.
[22] M. G. Moleirinho *et al.*, "Clinical-grade oncolytic adenovirus purification using polysorbate 20 as an alternative for cell lysis," *Curr. Gene Ther.*, vol. 18, no. 6, pp. 366–374, 2018.
[23] P. Guo, Y. El-Gohary, K. Prasadan, C. Shiota, X. Xiao, J. Wiersch, J. Paredes, S. Tulachan, G. K. Gittes, "Rapid and simplified purification of recombinant adeno-associated virus," *J. Virol. Methods* vol. 183, no. 2, pp. 139–146, 2013. doi: 10.1016/j.jviromet.2012.04.004
[24] S. Zhang, C. Thwin, Z. Wu, T. Cho, and S. Gallagher, "Method for the production and purification of adenoviral vectors," 2010.
[25] L. Besnard *et al.*, "Clarification of vaccines: An overview of filter based technology trends and best practices," *Biotechnol. Adv.*, vol. 34, no. 1, pp. 1–13, 2016.

[26] Y. Yigzaw, R. Piper, M. Tran, and A. A. Shukla, "Exploitation of the adsorptive properties of depth filters for host cell protein removal during monoclonal antibody purification," *Biotechnol. Prog.*, vol. 22, no. 1, pp. 288–296, 2006.
[27] M. Mellado and C. Peixoto, "Clarification of Adenovirus serotype 5: Robust protection of downstream purification steps application note," 2017.
[28] T. P. Pato *et al.*, "Development of a membrane adsorber based capture step for the purification of yellow fever virus," *Vaccine*, vol. 32, no. 24, pp. 2789–2793, 2014.
[29] T. Rodrigues, M. J. T. Carrondo, P. M. Alves, and P. E. Cruz, "Purification of retroviral vectors for clinical application: Biological implications and technological challenges," *J. Biotechnol.*, vol. 127, no. 3, pp. 520–541, Jan. 2007.
[30] S. B. Carvalho *et al.*, "Efficient filtration strategies for the clarification of influenza virus-like particles derived from insect cells," *Sep. Purif. Technol.*, vol. 218, pp. 81–88, 2019.
[31] A. Xenopoulos, "Production and purification of hepatitis C virus-like particles [Webinar]. EMD Millipore Webinar Series," 2015.
[32] B. Kalbfuss, Y. Genzel, M. Wolff, A. Zimmermann, R. Morenweiser, and U. Reichl, "Harvesting and concentration of human influenza A virus produced in serum-free Mammalian cell culture for the production of vaccines," *Biotechnol. Bioeng.*, vol. 97, no. 1, pp. 73–85, 2007.
[33] C. C. Liu *et al.*, "Purification and characterization of enterovirus 71 viral particles produced from vero cells grown in a serum-free microcarrier bioreactor system," *PLoS One*, vol. 6, no. 5, Article no. e20005, 2011.
[34] B. Zhang *et al.*, "Immunogenicity of a scalable inactivated rotavirus vaccine in mice," *Hum. Vaccin.*, vol. 7, no. 2, pp. 248–257, Feb. 2011.
[35] Y. E. Thomassen *et al.*, "Scale-down of the inactivated polio vaccine production process," *Biotechnol. Bioeng.*, vol. 110, no. 5, pp. 1354–1365, 2013.
[36] K. Trabelsi, M. Ben Zakour, and H. Kallel, "Purification of rabies virus produced in Vero cells grown in serum free medium," *Vaccine*, vol. 37, no. 47, pp. 7052–7060, 2019.
[37] R. J. S. Silva *et al.*, "A flow – Through chromatographic strategy for hepatitis C virus – Like particles purification,"*Processes*, vol. 8, pp. 1–13, 2020.
[38] Y. Cherradi, S. L. Merdy, L.-J.- Sim, T. Ito, P. Pattnaik, J. Haas , and A. Boumlic , *BioProcess Int.* "Filter-based clarification of viral vaccines and vectors," vol. 16, no. 4, 2018.
[39] J. Vellinga *et al.*, "Challenges in manufacturing adenoviral vectors for global vaccine product deployment," *Hum. Gene Ther.*, vol. 25, no. 4, pp. 318–327, 2014.
[40] Y. Genzel *et al.*, "High cell density cultivations by alternating tangential flow (ATF) perfusion for influenza A virus production using suspension cells," *Vaccine*, vol. 32, no. 24, pp. 2770–2781, 2014.
[41] D. Vázquez-Ramírez, I. Jordan, V. Sandig, Y. Genzel, and U. Reichl, "High titer MVA and influenza A virus production using a hybrid fed-batch/perfusion strategy with an ATF system," *Appl. Microbiol. Biotechnol.*, vol. 103, no. 7, pp. 3025–3035, 2019.
[42] B. Minow, F. Egner, F. Jonas, and B. Lagrange, "High-cell-density clarification by single-use diatomaceous earth filtration," *BioProcess Int*, vol. 12, no. 4, 2014.
[43] T. Williams *et al.*, "Lentiviral vector manufacturing process enhancement utilizing TFDF™ technology," *Cell Gene Ther. Insights*, vol. 6, no. 3, pp. 455–467, 2020.
[44] J. O. Konz, A. L. Lee, J. A. Lewis, and S. L. Sagar, "Development of a purification process for adenovirus: Controlling virus aggregation to improve the clearance of host cell DNA," *Biotechnol. Prog.*, vol. 21, no. 2, pp. 466–472, 2005.
[45] A. Ward, "Exploring a New Enzymatic Tool for AAV Production," *Genet. Eng. Biotechnol. News*, vol. 38, no. 3, 2018.

[46] T. Kröber, A. Knöchlein, K. Eisold, B. Kalbfuß-Zimmermann, and U. Reichl, "Dna depletion by precipitation in the purification of cell culture-derived influenza vaccines," *Chem. Eng. Technol.*, vol. 33, no. 6, pp. 941–959, 2010.

[47] A. T. Hanke and M. Ottens, "Purifying biopharmaceuticals: Knowledge-based chromatographic process development," *Trends Biotechnol.*, vol. 32, no. 4, pp. 210–220, 2014.

[48] M. Mellado and C. Peixoto, "Ultrafiltration and Diafiltration of Adenovirus serotype 5 with Sartocon® Slice cassettes installed within a SARTOFLOW ® Smart benchtop crossflow system application note," 2017.

[49] B. Carvalho *et al.*, "Membrane – Based Approach for the Downstream Processing of In fl uenza Virus – Like Particles,"*Biotechnol. J.* vol. 1800570, pp. 1–12, 2019.

[50] G. E. Healthcare and L. Sciences, "Concentration and diafiltration of cell-derived, live influenza virus using 750 C hollow fiber filter cartridge,"

[51] K. Marino and P. Levison, "Achieving Process Intensification with Single-Pass TFF," *Genet. Eng. Biotechnol. News*, vol. 37, no. 15, pp. 30–31, Aug. 2017.

[52] M. Zhao, M. Vandersluis, J. Stout, U. Haupts, M. Sanders, and R. Jacquemart, "Affinity chromatography for vaccines manufacturing: Finally ready for prime time?," *Vaccine*, vol. 37, no. 36, pp. 5491–5503, Apr. 2018.

[53] K. M. Łącki and F. J. Riske, "Affinity Chromatography: An Enabling Technology for Large-Scale Bioprocessing," *Biotechnol. J.*, vol. 15, no. 1, pp. 1–11, 2020.

[54] T. Weigel, T. Solomaier, A. Peuker, T. Pathapati, M. W. Wolff, and U. Reichl, "A flow-through chromatography process for influenza A and B virus purification," *J. Virol. Methods*, vol. 207, pp. 45–53, Oct. 2014.

[55] M. W. Wolf and U. Reichl, "Downstream processing of cell culture-derived virus particles," *Expert Rev. Vaccines,* vol. 10, no. 10, pp. 1451–1475, Oct. 2011.

[56] Sartorius Stedim Biotech GmbH, "Optimizing adenovirus purification processes," 2017.

[57] S. B. Carvalho *et al.*, "Purification of influenza virus-like particles using sulfated cellulose membrane adsorbers," *J. Chem. Technol. Biotechnol.*, vol. 93, no. 7, pp. 1988–1996, 2018.

[58] S. Tinch, K. Szczur, W. Swaney, L. Reeves, and S. R. Witting, "A scalable lentiviral vector production and purification method using mustang Q chromatography and tangential flow filtration," *Methods Mol. Biol.*, vol. 1937, pp. 135–153, 2019.

[59] D. Vincent *et al.*, "The development of a monolith-based purification process for Orthopoxvirus vaccinia virus Lister strain," *J. Chromatogr. A*, vol. 1524, pp. 87–100, 2017.

[60] L. M. Fischer, M. W. Wolff, and U. Reichl, "Purification of cell culture-derived influenza A virus via continuous anion exchange chromatography on monoliths," *Vaccine*, vol. 36, no. 22, pp. 3153–3160, 2018.

[61] N. K. Jain, N. Sahni, O. S. Kumru, S. B. Joshi, D. B. Volkin, and C. Russell Middaugh, "Formulation and stabilization of recombinant protein based virus-like particle vaccines," *Adv. Drug Deliv. Rev.*, vol. 93, pp. 42–55, 2015.

[62] T. J. Moyer *et al.*, *Engineered immunogen binding to alum adjuvant enhances humoral immunity*, vol. 26, no. 3. 2020.

[63] M. M. E. Sunay *et al.*, "Glucopyranosyl lipid adjuvant enhances immune response to Ebola virus-like particle vaccine in mice," *Vaccine*, vol. 37, no. 29, pp. 3902–3910, 2019.

[64] G.-A. Junter and L. Lebrun, "Polysaccharide-based chromatographic adsorbents for virus purification and viral clearance," *J. Pharm. Anal.*, vol. 10, no. 4, pp. 291–312,2020.

[65] Y. Kurosawa, P. Khandelwal, D. Yoshikawa, and M. Snyder, "Single-step influenza and dengue virus purification with mixed-mode CHT ceramic hydroxyapatite XT media," Bulletin 7115, Bio-Rad Laboratories, Inc. 2018. https://www.bio-rad.com/webroot/web/pdf/psd/literature/Bulletin_7115.pdf
[66] P. Fernandes, C. Peixoto, V. M. Santiago, E. J. Kremer, A. S. Coroadinha, and P. M. Alves, "Bioprocess development for canine adenovirus type 2 vectors," *Gene Ther.*, vol. 20, no. 4, pp. 353–360, 2013.
[67] S. Shoaebargh *et al.*, "Sterile filtration of oncolytic viruses: An analysis of effects of membrane morphology on fouling and product recovery," *J. Memb. Sci.*, vol. 548, no. November 2017, pp. 239–246, 2018.
[68] L. X. Yu *et al.*, "Understanding pharmaceutical quality by design," *AAPS J.*, vol. 16, no. 4, pp. 771–783, 2014.
[69] A. C. A. Roque *et al.*, "Anything but conventional chromatography approaches in bioseparation," *Proteomics*, vol. 8, pp. 1–25, 2020.
[70] M. G. Moleirinho, R. J. S. Silva, P. M. Alves, M. J. T. Carrondo, and C. Peixoto "Current challenges in biotherapeutic particles manufacturing," *Expert Opin. Biol. Ther.*, pp. 1–15, 2020.
[71] A. E. Ashcroft, "Mass spectrometry-based studies of virus assembly," *Curr. Opin. Virol.*, vol. 36, pp. 17–24, 2019.
[72] C. Uetrecht *et al.*, "High-resolution mass spectrometry of viral assemblies: Molecular composition and stability of dimorphic hepatitis B virus capsids," *Proc. Natl. Acad. Sci. USA.*, vol. 105, no. 27, pp. 9216–9220, 2008.
[73] S. A. Nass *et al.*, "Universal Method for the Purification of Recombinant AAV Vectors of Differing Serotypes," *Mol. Ther. - Methods Clin. Dev.*, vol. 9, no. June, pp. 33–46, 2018.
[74] S. A. Berkowitz and J. S. Philo, "Monitoring the homogeneity of adenovirus preparations (a gene therapy delivery system) using analytical ultracentrifugation.," *Anal. Biochem.*, vol. 362, no. 1, pp. 16–37, Mar. 2007.
[75] BioPhorum Group, "Improving the biomanufacturing facility lifecycle using a standardized, modular design, and construction approach," 2019.
[76] J. Markarian, "Flexible facilities for viral vector manufacturing," *BioPharm Int.*, vol. 33, no. 3, pp. 23–24, 2020.
[77] P. Walters, "Understanding the unique design and engineering needs for Gene therapy production," *Pharma's Almanac*, 2018.
[78] P. Nestola, "Integrated Technologies to Accelerate Process Intensification for Viral Vaccine Manufacturing," *Sartorius Stedim*, no. White paper, 2020.
[79] P. Abrecht, H. Pressac, G. Boulais, A., and Permanne, "Adenovirus downstream process intensification implementation of a membrane adsorber," *Bioprocess Int.*, vol. 17, no. 10, pp. 38–44, 2019.
[80] T. Weigel, T. Solomaier, S. Wehmeyer, A. Peuker, M. W. Wolff, and U. Reichl, "A membrane-based purification process for cell culture-derived influenza A virus," *J. Biotechnol.*, vol. 220, pp. 12–20, 2016.
[81] X. Gjoka, R. Gantier, and M. Schofield, "Platform for integrated continuous bioprocessing," *BioPharm Int.*, vol. 30, no. 7, p. 26, 2017.
[82] R. Patil and M. Zhao, *Downstream Process Intensification of Virus Purification Using Single-Use Membrane Chromatography*, vol. 38. 2019.
[83] A. S. Rathore, N. Kateja, and D. Kumar, "Process integration and control in continuous bioprocessing," *Curr. Opin. Chem. Eng.*, vol. 22, pp. 18–25, 2018.

8 Analytics and virus production processes

Emma Petiot
3d.FAB, Univ Lyon, Université Lyon1, CNRS, INSA, CPE-Lyon, ICBMS, UMR 5246, 43, Bd du 11 Novembre Villeurbanne cedex, France

CONTENTS

8.1 INTRODUCTION: REGULATORY CONTEXT FOR ANALYTICS DEVELOPMENT

For pharmaceutical products and especially viral vaccines, clear guidelines are issued by regulatory agencies (FDA, EMEA) to release production lots. This means that the analytical tools allowing for production lot characterization and validation also must be approved and validated upfront. This is to be considered while implementing a new analytical test for a process or replacing an "old-fashioned" analytical assay. Developing new analytical methods used in pharmaceutical production processes is managed through different methodologies described for example in the FDA guidance [1]. Ultimately, such quantification or detection methods must be validated. If the process is implemented in GMP facilities, fully validated methods are required with an extensive evaluation of the precision and robustness of each assay. Changes in the process (such as different matrix) need to be assessed as well as how these changes will affect the assay. The documentation to be submitted to the regulatory agencies must include a full validation protocol and extensive report. As described

DOI: 10.1201/9781003229797-8

in such guidance, "method development" involves optimizing the procedures and conditions involved with extracting and detecting the analyte. Method development includes the optimization of the following bioanalytical parameters:

- *Reference standards*
- *Critical reagents*
- *Calibration curve and linearity range*
- *Quality control samples (QCs)*
- *Selectivity and specificity, the ability to assess the analyte when it is in the presence of other components, such as a matrix.*
- *Sensitivity, related to limit of detection and quantification*
- *Accuracy, the assessment of the difference between the measured value and the real value. That could be quite difficult to evaluate if you have not yet a reference standard.*
- *Precision, the measure of the agreement of multiple measurements on the same sample*
- *Recovery, if you have sample pretreatment before the analysis*
- *Stability of the analyte in the matrix*

Also, certain considerations should be envisioned upfront. The evaluation of the limits of detection or quantification is not the same. The **limit of detection** is the lowest amount of analyte that can be detected while the **limit of quantification** is the lowest amount of analyte that could be determined accurately. Thus, an assay can give a result at a very low level of analytes, but the precision of the test is not enough to reach proper quantification. The linearity range is also important to determine. The linearity is the range in which there is no saturation of the detection signal used for the assay. The robustness of the assay regarding changes in the process is also of importance if a tool is implemented in the development stages of the production process. As an assay could be strongly impacted by the composition of the sample matrix, it is important to select appropriate analytical methods if the process has been designed to modify this matrix. Analytical tools can be affected by factors such as the manufacturing process, change in formulation, or equipment used to synthesize the product. Overall, any test needs to be validated; consequently, the way the assay is performed must be enough robust to allow validation of the test.

8.2 ANALYSIS OF VIRAL PRODUCTS

Viral production processes are various, and consequently, the type of biological material composition to characterize and quantify could involve a wide variety of quality attributes. Indeed, depending on the end product, if the production process for example, aims to produce viral vaccines or viral vectors, manufacturers will not target the same analytes.

8.2.1 Biological Attributes of Virus-Based Products

To start with definitions, **biological activity is a critical quality attribute (CQA)**, which means that biological activity is qualifying the viral product in terms of efficiency. Different types of assays are applied to evaluate the biological activity of virus production. Most of the assays are biochemical or cell-culture-based assays. Molecular biology assays were extensively developed in the last twenty years to identify more effectively viral variants and strains. **Viral potency** is the quantitative measurement of the biological activity of a viral product. Thus, the potency of a viral product refers to the comprehension of the relation between the product activity and its biological quantity. As an example, a potency assay could be quantifying the amount of protein needed to give a specific activity of a vaccine, such as protection of a patient. The viral potency is thus dependent on the targeted molecule's affinity and its efficacy. For vaccines, the main target is to evaluate the **product immunogenicity**. In such cases, the quantification techniques aim to describe the amount of antigen (or antigen epitopes) which are necessary for the onset of an immune response in-vivo, either on animals or in human patients. The protective effect of such induced immune response is then further evaluated. The immune response quality will be evaluated by the specific quantification of both B-cell humoral and specific antibodies release or T-cell cytotoxic response to protect against the infectious disease.

In the case of a vaccine candidate, the major critical quality attribute (CQA) is the antigen content and its bioactivity. An antigen can be of different forms depending on the type of vaccine. Thus, when the vaccine candidate is an attenuated vaccine, the amount of infectious and complete viral particles is of importance for the product quality assessment. Whereas if the vaccine candidate is an inactivated split vaccine, then only the antigen protein content is assessed. For other types of viral products, like viral vectors aimed to be integrated within gene therapies or cell therapies treatment strategies, the viral potency will here only target the product capacity to infect naïve cells. Thus, the amount of infectious viral particles will be the main read-out. These considerations will allow for the selection of the appropriate analytical tools and methodologies. In the case of a viral vector, the product efficacy corresponds to the virus capacity to target specific cells, its entry in the cells, and capacity to deliver a modified genome. Thus, here the main attribute of the product is the infectivity and infectious dose of the product. Table 8.1 presents the type of assay to target depending on the viral product application.

8.2.2 Implications of Process Phase on Analytics Choices

The second element to consider for the selection of appropriate analytics is the process sequence step. Analytics is not only implemented at the final step of the manufacturing process for product qualification and lot-release. Many analytical tools were developed to allow for the evaluation of process consistency or process improvement across the different steps of the manufacturing process. Consequently, analytical tools will be subject to different conditions with different challenges depending on the production stage at which they are implemented.

After fill-and-finish process step, product quality and potency are assessed on highly purified material. On the contrary, analytics implemented along the upstream and downstream process need to cope with a wide range of purity. Thus, the **matrix effect**, namely the impact of the solution in which the product is suspended, must be considered. The composition of such a matrix might be ranging from spent media, purification saline

TABLE 8.1
Description of characteristics targeted by analytical tools for viral product description and processes monitoring

	Product Characteristics	Products Application
WHOLE ACTIVE VIRUS	*infectious particles* *complete replicative viral particles* *viral genome* *bioactive viral proteins*	viral vectors Attenuated viral vaccines
WHOLE INACTIVATED VIRUS (heat or chemically inactivated)	***non-infectious particles*** *uncomplete viral particles* *viral genome* *bioactive viral proteins*	inactivated vaccines
VIRUS-LIKE PARTICLES (VLP)	***non-infectious particles*** *Uncomplete viral particles* *bioactive viral proteins* ***no viral genome***	vaccines
VIRUS SUB-UNITS (purified antigens or split viral particles)	***non-infectious particles*** *bioactive viral proteins* ***no viral genome***	vaccines

buffer, to formulation solution containing stabilizer and adjuvants. All these matrices could have a strong effect on cell-based or biochemical assays and will complexify the extraction of meaningful data from the assays. A strong matrix effect to highlight, especially for viral particle counting tools, is the co-secretion of extracellular vesicles (EVs) by cultivated and infected cells while they are undergoing viral replication. Such EVs, could be part of the host-cell response to viral infection. The main drawback of such nanoparticle contamination is due to their physical characteristics that are very close to the viral particles themselves. EVs are spherical particles with mean diameters ranging from 50 nm for exosomes up to 1 μm for micro-vesicles (see Figure 8.2).

Similarly, cells following infection undergo lysis, most commonly releasing their intracellular content in the culture supernatant. In such a case, non-assembled free viral antigens, and viral DNA, but also host-cell DNA and host-cell proteins are released. The viral replication process being in principle not fully efficient, it is very common that culture broth will contain both complete viruses that are infectious and efficient to replicate but also defective viral particles (DIPs), which will be part of a global viral particle population. For example, the number of infectious particles compared to defective particles released by cells could be very low. A ratio between the infectious viral particles and the total viral particles might be as low as 1% for an influenza viral strain (see Figure 8.4). This should also be considered as a matrix effect for the infectious assays for example (see the section of infectious particles assays). The ratio between the total viral particles and the infectious particles is strongly dependent on the virus type, the virus strain, the process, or the cell production platform used [2]. The notion of the ratio between the total viral particle and the infectious ones is very important for monitoring and qualifying viral production processes' efficiency. In the last 10 years, the scientific community has developed more and more quantification tools to describe and better understand the viral replication mechanisms associated with these observations. Nevertheless, this concept is not yet fully understood, as fundamental virology has for long not been sufficiently quantitative with regards to incomplete processing of the viral particles leading to DIPs. As an example, it is not rare that the ratio between the infectious particles and the total particles generated by an identical production process changes with the viral strain as it was demonstrated for influenza viruses [3].

Regarding the selection of assays, further guidelines will be discussed depending on the information that must be provided to the regulatory agencies. Besides, several critical quality attributes will have to be evaluated at several steps of the process. In general, for viral production processes, three main attributes of the product are targeted (i) the infectious viral particles (IVP), (ii) the total viral particle (VP), and (iii) the total antigen content.

8.3 VIRAL QUANTIFICATION METHODS

As described earlier, several biochemical, cell-based, or molecular biology methods are applied for the quantification of viral samples. Such methods can be categorized into four main sections: infectivity assays, protein content or nucleic acid content (genome) assessment, and particle counts. Assays will be further presented in the following sections.

The following section describes the main assays implemented for viral production process monitoring. Assays presently in use have several limiting drawbacks for both the qualification of virus lots and the monitoring of in-process analysis of product quality. Indeed, most of these viral quantification methods are time consuming and costly experiments. They also present a lot of variabilities, either operator-dependent, standard-dependent, or matrix-dependent. This limits the use of a unique assay throughout the entire process. They are not extremely specific, and the limits of detection and quantification are high. Thus, for now, the scientific community does not have access to highly satisfactory assays for the evaluation of their processes. Therefore, different orthogonal assays are combined to determine the consistency of the process and its productivity.

8.3.1 INFECTIOUS PARTICLE QUANTIFICATION

For the evaluation of infectious viral particles, methods test the capacity of the virus to infect, to replicate, and kill plated cultivated cells (Figure 8.1). Such cell-based assays have high variability and could take from days up to weeks to provide results. They could fall in two categories: **plaque assays** and **tissue culture infectious doses (TCID50).**

Plaque assays are one of the oldest assays for detecting viruses and quantifying the number of infectious viral particles. Plaque assay is based on putting in contact different dilutions of virus suspension on confluent cells for a short period (commonly below 30 min). Plated cells are then covered with a hydrogel commonly composed of soft agarose prepared within a culture medium (see Figure 8.1). This

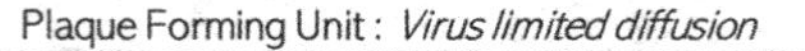

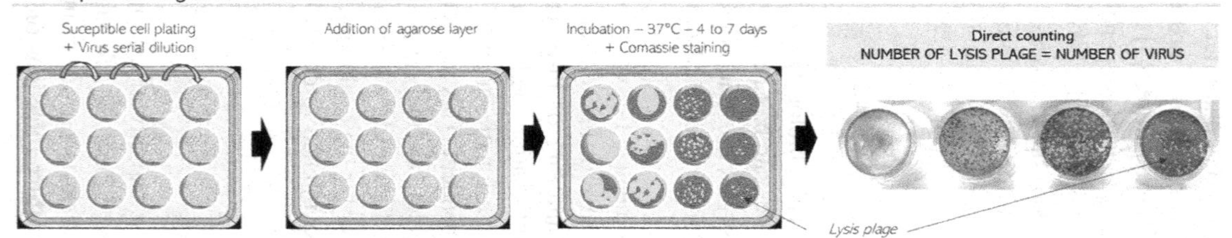

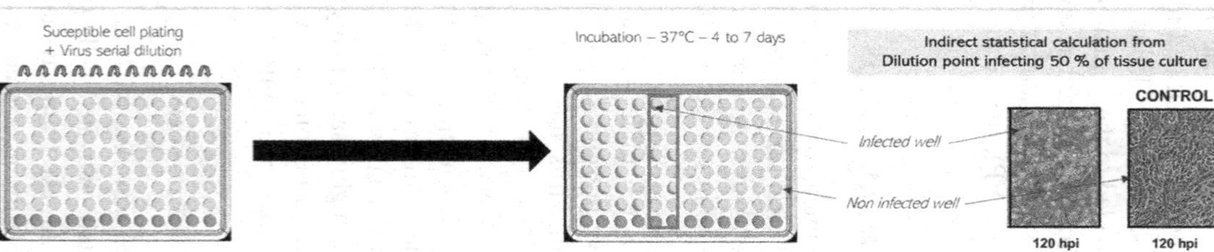

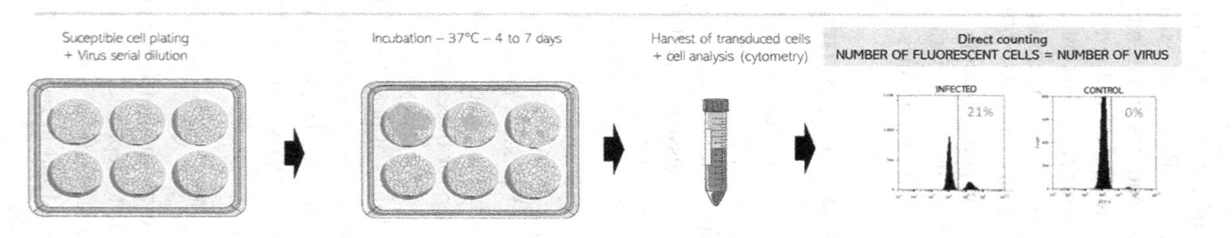

FIGURE 8.1 Presentation of infectivity assays to quantify infectious viral particles i.e., plaque forming unit (pfu) assay, Tissue Culture Infectious Dose 50 assay, and Transducing units.

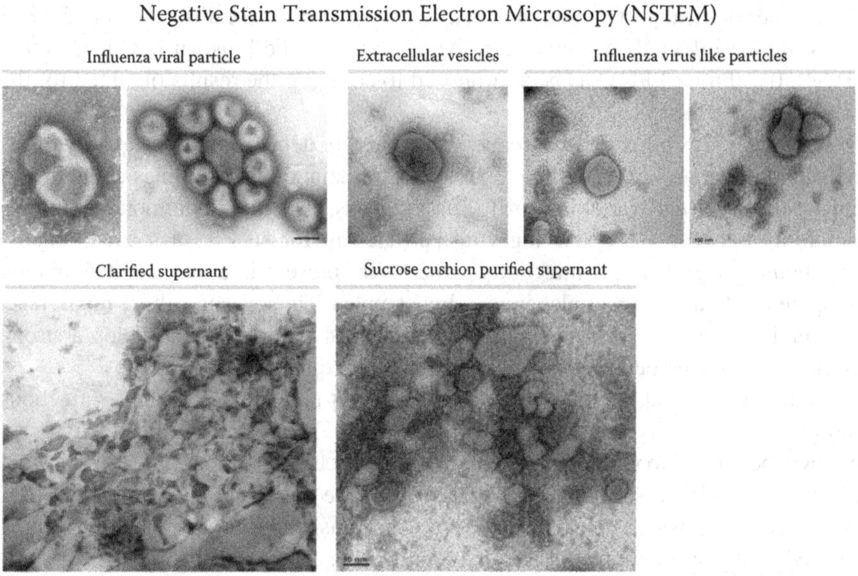

Negative Stain Transmission Electron Microscopy (NSTEM)

Influenza viral particle | Extracellular vesicles | Influenza virus like particles

Clarified supernant | Sucrose cushion purified supernant

FIGURE 8.2 Negative Staining Transmission Electron Microscopy (NSTEM) observation of viral preparations. A – Sucrose cushion and sucrose gradient purified preparation of influenza virus, extracellular vesicles, or influenza virus like particles. B – Immunogold labeling of sucrose cushion and sucrose gradient purified preparation of influenza virus. C – Several purification steps of culture supernatant.

will allow a limit to the dissemination of the newborn viruses onto the tissue cell culture and to have a transmission of the infection only to the closest cells. This way of viral expansion led to the creation of lysis plaque within the tissue culture, which ultimately will allow for counting the initial number of viruses applied in the suspension. This is possible for a viral concentration, which allows for the visual or microscopic counting of the lysis plaque.

The method assumes that only one virus infects one cell which might introduce a bias. The case of several defective incomplete viruses entering the same cell and self-complementing cannot be ruled out. Therefore, false positives are possible for such assays.

$TCID_{50}$- Tissue Culture Infectious Dose 50 uses a similar principle of applying a suspension of the virus at different concentrations onto a plated tissue culture. In this case, the diffusion of the newborn viruses is not limited to inducing several cycles or replications on the same plated culture. Ultimately the whole culture should be infected and lysed. Therefore, the read-out of such assays is evaluating the *cytopathic effect* of a virus suspension. Such an assay lasts between 4 to 7 days [4]. The test is commonly performed on 96-well plates with replicates to evaluate the percent of tissue cultures that have been infected after several potential replication cycles. Thus, the wells that are infected are counted, among the wells where the tissue culture is healthy. A statistical analysis is performed to determine the number of viral particles in this assay by targeting the conditions where 50% of the wells are infected. A conversion factor based on the Poisson law distribution (TCDI50 = 0.69 × PFU) has been proposed to compare the results obtained for the two infectivity assays.

Both assays, as other cell-based assays, are highly dependent on the quality of the cells used. Operator effect and methodologies to maintain the cell lines could highly impact the level and variability of the assay. Thus, it is quite common to obtain a variability between 0.5 and 1 log viral particles/ml for such assays. Additionally, one should remember that infectious particles are present in a very limited amount compared to total viral particles in the viral samples. Thus, tissue culture might react differently to samples carrying a high level of viral particles and develop a strong innate cell defense mechanism that will impact the quantification assay.

On the other hand, in some processes, the target analyte is not a full "infectious particle." For virus-like particles (VLPs) or defective viral vectors, the infectivity assays will be limited to viral particles entering the target cells without further replicating and eventually expression of the transgene in the case of defective viral vectors. Here, the virus can get inside a cell, and just express the transgene without replication. This is referred to as transduction with the additional notion of transducing units.

Transducing units correspond to the number of viral particles capable to transfer a gene of interest into a cell. This is commonly evaluated using a reporter gene as fluorescent GFP or RFP proteins. Transducing particles could either be quantified with an assay adapted from the TCID50 assay presented previously, or with flow cytometry. Indeed, the principle of applying the suspension of virus onto a cell culture is maintained and the read-out is no longer the cytopathic effect of the suspension but the expression of the reporter gene. Positive wells are then numbered based on the presence or absence of fluorescent gene expression.

Additional protocols based on flow cytometry were developed to achieve more statistically significant results and to develop high throughput assays. After sufficient cultivation duration following cells contact with the virus suspensions at different dilutions, cells are resuspended to be analyzed by flow cytometry.

8.3.2 Total Viral Particles Quantification

As described in the first section, the number of total particles including defective particles is of interest for qualification of some process steps. Several tools were proposed to capture such information. In the 1960s, early analysis was based on microscopic observation using electron microscopy [5]. In this case, the method has been used satisfactorily with poliovirus, bacteriophage, and vaccinia virus, three viruses differing greatly in size and morphology.

Negative staining transmission electron microscopy (NSTEM) is the only assay allowing for direct observation of the viruses. Such a tool is fully adapted to observe viruses that have a range of sizes between 30 nm to 300 nm. It gives the possibility to describe viral particles morphology, structure, and composition. To confirm the presence of viral proteins, it is possible to label specific antibodies with gold particles (see Figure 8.2). The negative staining technic principle holds on viral suspension staining with electron-dense stain prepared from heavy metals after its fixation. The heavy metals then coat the surface of the virus particles to reveal their internal and external structures [6]. The stained samples are then deposited and dried on dedicated grids. The quantitative evaluation of viral particles is feasible but with high variability as the analysis is performed by visual counting on NSTEM grids.

Such tools also permit to qualify the viral suspension quality as it will be possible to observe other nanoparticles like exosomes or micro-vesicles. However, such assays are very complex to implement as a quantification assay. NSTEM is very long and costly. Consequently it can not be used for screening. It is most of the time applied on highly purified samples as protein and cell debris will also have a strong response to electron-dense stain.

For process optimization and development, electron microscopy is not the appropriate tool as it will be necessary to screen production conditions with impure material. Thus, other types of assays allowing the quantification of parts of the virus (genome, viral protein) are preferred for high-throughput analyses. In such a case, the evaluation of the number of total viral particles is performed by calculating the theoretical amount of genome or protein within a viral particle. Such indirect quantification methods imply that an external standard is used to calibrate the amount of total viral particles. Thus, biochemical or molecular biology assays are qualified with electron microscopy viral particle counting.

Viral genome quantification by **quantitative qPCR and RT-qPCR** was in the last 10 years the method of choice to determine total viral particles amount. Indeed, the assumption is here that each viral particle carries a single copy of its genome. Thus, quantifying the number of viral genomes within a sample gives access to the number of particles. Such tools based on molecular biology techniques allow for the design of probes that could have nucleotide sequences highly specific to a virus or a viral strain. It was particularly exploited for the presence of viral adventitious agents within pharmaceutical products as it could allow screening for several viruses at the same time. As preliminary steps, this assay necessitates extracting the viral genome from the particles and eventually convert RNA in cDNA if the virus strain is an RNA virus. These two steps include purification and sample handling steps, which might affect again the variability of the assay. The second step consists of the incubation of the viral genomes with a specific probe carrying both a fluorophore and its quencher to allow for fixation of the probes on the nucleic acid sequences. Release of the fluorophore then happens while the DNA polymerase degrades the specific probe. Correlation between the number of fluorophores released and the number of viral particles then must be established upfront. This means that a reference material with its associate reference qualification assay is then necessary. In the last decade, high-throughput droplet-based digital PCR (ddPCR) has been developed as an improvement of the conventional polymerase chain reaction (PCR) methods. In ddPCR, DNA/RNA is encapsulated inside reaction chambers formed of microdroplets. This was render possible thanks to the improvement reached in microfluidic science working on mixture between immiscible fluid (i.e., the so-called dispersed fluid) and a continuous fluid to generate submicroliter droplets at kilohertz rates [7]. Thus, the reaction chamber contains one or fewer copies of the DNA or RNA. Some of the main features of ddPCR include high sensitivity and specificity, absolute quantification without a standard curve, high reproducibility, good tolerance to PCR inhibitor, and high efficacy compared to conventional molecular methods. This tool is nowadays deployed on a variety of viruses including influenza, cytomegalovirus, HIV, hepatitis, and was extensively evaluated during the SARS-CoV-2 outbreak. Indeed, some results reveal that ddPCR was 500 times more sensitive to SARS-CoV-2 than RT-PCR in low-viral throat swabs. Nevertheless, as the pandemic is still underway, much of the findings concerning SARS-CoV-2 detection should be taken with causion [7] (Table 8.2).

Viral protein quantifications could be performed using several types of assays. Here diagnosis and biochemistry tools have reached years of development, allowing one to choose between biochemical assays (protein activity or total protein quantification assays), immuno-based assays (ELISA, SRID), biosensors (SPR), chromatography (UPLC, HPLC), or mass spectrometry following liquid chromatography separation (LC-MS). Most of the quantification tools developed for viral proteins often target the dominant viral antigens or external proteins. Indeed, their amount or activity is one of the main quality attribute of the viral-based products. Thus, these tools could either be exploited as the evaluation of critical quality attributes of a viral-based product or to access the number of total virus particles for process optimization. The first assays presented are historical virology assays allowing for viral activity evaluation and description. Nevertheless, most of them remain reference methods to evaluate the quality of the viral product for product lot release.

The **hemagglutination assay** is still in use in many labs for different types of viruses. This includes the viral families of *orthomyxoviridae, paramyxoviridae, togaviridae, reoviridae, adenoviridae* with for example *influenza, measles,* or *rubella* viruses [21]. The hemagglutination assay detects the interaction between the virus and red blood cells. Virus suspensions are incubated with red blood cells (RBCs) to allow for attachment of viral antigens with RBC specific receptors. In highly concentrated viral suspension, RBC and virus will then form a network blocking the RBC sedimentation. When performed within a conical bottom well plate, it is thus easy to visually distinguish a condition where the network has been formed (no sedimentation) from a condition where not enough viruses were present to form a network (sedimentation observed with a red dot). Hemagglutination assay is highly dependent on the purity of the viral material tested and on the RBC quality and origin. Mostly RBCs are used from chicken blood, but guinea pigs and other types of poultry can also be used. RBCs are used fresh, ideally collected the day before the assay which renders the analysis process complex to plan. Donor-to-donor animal variability could strongly impact the results; therefore, standard reference samples are required. The assay sensitivity is also quite poor compared to further detailed immune-based assay. Nevertheless, this essay is simple to perform, rapid, easy to read by visual evaluation,and allows for the comparison of many conditions. Such assay has already been used to quantify total viral particles. Indeed, in 1954, Donald and Issacs established a quantification method of viral particles based on the hypothesis that there is approximately one influenza virus for each red blood cell at the end point of agglutination [22]. Viruses preparations were quantified by both electron microscopy and red blood cells assay to establish such correlation. Nevertheless, because of its high degree of variability and dependency on RBC origin or operator reading, hemagglutination assay should be used with

TABLE 8.2
Listing of analytical methods applied to qualify and quantify viral preparations

Method		Duration	Principle	Necessitate labeling	Analytical Range related to total particle count	Reproducibility % CV	REF
Infectivity assay	Plaque assays	days	Cytopathic effect	no	>10°	7%	[8]
	Tissue culture infectious doses (TCID50)		Cytopathic effect	no	>10°	27%	[8]
	Fluorescence focusing assay (FFA)		Fluorescent marker detection	yes	>10°	N.D.	[9]
Activity assay	Hemagglutination assay	hours	Hemagglutination i.e., attachment to syalic acid residues of Red blood cells	no	$>10^{6}$	30–40%	[10]
Immuno-based assays	Single Radial Immunodiffusion (SRID) assay	days	Specific Immunoprecipitation for viral antigens in a gel	no	N.A.	N.A.	[11]
	ELISA	hours-days	Capture of viral antigens by specific antibodies	yes	N.A.	N.A.	[12]
Genome quantification	Quantitative PCR (qPCR and RT-qPCR)	hours	Viral DNA or RNA amplification	yes	>10°	45–130%	[13]
	ddPCR and dd RTPCR	hours		yes	>10°	N.A.	[7]
Total particles counting	Individual particle counting: Negative Stain Transmission Electron Microscopy (NSTEM) Atomic Force Microscopy (AFM)	days	Visual counting	no	N.D.	N.D.	[4]

(Continued)

TABLE 8.2 (Continued)
Listing of analytical methods applied to qualify and quantify viral preparations

Method			Duration	Principle	Necessitate labeling	Analytical Range related to total particle count	Reproducibility % CV	REF
		Flow virometry	Minutes/ hours	Fluorescent labeling of viral components	yes	$10^5 - 10^{12}$	44% [14] 9–31% [15]	[14], [15]
		Tunable resistive pulse sensing (TRPS)		Charge and size determination thanks to coulter effect	no	$10^5 - 10^{12}$	6% [16] 24–52% [17]	[16], [17]
		Nanoparticle Tracking Analysis (NTA)		Size measurement thanks to detection of Brownian motion induced by particles fluorescence or light diffusion	no	$10^7 - 10^9$	14%	[18]
	Particle Population counting	*Flow field fractionation – Multiple angle light scattering* (FFF-MALLS)		Light scattering	yes	$>10^6$	2%	[13], [19]
		Ion-exchange Liquid Chromatography (HPLC)		Separation of virus particles by chromatography on monolith column thanks to charge and size	no	$10^8 - 10^{11}$	6%	[3], [20]
		Surface plasmon resonance (SPR)		Measure of the local refraction variation index	no	$>10^6$	< 3%	[11]

care to compare production levels between conditions, and even more between laboratories.

8.3.2.1 Immuno-Based Assay

The **single-radial immunodiffusion assay (SRID)** is another method that is often used for viral quantification. It is also used for the quantification of other kinds of proteins. The principle holds on the embedment of specific antibodies in agarose gel in which target proteins will later diffuse (Figure 8.3). In the case of viral preparation, polyclonal antibodies are preferred to enhance the potential interaction between viruses and antibodies through their attachment to different epitopes. Holes of calibrated size (commonly 3 to 4 mm diameter) are punched in the agarose where the viral suspension is deposited at various dilutions. Viral components migration stops its diffusion while forming a precipitation ring when the network of antibodies/virus are too large to pass through the agarose mesh. The quantification assay is then based on a calibrated reagent with a known content of viral antigens while the amount of antigen is correlated to the size of the precipitation rings. As for hemagglutination assay, such quantification method is highly variable due to operator experimentation (holes preparation, measurement of precipitation rings) and to reference reagent lot to lot variability. Nevertheless, in some cases (influenza vaccine), the SRID assay stays the only assay validated by the FDA to release vaccine lots. In this specific case, great effort has been dedicated by research teams to replace this assay with alternative quantification approaches, which could overcome the problematics of reference reagent supply.

The ***ELISA*, or enzyme-linked immunosorbent assay**, uses a similar principle with improved control of the antigen/antibody's interactions. In that case, the antibodies or the antigens are coated on a planar surface, generally in a 96-well plate allowing for high throughput analysis. The read-out is the presence of attachment/interaction between a specific antibody and the protein of interest. Indeed, non-attached analytes will be washed out by several buffer rinses. ELISA was developed for several kinds of viruses and generally allows to reach a lower limit of detection and quantifications than SRID assays. The reproducibility is also much better than the SRID. Nevertheless, for now, it is not sufficient to replace the SRID. Indeed, as an example, for the influenza vaccine lot qualification, the use of polyclonal antibodies raised in sheep after inoculation of influenza viruses demonstrate the specificity of the antigens towards a global immune response. Whereas, in the case of ELISA, the most common antibodies used are monoclonal antibodies produced at a high rate in cell culture and which target a single epitope of the viruses [12] or receptor-based assays [23].

Biosensors have been developed for multiple applications in the diagnosis field. Some of these technologies were used in the bioproduction field. One of these techniques for example is the **surface plasmon resonance (SPR).** It was originally applied to investigate virus/host interactions [24]. In the viral quantification field, it is to be compared/included in the immuno-based assay category as it was first applied by antibodies grafting onto sensor surfaces. SPR does present some advantages over ELISA, as it does not required for detection of a fluorescence or HRP label (and thus a secondary antibody). The interaction between the analyte (here viruses or antigens) and the ligand (here the antibodies) is taking place within a liquid solution on the biosensor surface. The detection of the interaction is based on the physical

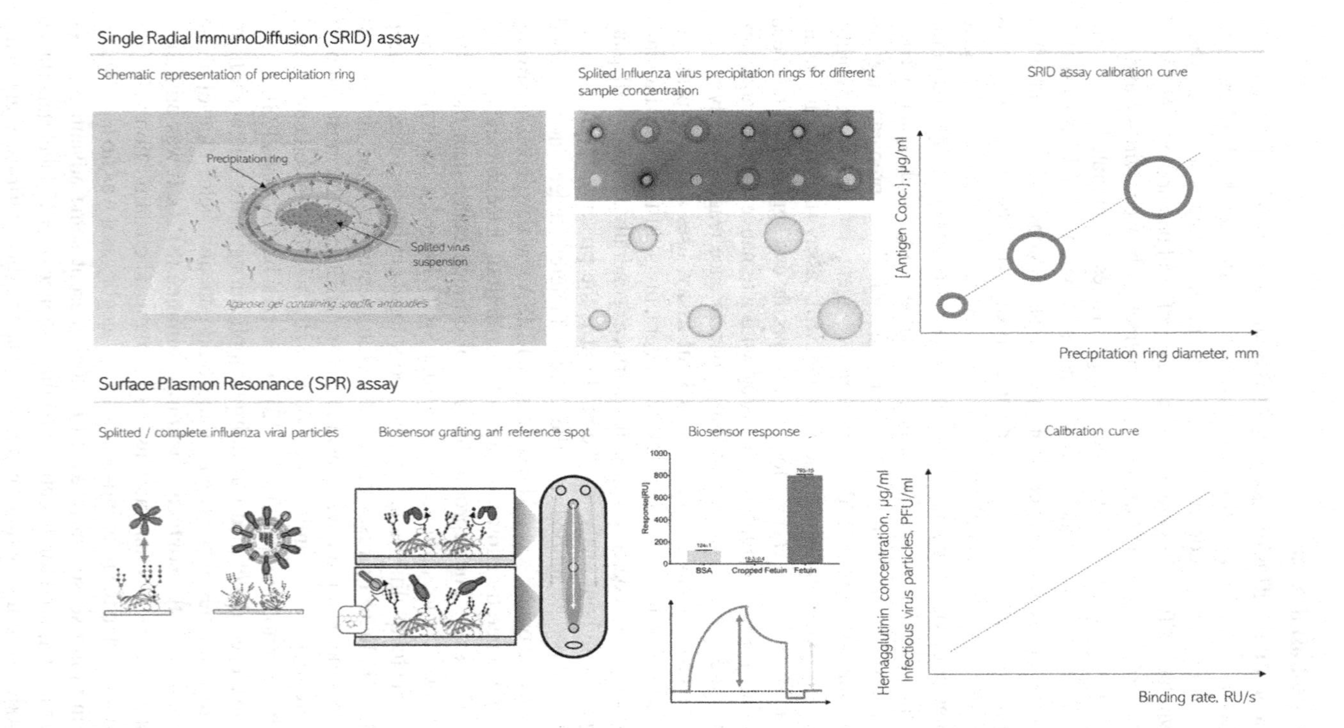

FIGURE 8.3 Negative Staining Transmission Electron Microscopy (NSTEM) observation of viral preparations. A – Sucrose cushion and sucrose gradient purified preparation of influenza virus, extracellular vesicles or influenza virus like particles. B – Immunogold labeling of sucrose cushion and sucrose gradient purified preparation of influenza virus. C – Several purification steps of culture supernatant.

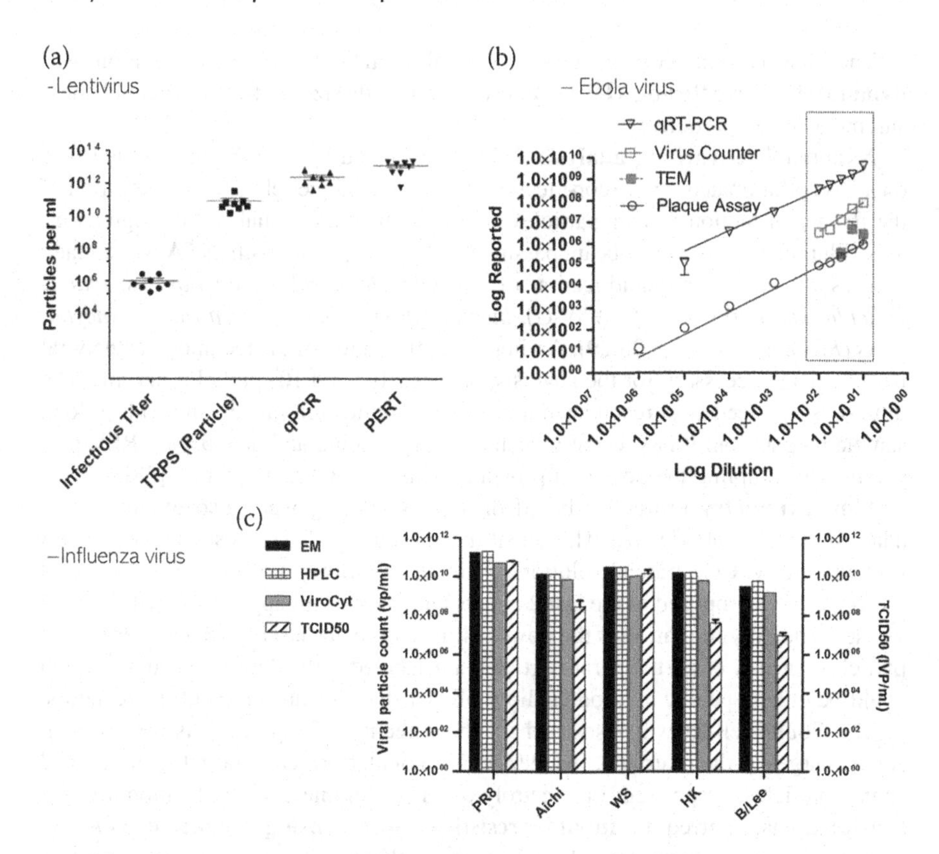

FIGURE 8.4 Comparative studies of quantification strategies on viral preparations. a – Lentivirus quantifications (data extracted from Heider et al. 2014 [4]). b – Ebolavirus quantifications (data extracted from Rossi et al. 2015 [27]). c – Influenza virus of various strain quantifications (data extracted from Transfiguracion et al. 2015 [3]).

measurement of the modification of the sensor surface refractive index. The surface of such biosensors is commonly in metals (gold or platinum) and ligands grafted could be of various origins (antibodies, cell receptors, glycans, aptamers). Antibody grafting was the first applied for virus detection [25], but antibody independent assays were soon proposed based on the grafting of viral cell receptors or target as for the glycans in the case of influenza viruses [11], [26]. The limit of detection of such tools is highly competitive and it was recently proven to be the first tool to allow for both viral activity validation and complete viral particle counting [11] (see Figure 8.3).

8.3.2.2 Particle Counters

Viral particle counting technologies have been developed in the last twenty years and could be referred to as **particle counters** for most of them. Such technologies have been developed or adapted from other fields for the real-time quantification of the total viral particles. They rely on various technologies involving either light scattering, Coulter effect, or fluorescence in-flow detection [4]. Examples are given in Figure 8.4.

The first particle counter used for viral particle counting in solution was **dynamic light scattering (DLS),** allowing for both size and zeta potential measurement of the particles.

Nanoparticle tracking analysis (NTA), developed by NanoSight Ltd., is based on a laser-illuminated microscope technique. The technique relies on the detection of the Brownian motion of nanoparticles in liquid which are visualized through an intense illumination by a laser beam passing through an optical prism. **NTA** was already successfully used for quantification of *vaccinia virus, rabies, adenovirus, phage, vesicular stomatitis virus G (VSV-G), human cytomegalovirus, respiratory syncytial virus (RSV)*, and *HIV* [18,28,29]. An important limitation of the technique is the viral particle range necessary for the analysis, between 10^7 and 10^9 particles per mL. The dilution steps needed to reach such a narrow quantification range can result in long sample preparation. It has recently been applied to sizing and quantifying RSV preparation by coupling the detection principle with a fluorescent aptamer [28].

Flow cytometry principles based on light scattering were also exploited and adapted to virus particle size. The scientific community also calls such technologies flow virometry. Commercial solutions (Virocyt®– Sartorius) or in-house developed protocols were proposed in the last 20 years with successful application with several viruses. The only limitation is the power of the cytometer lasers to detect very small particles. In this case, the viral particles are labeled with fluorescent dyes which could be either specific antibodies, lipid bilayer dyes, or nucleic acid intercalators.

The Coulter effect was also used for the detection of a particle as they present some charges at their surface. Indeed, coulter counter principle target counting and sizing particles suspended in electrolytes. The commercial application of this technology is referred to **tunable resistive pulse sensing technique (TRPS).** Viruses covered with proteins have a specific charge that is commonly negative. The reference commercial equipment for virus particle counting is the qNano® equipment (IZON). Here the particles are pushed through pores of different sizes by electric current between two electrodes. The number of charges present at the surface of the particles will induce differences in the speeds rate of the particles through the holes. This way, particles are discriminated into populations based on their particle's diameter but also their zeta-potential (charges). The Coulter effect is also now referenced in literature for the characterization and quantification of several virus types (adenovirus, VSV, influenza) [11,30].

Charges of viral particles were also exploited for virus separation and quantification by liquid chromatography. **Ion-exchange chromatography IEX-HPLC** was applied to quantify several viruses and viral vectors (influenza virus, baculovirus, adenovirus, adeno-associated virus) [3], [31–33]. In this case, the charges of the virus particles are used to create interaction with cationic support and to allow for separation from low negatively charged free proteins and strongly negative DNA/RNA molecules. Separation is commonly realized by saline buffer allowing to conserve structural integrity of the viral particles. Detection of the viral particle peaks is performed on native absorbance or fluorescence of the proteins with respectively either UV detectors at 280 nm (protein main absorbance) or fluorescence detectors excitation at 290 nm and emission at 335 nm (specific fluorescence of tryptophan residues).

The limitation of all these technologies comes from their capacity to specifically quantify viral particles. Although most of them can handle non-purified material at the first stages of the production process. None of them can discriminate viral particles from exosomes or extracellular vesicles (EVs). Indeed, most of these technologies have been applied for the quantification and characterization of EVs. This is a major limitation if such technologies are to be applied at the very first stages of the process while viral production is occurring upstream. Indeed, most of the cell factories are secreting exosomes and EVs during the viral replication process. In that case, technologies allowing for specific labeling with antibodies (virometry, HPLC) are still preferred to confirm the identity of the particles.

The selection of dedicated analytics for a specific application and process phase is highly complex. Most of the development of new quantification technologies has necessitated correlation/comparison with reference assays. Several comparison studies show high discrepancies between the different tools used for quantification thus underlying a strong particularity of the viral samples. First, while comparing infectious titration assays (either TCID50 of pfu) and total particle counting tools (TPRS, HPLC, flow virometry), a difference of 1 to 3 log between infectious and total particles is observed ([3], [4], and [27]). On the opposite, the total viral genome quantified (qPCR, qRT-PCR, ddPCR) are always providing ranges above the total particle counting methods. Indeed, while the producing cells are dying in the dynamic production process, as the infection process is not perfect and complete, several viral genomes not encapsulated in particles are released in the medium. If one relies only on the qPCR to count the viral particles, this assay might overestimate the production yield. Consequently, to develop or assess the performances of a viral production process, at least two or three orthogonal quantification methods should be implemented to generate consistent information about the production. Ideally, the selection of assays targeting (i) the genome, (ii) the particle count, (iii) the infectivity assays, or the protein content should be mandatory, and these assays should be indicating the same trend of viral product accumulation over time.

8.4 PROCESS ANALYTICAL TECHNOLOGIES AND IN-LINE ANALYTICS FOR VIRAL PRODUCTION PROCESSES

Process analytical technology (PAT) is a major part of the 2004 FDA guidance establishing the concept of **quality by design (QbD)**. The aim of quality by design is to integrate quality assessment through all the processes and not only at the late stage while assaying the final quality of the pharmaceutical product. Thus, regulatory agencies supported the concept to develop products where the quality will be better described along the different process steps. The broad definition of PAT includes all the tools that could be implemented to gain an improved understanding and monitoring of the processes and the quality of the product. PAT tools are thus applied for the monitoring and the detection of key parameters for a viral production process, ranging from cell biomass quantification, evaluation of metabolite concentrations, physicochemical cell environment, etc. These tools can be classified into four categories depending on their interaction with the process operations. Thus, you can range analytical tools from *off-line*, *at-line*, *on-line*, and *in-line* where the latter is in direct contact with the product within the production process. The final goal of implementing such technologies is to achieve methodologies allowing the assessment of the product quality across the different steps of the process thus allowing for a real-time lot release. It is also intended to improve the assessment of the process variability, to facilitate continuous improvement and automation of the process while reducing duration and production costs.

8.4.1 Off-line, At-line, On-line, and In-line Definitions

Most of the techniques that are applied for viral production and quantification are **off-line**. This is the case of all quantification techniques that necessitate sampling the production process and processing the analytical steps externally, the analytics being located outside the production zone. Analyses are commonly long to perform and necessitate a dedicated environment for example for the infectivity assays (see section before). The limitations of off-line analyses is that they do not allow for retro-control on the production process in real-time. Indeed, the durations to complete such analyses are longer than the production steps (from 2 hours up to 1 week) and will generally be performed after sample storage.

On the contrary, **at-line** analyses are performed within the production zone close to the production process. They still require process sampling but here, the assays are much faster (30 min to 2 hours). This allows for a cell-based process to have in-process monitoring, meaning informative quantification in reasonable time-frequency to track the targeted biological event. As examples for targeted biological events, common cell doubling times are of 20 to 30 hours and viral cycles are of 6 to 30 hours. This allows the operators to exploit the at-line analytical results to correct process deviations or to start new process phases based on such external quantifications.

To go further and reach automation of the production process, PAT targeted the development of **on-line** and **in-line** tools. The distinction between those two types of analytical tools is not always clearly set in the literature, therefore they are be re-defined herein. Both are analytical tools giving access to information in real-time. This means that the acquisition and analysis results are available in a time frame and frequency much higher than for at-line tools. We refer here to measurements lasting seconds or minutes. Such types of equipment are connected to the production process and could be integrated into feedback control loops. Operators are no longer involved either in sampling or changing the operating parameters on the process. Probes or analytical equipment could be part of the regulation loops.

In-line defines for example a probe that is directly in contact with the analyte inside the production process. Such analyses are non-destructive and commonly provide results in a time frame of seconds. On-line, on the contrary, is not implemented within the process but connected on a derivation thanks to a measurement chamber. The analysis is also non-destructive.

8.4.2 Process Analytical Technology for Viral Production Processes

All the PAT tools applied for cell-based culture processes associated with recombinant protein production could also be applied in the present case. Thus, all the in-line electrochemistry probes applied for the monitoring of the cell culture environment with the pH, pO_2, and pCO_2 probes are used. This allows for close monitoring of the cellular respiration and metabolic activity (lactic acid secretion) within the cultivation broth. Eventually some at-line tools allowing for cell counting like automated cell counters might be used. Cell metabolites secreted or consumed by cultivated cells could also be monitored using several technologies. To replace the old off-line biochemical assays, novel automated metabolites analyzers were developed and implemented at-line thanks to automated sterile sampling devices in the last 20 years. This now allows discrete sampling of the production process to quantify all the main cellular metabolites as well as complementary information like trace elements or osmolarity.

Additionally, several developments were dedicated in the last 20 years to achieving inline monitoring of viral production processes key parameters. The technologies evaluated were mostly spectroscopic (infrared, near infrared, fluorescence, raman, dielectric spectroscopies) [34–37].

More specifically, if we focus on the viral production processes, we will distinguish two types of analytical tools. First, the technologies aiming at quantifying the final product and which could be implemented at-line. Here we find again the viral particle counter already described in section 8.3.2.2. Indeed, most viral counters allow for viral particles count within a wide range of sample matrix from culture supernatant up to highly purified solutions. Such assays allow for better process understanding and monitoring of the biological events occurring in the production phase (viral release, viral cycle length, etc.). The only information that is not yet reached with such a technology is the activity and purity of the viral particles. Also, information is available only at sampling times.

The second kind of technology is the one targeting the production cells. Indeed, cells are here the production units modifications of the total cell biomass available for production in terms of quantity or quality can strongly affect the production efficiency. Therefore, from the late 1990s, analytical tools were developed to allow for live biomass monitoring in cell-based production. Two examples of technologies should be emphasized here, the Raman and the capacitance spectroscopy, besides other technologies that have also been explored.

Raman spectroscopy was originally assessed to monitor the composition of the cultivation media over the cultivation process in cell-based processes. Indeed, this vibrational spectroscopy tool allows for the detection of all the organic molecules. It will record spectral fingerprints of the molecules mixture thus necessitating an additional chemometric spectral post-treatment to extract information on the targeted analytes. It was successfully applied for the monitoring of several cell metabolites, either essential nutrients (glucose, amino acids) or metabolic cell by-products (lactate, other acids), in cell-based production of recombinant protein production [37–40]. Also, as cells are composed of several specific organic molecules, Raman spectroscopy was concomitantly evaluated for its capacity to monitor cell productive biomass. It was successfully applied for CHO cell culture which is the gold standard technology for recombinant protein production [39], [41], [42]. Raman probes are now commercialized by several vendors and applied either on upstream or downstream cell-based processes. They are currently explored to directly access the product levels and quality within the production steps as demonstrated by the two studies published recently for monoclonal antibody titer and glycosylation monitoring [43–45]. Such tools are highly promising if applied to viral production processes for the in-line detection of total viral particles.

Capacitance spectroscopy aims at monitoring live biomass based on the principle of dielectric spectroscopy. Capacitance probes consist of small electrodes immersed in a cell suspension which will allow for its polarization under specific frequencies. The polarization occurs in a volume of 1 cm^3 around the probe thus implying a good homogeneity of the suspension. This technology then measures the charge and discharge of the viable and intact cells which then behave like small capacitors. Thus, it can detect all the cells present within the cell suspension culture volume, allowing to use this technology as a cell counting probe or a biomass quantification tool. This is why such technology was successfully applied to cell growth and death monitoring of a wide range of cell types used in manufacturing processes including both suspension cells and adhered cells on suspended microcarriers [46–50]. The acquisition time being below seconds allows the integration in feedback control loops as for fed-batch feeding strategies [51], [52].

Such technology that analyses the specific dielectric properties of cells also allows access to several cell biological properties indirectly. Namely, cell size and volume, cell membrane thickness and villosities, and cell cytoplasm composition would impact the dielectric measurement. This is of great interest in the context of viral production processes. Indeed, it has repeatedly been proven that viral replication within a cell affects their biological properties including cytoplasm composition with the accumulation of viral particles components (antigens or capsids) or the membrane structure (effect of viral budding). Thus, this technology was also exploited to monitor indirectly viral cycle biological events within the production phase. For several virus and cell pairs, capacitance spectroscopy was proven to be a valuable tool, allowing for inline monitoring of viral production phases [47], [53–55].

Process analytical technology tools available for viral production processes are not yet at the stage to provide information directly on the viral product in-line or within the production process equipment. Nevertheless, regarding the advances performed for similar cell-based processes for large-scale recombinant protein production, we can expect that new tools will be soon available to reach this stage. Spectroscopic tools are for now some of the most promising analytical equipment to fulfill such needs but the connection of at-line analytics using automated and sterile sample loops could allow reaching faster the target proposed by regulatory agencies, the holy grail being access to a direct quantification of active viral particles.

REFERENCES

[1] FDA, "Bioanalytical method validation guidance for industry," 2018.

[2] P. J. Klasse, "Molecular determinants of the ratio of inert to infectious virus particles," *Prog. Mol. Biol. Transl. Sci.*, vol. 129, pp. 285–326, 2015.

[3] J. Transfiguracion, A. P. A. P. Manceur, E. Petiot, C. M. C. M. C. M. Thompson, and A. A. A. Kamen, "Particle quantification of influenza viruses by high performance liquid chromatography," *Vaccine*, vol. 33, no. 1, pp. 78–84, Jan. 2015, doi: 10.1016/j.vaccine.2014.11.027
[4] S. Heider and C. Metzner, "Quantitative real-time single particle analysis of virions," *Virology*, vol. 462-463C, no. 1, pp. 199–206, Jul. 2014, doi: 10.1016/j.virol.2014.06.005
[5] K. O. Smith and J. L. Melnick, "Electron microscopic counting of virus particles by sedimentation on aluminized grids," *J. Immunol.*, vol. 89, no. 2, pp. 279 LP–284, Aug. 1962.
[6] S. Payne, "Methods to study viruses," *Viruses*, pp. 37–52, 2017, doi: 10.1016/b978-0-12-803109-4.00004-0
[7] A. A. Kojabad *et al.*, "Droplet digital PCR of viral DNA/RNA, current progress, challenges, and future perspectives," *J. Med. Virol.*, vol. 93, no. 7, pp. 4182–4197, 2021, doi: 10.1002/jmv.26846
[8] A. Roldão, R. Oliveira, M. J. T. Carrondo, and P. M. Alves, "Error assessment in recombinant baculovirus titration: Evaluation of different methods," *J. Virol. Methods*, vol. 159, no. 1, pp. 69–80, 2009, doi: 10.1016/j.jviromet.2009.03.007
[9] A. L. Masci, E. B. Menesale, W.-C. Chen, C. Co, X. Lu, and S. Bergelson, "Integration of fluorescence detection and image-based automated counting increases speed, sensitivity, and robustness of plaque assays," *Mol. Ther. – Methods Clin. Dev.*, vol. 14, pp. 270–274, Sep. 2019, doi: 10.1016/j.omtm.2019.07.007
[10] B. Y. G. K. Hirst, "The quantitative determination of influenza virus and antibodies by means of red cell agglutination," *J. Exp. Med.*, vol. 75, no. 1, pp. 49–64, 1942.
[11] L. Durous *et al.*, "SPRi-based hemagglutinin quantitative assay for influenza vaccine production monitoring," *Vaccine*, vol. 37, no. 12, pp. 1614–1621, Mar. 2019, doi: 10.1016/j.vaccine.2019.01.083
[12] F. Schmeisser, A. Vasudevan, J. Soto, A. Kumar, O. Williams, and J. P. Weir, "A monoclonal antibody-based immunoassay for measuring the potency of 2009 pandemic influenza H1N1 vaccines," *Influenza Respi. Viruses*, vol. 8, no. 5, pp. 587–595, 2014, doi: 10.1111/irv.12272
[13] T. Bousse *et al.*, "Quantitation of influenza virus using field flow fractionation and multi-angle light scattering for quantifying influenza A particles," *J. Virol. Methods*, vol. 193, no. 2, pp. 589–596, Nov. 2013, doi: 10.1016/j.jviromet.2013.07.026
[14] P. C. Stepp, K. A. Ranno, E. D. Dawson, K. L. Rowlen, and M. M. Ferris, "Comparing H1N1 Virus quantification with unique flow cytometer and quantitative PCR," *Bioprocess Int.*, vol. 9, no. 8, pp. 50–56, 2011.
[15] C. A. Rossi *et al.*, "Evaluation of ViroCyt®Virus counter for rapid filovirus quantitation," *Viruses*, vol. 7, no. 3, pp. 857–872, 2015, doi: 10.3390/v7030857
[16] R. Vogel *et al.*, "Quantitative sizing of nano/microparticles with a tunable elastomeric pore sensor," *Anal. Chem.*, vol. 83, no. 9, pp. 3499–3506, 2011, doi: 10.1021/ac200195n
[17] R. Vogel *et al.*, "A standardized method to determine the concentration of extracellular vesicles using tunable resistive pulse sensing," *J. Extracell. Vesicles*, vol. 5, no. 1, 2016, doi: 10.3402/jev.v5.31242
[18] P. Kramberger, M. Ciringer, A. Štrancar, and M. Peterka, "Evaluation of nanoparticle tracking analysis for total virus particle determination," *Virol. J.*, vol. 9, p. 265, Jan. 2012, doi: 10.1186/1743-422X-9-265
[19] Z. Wei *et al.*, "Biophysical characterization of influenza virus subpopulations using field flow fractionation and multiangle light scattering: correlation of particle counts, size distribution and infectivity," *J. Virol. Methods*, vol. 144, no. 1–2, pp. 122–132, Sep. 2007, doi: 10.1016/j.jviromet.2007.04.008

[20] J. Transfiguracion, A. Bernier, R. Voyer, H. Coelho, M. Coffey, and A. Kamen, "Rapid and reliable quantification of reovirus type 3 by high performance liquid chromatography during manufacturing of Reolysin," *J. Pharm. Biomed. Anal.*, vol. 48, no. 3, pp. 598–605, Nov. 2008, doi: 10.1016/j.jpba.2008.05.038

[21] K. A. Brogden, "The sweet spot: defining virus–sialic acid interactions Jennifer," *Nat. Rev. Microbiol.*, vol. 3, no. 11, pp. 238–250, 2005, doi: 10.1038/nrmicro3346.

[22] H. B. Donald and a Isaacs, "Counts of influenza virus particles," *J. Gen. Microbiol.*, vol. 10, no. 3, pp. 457–464, Jun. 1954.

[23] A. M. Hashem *et al.*, "A novel synthetic receptor-based immunoassay for influenza vaccine quantification," *PLoS One*, vol. 8, no. 2, p. e55428, Jan. 2013, doi: 10.1371/journal.pone.0055428

[24] M. Rusnati, P. Chiodelli, A. Bugatti, and C. Urbinati, "Bridging the past and the future of virology: Surface plasmon resonance as a powerful tool to investigate virus/host interactions," *Crit. Rev. Microbiol.*, vol. 7828, no. February, pp. 1–23, 2013, doi: 10.3109/1040841X.2013.826177

[25] Y. F. Chang *et al.*, "Simple strategy for rapid and sensitive detection of avian influenza A H7N9 virus based on intensity-modulated SPR biosensor and new generated antibody," *Anal. Chem.*, vol. 90, no. 3, pp. 1861–1869, 2018, doi: 10.1021/acs.analchem.7b03934

[26] S. Khurana, L. R. King, J. Manischewitz, E. M. Coyle, and H. Golding, "Novel antibody-independent receptor-binding SPR-based assay for rapid measurement of influenza vaccine potency," *Vaccine*, vol. 32, no. 19, pp. 2188–2197, 2014, doi: 10.1016/j.vaccine.2014.02.049

[27] C. A. Rossi *et al.*, "Evaluation of ViroCyt® Virus counter for rapid filovirus quantitation," *Viruses*, vol. 7, no. 3, pp. 857–872, 2015, doi: 10.3390/v7030857

[28] Z. Szakács, T. Mészáros, M. I. De Jonge, and R. E. Gyurcsányi, "Selective counting and sizing of single virus particles using fluorescent aptamer-based nanoparticle tracking analysis," *Nanoscale*, vol. 10, no. 29, pp. 13942–13948, 2018, doi: 10.1039/c8nr01310a

[29] N. Sanchez *et al.*, "Rabies vaccine characterization by nanoparticle tracking analysis," *Sci. Rep.*, vol. 10, no. 1, pp. 1–8, 2020, doi: 10.1038/s41598-020-64572-6

[30] L. Yang and T. Yamamoto, "Quantification of virus particles using nanopore-based resistive-pulse sensing techniques," *Front. Microbiol.*, vol. 7, no. SEP, Article 1500, 2016, doi: 10.3389/fmicb.2016.01500

[31] B. Lorbetskie *et al.*, "Optimization and qualification of a quantitative reversed-phase HPLC method for hemagglutinin in influenza preparations and its comparative evaluation with biochemical assays," *Vaccine*, vol. 29, no. 18, pp. 3377–3389, Apr. 2011, doi: 10.1016/j.vaccine.2011.02.090

[32] J. Transfiguracion, H. Coelho, and A. Kamen, "High-performance liquid chromatographic total particles quantification of retroviral vectors pseudotyped with vesicular stomatitis virus-G glycoprotein," *J. Chromatogr. B*, vol. 813, no. 1–2, pp. 167–173, 2004.

[33] P. S. Chahal, J. Transfiguracion, A. Bernier, R. Voyer, M. Coffey, and A. Kamen, "Validation of a high-performance liquid chromatographic assay for the quantification of Reovirus particles type 3," *J. Pharm. Biomed. Anal.*, vol. 45, no. 3, pp. 417–421, Nov. 2007, doi: 10.1016/j.jpba.2007.06.025

[34] M. B. Haack, A. E. Lantz, P. P. Mortensen, L. Olsson, and A. E. L. P. P. M. L. O. Martin B. Haack, "Chemometric analysis of in-line multi-wavelength fluorescence measurements obtained during cultivations with a lipase producing *Aspergillus oryzae* strain," *Biotechnol. Bioeng.*, vol. 9999, no. 9999, 2006, doi: 10.1002/bit

[35] C. Schaefer, D. Clicq, C. Lecomte, A. Merschaert, E. Norrant, and F. Fotiadu, "A Process Analytical Technology (PAT) approach to control a new API manufacturing

process: Development, validation and implementation," *Talanta*, vol. 120, pp. 114–125, Mar. 2014, doi: 10.1016/j.talanta.2013.11.072
[36] S. M. Mercier, P. M. Rouel, P. Lebrun, B. Diepenbroek, R. H. Wijffels, and M. Streefland, "Process analytical technology tools for perfusion cell culture," *Eng. Life Sci.*, vol. 16, no. 1, pp. 25–35, 2016, doi: 10.1002/elsc.201500035
[37] H. Mehdizadeh, D. Lauri, K. M. Karry, M. Moshgbar, R. Procopio-Melino, and D. Drapeau, "Generic Raman-based calibration models enabling real-time monitoring of cell culture bioreactors," *Biotechnol. Prog.*, vol. 31, no. 4, pp. 1004–1013, 2015, doi: 10.1002/btpr.2079
[38] S. Craven, J. Whelan, and B. Glennon, "Glucose concentration control of a fed-batch mammalian cell bioprocess using a nonlinear model predictive controller," *J. Process Control*, vol. 24, no. 4, pp. 344–357, 2014, doi: 10.1016/j.jprocont.2014.02.007
[39] B. Berry, J. Moretto, T. Matthews, J. Smelko, and K. Wiltberger, "Cross-scale predictive modeling of CHO cell culture growth and metabolites using Raman spectroscopy and multivariate analysis," *Biotechnol. Prog.*, vol. 31, no. 2, pp. 566–577, 2015, doi: 10.1002/btpr.2035
[40] R. M. Santos, P. Kaiser, J. C. Menezes, and A. Peinado, "Talanta Improving reliability of Raman spectroscopy for mAb production by upstream processes during bioprocess development stages," *Talanta*, vol. 199, no. November 2018, pp. 396–406, 2019, doi: 10.1016/j.talanta.2019.02.088
[41] J. Whelan, S. Craven, and B. Glennon, "In situ Raman spectroscopy for simultaneous monitoring of multiple process parameters in mammalian cell culture bioreactors," *Biotechnol. Prog.*, vol. 28, no. 5, pp. 1355–1362, 2012, doi: 10.1002/btpr.1590
[42] N. R. Abu-Absi *et al.*, "Real time monitoring of multiple parameters in mammalian cell culture bioreactors using an in-line Raman spectroscopy probe," *Biotechnol. Bioeng.*, vol. 108, no. 5, pp. 1215–1221, May 2011, doi: 10.1002/bit.23023
[43] L. Saint *et al.*, "In-line and real-time prediction of recombinant antibody titer by in situ Raman spectroscopy," *Anal. Chim. Acta*, vol. 892, pp. 148–152, 2015, doi: 10.1016/j.aca.2015.08.050
[44] M. Li, B. Ebel, F. Chauchard, E. Guédon, and A. Marc, "Parallel comparison of in situ Raman and NIR spectroscopies to simultaneously measure multiple variables toward real-time monitoring of CHO cell bioreactor cultures," *Biochem. Eng. J.*, vol. 137, pp. 205–213, 2018, doi: 10.1016/j.bej.2018.06.005
[45] M. Li, B. Ebel, F. Chauchard, E. Guedon, and A. Marc, "Real-time monitoring of antibody glycosylation site occupancy by in situ raman spectroscopy during bioreactor CHO cell cultures,"Biotechnol. Prog., vol. 34, no. 2, 486–493, 2018, doi: 10.1002/btpr.2604
[46] S. Metze *et al.*, "Multivariate data analysis of capacitance frequency scanning for online monitoring of viable cell concentrations in small-scale bioreactors," *Anal. Bioanal. Chem.*, vol. 412, no. 9, pp. 2089–2102, 2020, doi: 10.1007/s00216-019-02096-3
[47] E. Petiot and A. A. Kamen, "Real-time monitoring of influenza virus production kinetics in HEK293 cell cultures," *Biotechnol. Prog.*, vol. 29, no. 1, pp. 275–284, Jul. 2012, doi: 10.1002/btpr.1601
[48] J. P. Carvell and J. E. Dowd, "On-line Measurements and Control of Viable Cell Density in Cell Culture Manufacturing Processes using Radio-frequency Impedance," *Cytotechnology*, vol. 50, no. 1–3, pp. 35–48, Mar. 2006, doi: 10.1007/s10616-005-3974-x
[49] K. Braasch *et al.*, "The changing dielectric properties of CHO cells can be used to determine early apoptotic events in a bioprocess," *Biotechnol. Bioeng.*, vol. 110, no. 11, pp. 2902–2914, 2013, doi: 10.1002/bit.24976

[50] E. Petiot *et al.*, "Real-time monitoring of adherent Vero cell density and apoptosis in bioreactor processes," *Cytotechnology*, vol. 64, no. 4, pp. 429–441, Feb. 2012, doi: 10.1007/s10616-011-9421-2

[51] L. Párta, D. Zalai, S. Borbély, and Á. Putics, "Application of dielectric spectroscopy for monitoring high cell density in monoclonal antibody producing CHO cell cultivations," *Bioprocess Biosyst. Eng.*, vol. 37, no. 2, pp. 311–323, 2013, doi: 10.1007/s00449-013-0998-z

[52] A. Zhang *et al.*, "Advanced process monitoring and feedback control to enhance cell culture process production and robustness," *Biotechnol. Bioeng.*, vol. 112, no. 12, pp. 2495–2504, 2015, doi: 10.1002/bit.25684

[53] S. Ansorge, G. Esteban, and G. Schmid, "On-line monitoring of infected Sf-9 insect cell cultures by scanning permittivity measurements and comparison with off-line biovolume measurements," *Cytotechnology*, vol. 55, no. 2–3, pp. 115–124, Dec. 2007, doi: 10.1007/s10616-007-9093-0

[54] E. Petiot, S. Ansorge, M. Rosa-Calatrava, and A. Kamen, "Critical phases of viral production processes monitored by capacitance," *J. Biotechnol.*, vol. 242, pp. 19–29, Jan. 2017, doi: 10.1016/j.jbiotec.2016.11.010

[55] S. Ansorge, S. Lanthier, J. Transfiguracion, O. Henry, and A. Kamen, "Monitoring lentiviral vector production kinetics using online permittivity measurements," *Biochem. Eng. J.*, vol. 54, no. 1, pp. 16–25, Mar. 2011, doi: 10.1016/j.bej.2011.01.002

LIST OF ABBREVIATIONS

CHO	Chinese hamster ovary
CQA	Critical quality attribute
DLS	Dynamic light scattering
ELISA	Enzyme-linked immunosorbent assay
EMA	European Medicament Agency
FDA	Food and Drug Administration
IEX-HPLC	Ion-exchange high performance liquid chromatography
IVP	Infectious viral particle
LC-MS	Mass spectrometry following liquid chromatography separation
NSTEM	Negative staining transmission electron microscopy
NTA	Nanoparticle tracking analysis
PAT	Process analytical technology
Pfu	Plaque forming unit
QbD	Quality by design
RBC	Red blood cells
SPR	Surface plasmon resonance
SRID	Single radial immuno diffusion assay
TCID50	Tissue culture infectious Dose 50
TRPS	Tunable resistive pulse sensing technique
TU	Transducing unit
VLP	Virus-like particle
VP	Viral particle

9 Manufacturing of seasonal and pandemic influenza vaccines–A case study

Cristina A. T. Silva
Department of Chemical Engineering, Polytechnique Montréal, Montréal, Canada/Department of Bioengineering, McGill University, Montréal, Canada

Shantoshini Dash and Amine Kamen
Viral Vectors and Vaccines Bioprocessing Group, Department of Bioengineering, McGill University, Montréal, Canada

CONTENTS

9.1 INTRODUCTION

The influenza disease is an upper respiratory tract infection caused by the influenza virus, usually characterized by mild symptoms, such as cough, fever, sore throat, and muscle pain, sometimes evolving to more severe and even lethal pneumonia [1]. Influenza outbreaks are believed to have occured since at least the Middle Ages, possibly before that, affecting humanity as a whole in the form of localized outbreaks,

DOI: 10.1201/9781003229797-9

seasonal epidemics, and occasional pandemics [2]. Every year, between 290,000–645,000 people die of complications from seasonal influenza, particularly young infants, and adults older than 65 [3]. While it is possible to estimate when seasonal influenza epidemics will occur around the world, predicting their magnitude and severity, as well as the exact composition of the infectious agent, is a challenging task for public health authorities. Additionally, since vaccination is the most effective way of protection, defining vaccine composition based on circulating strains and organizing worldwide manufacturing and supply of vaccines is a formidable effort. Influenza pandemics, on the other hand, occur every 10–40 years as a result of the introduction of a new and antigenically distinct influenza strain, that usually has zoonotic origins [4]. Whereas it is currently impossible to predict exactly when or how it will occur, understanding the genetic and epidemiological characteristics of past pandemics is critical to develop and optimize influenza surveillance tools [5]. Since the beginning of the 20th century, humanity has seen four global influenza pandemics: in 1918–1920, the Spanish flu is estimated to have killed up to 50 million people around the globe; between 1957–1958, the Asian flu resulted in the death of 1.5 million people; the Hong Kong flu, was responsible for 1 million deaths from 1968–1969; finally, the swine flu, a milder and more recent pandemic, is estimated to have caused around 200,000 deaths globally in its first year of circulation [4,6].

9.2 THE INFLUENZA VIRUS

Influenza viruses are members of the Orthomyxoviridae family, being single-stranded negative-sense enveloped RNA viruses that present a segmented genome composed of seven to eight segments. Although wild virions are pleomorphic, virions that were subjected to multiple passages acquire a roughly spheroidal shape, approximately 100 nm in diameter [7,8]. While influenza A and B viruses (IAVs and IBVs, respectively) are responsible for worldwide recurrent influenza outbreaks and have a similar structure (both presenting eight genome segments), influenza C viruses are more divergent (with only seven genome segments) and cause a less severe disease in humans [9]. More recently, influenza D virus (IDV), a novel influenza C-like virus, was isolated from pigs. Although rare, serological surveys revealed the presence of antibodies against IDV's hemagglutinin in humans working with cattle, species considered as the natural reservoir for IDVs [10]. As ICVs and IDVs account for less severe disease in humans, those will not be further discussed in this chapter.

The influenza viral envelope is composed of a host-derived lipid bilayer that presents three different transmembrane proteins, HA (hemagglutinin), NA (neuraminidase), and M2 (matrix 2). HA, a glycosylated integral membrane protein, is the most abundant on the virus surface (around 80%), being responsible for the initial attachment of the virus to host cell receptors (bearing a terminal sialic acid) and later merging of viral envelope and host cell membrane. NA, which represents approximately 17% of viral surface proteins, cleaves the sialic acid residues in host cell receptors to release new virions, allowing for the spread of the virus [4,7]. Protective immunity against influenza viruses is mediated mainly by neutralizing antibodies against these two surface proteins, which prevents the infection and spread of the virus in host cells [11]. M2, present in minor quantities on the membrane, is an ion channel that has an important role in early phases of the infection. Inside the virus, each RNA segment is wrapped around nucleoprotein (NP) monomers, forming viral ribonucleoprotein (RNP) complexes alongside with the viral polymerases PB1, PB2 and PA. Attached to the inside of the membrane, M1 (matrix) protein interacts with RNP complexes [8].

As mentioned before, IAVs and IBVs are responsible for influenza outbreaks around the world. While both viruses result in similar burdens on public health for seasonal epidemics [12], IAVs are also responsible for sporadic global pandemics due to their greater genetic diversity and host range (infecting domestic animals, pigs, poultry, and wild birds, while IBVs infect almost exclusively humans): all known influenza pandemics were caused by IAVs, usually from zoonotic origin [4]. To this date, 18 HA and 11 NA subtypes were identified for IAVs, while only two antigenic lineages are reported for IBVs. HA plays a critical role in species restriction: human influenza viruses have HAs specific to α2–6 terminal sialic acid (SA) while avian HAs bind to α2–3 SA [2]. Influenza viruses undergo frequent mutations as viral polymerase lacks proofreading activity, around one introduced error per replicated genome [13]. When these mutations occur in the antigenic portions of the HA and NA surface proteins, it may result in advantages that allow some strains to escape pre-existing immunity, a process called antigenic drift. Additionally, as the influenza genome is composed of discrete segments, coinfection of a same host cell with two different viruses may result in virions that contain genome segments from both parental strains. This process, named antigenic shift, is critical in pandemic formation as one single event is able to result in significant antigenic modifications that allows, for example, the virus to jump from one species to another eventually infecting humans [7]. A pandemic occurs when a new influenza strain, containing antigens that are significantly different from previous viruses and to which humans have no or little immunity, appears and spreads, being able to contaminate a significant percentage of the population [2,4].

9.3 THE ANNUAL CYCLE FOR INFLUENZA VACCINE MANUFACTURING

Due to the high mutation rate of influenza viruses, a continuous surveillance of worldwide influenza activity must be conducted in order to assure a good match between seasonal vaccines and circulating strains, as well as to increase preparedness for future pandemics. The Global Influenza Surveillance and Response System (GISRS), coordinated by the World Health Organization since 1952, is responsible for monitoring, isolation, and identification of influenza viruses causing localized epidemics throughout the year. Samples from influenza patients collected from around the world are sent to one of the more than 150 National Influenza Centers for strain isolation and identification. Once a new strain is detected, it is sent to one of the WHO's reference laboratories for further antigenic and molecular analysis. Twice a year, the WHO will carefully review the data generated and announce the four influenza strains that are likely to circulate in the next flu season for Northern and Southern Hemispheres [14]. From the time of this announcement, manufacturers have a 6-month window to produce and provide vaccines for vaccination campaigns around the world [15,16]. Numerous factors may challenge this production timeline: the generation of egg-adapted virus seed stocks and a continuous egg supply, as ~80% of influenza vaccines are produced in embryonated hens' eggs [17]; the lack of specific reagents for vaccine quantification; the need for full characterization of the product and the fulfilment of regulatory aspects [18,19]. In spite of WHO's extensive efforts to increase global influenza vaccine production to respond to the threat of a global pandemic, manufacturing capacity is still not sufficient. According to Sparrow, et al. [17], an estimated 8.31 billion doses could be produced in a 12-month period by current active manufacturers, an estimation made based on various assumptions such as a sufficient supply of eggs and other reagents, which might not be the case if manufacturers are in the middle of seasonal vaccine production. Furthermore, multiple doses might be required for vaccine effectiveness, and some pandemic strains, such as the H5N1, have already shown suboptimal production in egg-based systems [20].

9.4 INFLUENZA VACCINES

Vaccination is the most cost-effective and efficient way to minimize the impact of flu outbreaks in society [21]. However, the influenza disease can also be treated with the use of antiviral drugs, although resistance to some of these drugs has already been reported [22]. Tables 9.1 and 9.2 summarize the different types of vaccines and antiviral drugs, respectively, available for influenza prevention and treatment.

Three types of vaccines are currently approved and available on the market for influenza: inactivated influenza vaccine (IIV), live attenuated influenza vaccine (LAIV), and recombinant vaccine (RV). Manufacture of IIVs and LAIVs, which together correspond to almost 90% of production capacity [17], relies on the generation of master viral seed stocks. As some influenza strains present limited yields when produced in vitro, master virus stocks are composed of high growth reassortants (HGR), a modified virus expressing the antigenic proteins HA and NA from the targeted strain within the backbone of a high growth strain such as the A/PR/8/34 H1N1. While critical for egg-based influenza vaccine production, it is yet not clear if HGRs are necessary for cell-culture–based vaccine manufacturing, with evidence suggesting no significant improvement in virus yields for cell-base

TABLE 9.1
Different types of Influenza vaccines available in the market

Vaccine type	Composition	% of global capacity (seasonal) [17]
Inactivated influenza vaccine (IIV)	trivalent, quadrivalent, adjuvanted	89.6%
Live attenuated vaccine (LAIV)	trivalent, quadrivalent	5.0%
Recombinant vaccine (RV)	trivalent, quadrivalent	5.4%

TABLE 9.2
Approved antiviral drugs for the influenza disease

Mechanism of action	Drug
Neuraminidase inhibitor (blocks particle release)	Rapivab (Peramivir), Relenza (Zanamivir), Tamiflu (Oseltamivir phosphate)
Inhibitor of viral polymerase activity (inhibits viral replication)	Xofluza (Baloxavir marboxil)

systems seeded with HGRs in comparison to wild-type viruses [23]. HGR generation, performed only by a few laboratories in the world, consists of coinfecting the same egg with both viruses and screening the viral progeny for a reassortant with the desired characteristics, a process that takes between 4–6 weeks to be completed [17,18]. To reduce the time required for HGR generation and increase responsiveness to potential pandemics, the use of reverse genetics for viral seed stock production have been studied. This technique consists of generating live viruses by co-transfecting cells with a set of plasmid-cloned cDNA encoding the influenza viral genome [24]. Rapid production of influenza virus seed stocks using reverse genetics have been demonstrated for different cell lines [25–27].

9.4.1 Inactivated Influenza Vaccine (IIV)

IIVs, which represent almost 90% of influenza vaccines produced globally, may be composed of virus sub-unit, whole or split virus. Inactivated vaccines are safe, owing to the inability of the virus to replicate, and are highly effective, promoting mostly humoral immune response [16,28]. Cellular immune response may be increased if viral structures are successfully preserved during the inactivation process. While virus inactivation may be achieved through chemical (formaldehyde, β-propiolactone) or physical processes (UV, gamma irradiation, USP laser), each technique results in a different product and, consequently, may trigger slightly different immunological responses [29].

The majority of IIVs, around 80%, is produced using embryonated hens' eggs, a technology established in 1940s where fertilized eggs are used for virus propagation [17]. Yet, driven by advances in large-scale cell culture for recombinant protein production [30,31], the last decades have seen an increase in the use of cell culture platforms as a faster alternative for IIV production [32–37]. In short, cells are cultivated until sufficiently high cell densities are achieved, followed by infection with the desired virus strain. Currently, five IIVs produced using cell-culture systems are approved for use, and only two of them are in production for commercial use (Table 9.3). Advantages of cell-culture–based systems over traditional egg production include: *i)* scalability, notably with the use of suspension cell lines [30]; *ii)* greater process control, which results in a more reliable product [38]; *iii)* a better match between vaccine and circulating strains, as egg-growth adaptation may generate undesirable antigenic modification [39,40]; *iv)* shorter production cycles, and consequently faster responses to

TABLE 9.3
Licensed cell-culture–based IIVs

Commercial name	Manufacturer	Composition	Cell platform	Approval	Current production status
Preflucel	Baxter	trivalent	Vero	EU	No longer in production
Celvapan	Baxter	monovalent (H1N1)	Vero	EU	No longer in production
Flucelvax	Seqirus	trivalent and quadrivalent	MDCK	FDA/EU	In production
Audenz	Seqirus	monovalent (H5N1)	MDCK	FDA	Not in production
SKYCellflu	SK Bioscience	quadrivalent	MDCK	MFDS	In production

Source: Adapted from Silva, et al. [32].

pandemic situations, as cell supply can be accelerated with the use of high cell density frozen seed stocks, unlike the longer timelines for egg supply [41,42]. Indeed, egg supply was shown to be a major bottleneck for vaccine production for a significant part of active manufacturers [17]. However, in order to encourage influenza vaccine producers to shift from the cost-effective and well-established egg-based system to cell-culture platforms, a number of challenges have yet to be overcome: *i)* the screening and development of new host cell lines, with characteristics such as high cell density suspension growth and the ability to grow in media free from animal-based supplements; *ii)* the development of cost-effective and scalable high cell density processes, that maximize productivities; *iii)* the improvement of culture media and media supplementation specific for virus production; *iv)* the development of online monitoring techniques to increase product quality and minimize product losses. Cell-culture processes for influenza vaccine manufacturing, as well as its challenges and opportunities have been reviewed elsewhere [32].

9.4.2 Live Attenuated Influenza Vaccines (LAIVs)

LAIVs represent around 5% of total influenza vaccine production, being made from a cold adapted (*ca*), temperature sensitive (*ts*), or attenuated (*att*) virus backbone containing the relevant antigenic proteins (HA and NA) from recommended seasonal strains. The resulting viruses are incapable of replicating in the warmer temperatures of the lower respiratory tract and, being restricted to the nasal passage, stimulate local humoral and cellular immune responses [19,28,43]. The vaccine, administered intranasally, have shown to be effective in young children but, because of the presence of live viruses, is not recommended for immunosuppressed individuals [28]. Vaccine strains are generated either by reassortment in eggs or, more recently, by reverse genetics recapitulating *ca, ts,* and *att* mutations [19]. LAIV antigen content is defined as 6.5–7.5 log fluorescent focus units (FFUs) for vaccines containing the Ann Arbor backbone and at least 6.5–7 log 50% egg-infected dose (EID50) for vaccines derived from the Leningrad backbone [17]. While all of the commercial LAIVs are currently produced in embryonate hens' eggs, its production using cell-based systems have been demonstrated for different adherent cell lines, including MDCK [44–46] and Vero cells [47].

9.4.3 Recombinant Vaccines

Recombinant influenza vaccines are composed of one or more viral proteins, in its free form or assembled in virus-like particles (VLPs), produced using recombinant expression systems [48] and represent 5.4% of global production capacity [17]. Because the production of this type of vaccine does not require adaptation of the influenza strain, a good antigenic match between vaccine and circulating strains is obtained. Additionally, unlike IIVs and LAIVs, recombinant vaccines do not require high-level biocontainment facilities, and can be produced in shorter periods of time [18]. Flublok (Sanofi-Pasteur), the first RV for influenza to be approved by the FDA, is composed of recombinant HA protein (full length, comprising the transmembrane domain, HA1 and HA2 regions) produced in sf9-derived insect cells using the baculovirus expression system [49,50]. Baculoviruses are DNA viruses that infect insect cells and induce the production of large amounts of the viral protein polyhedrin. Recombinant baculoviruses, in which a gene of interest replaces the polyhedrin gene, can then be used for the expression of large quantities of a foreign protein of interest [19,48]. Whereas this vaccine requires larger doses of HA protein (45 μg per strain, 180 μg for QIVs), the antibody response that is induced was shown to be comparable to that of IIVs or LAIVs [28,49]. Cadiflu-S (CPL Biologicals), a recombinant vaccine recently licenced in India, contains three immunogenic proteins from influenza (HA, NA, and M1) incorporated into a virus-like particle, also produced using the baculovirus expression system [17].

9.4.4 Emerging Technologies for Influenza Vaccine Production

While IIVs, LAIVs, and RVs represent all the influenza vaccines currently on the market, other emerging vaccine technologies, such as DNA, mRNA, and viral vectored vaccines, are presently in clinical trials (Table 9.4). These novel technologies that gained ground during the COVID-19 pandemic, are potentially faster to produce, could increase vaccine efficacy and, therefore, are likely to impact global influenza vaccine manufacture [51].

9.5 INFLUENZA VIRUS QUANTIFICATION

While efforts in the improvement of upstream processes for influenza production are crucial for the development of cell-culture–based influenza vaccines, the quantification of influenza viruses remain one of the greatest bottlenecks for both traditional and cell-based vaccine manufacturing. A number of different techniques are available for the quantification of either HA content, total viral particles, or total infectious particles, with different stages of the vaccine process development

TABLE 9.4
DNA, mRNA, and viral vectored influenza vaccines currently in clinical trials

Company	Vaccine type	Phase
Vaccitech	Viral vectored – Modified Vaccinia Ankara (MVA)	Phase II
Vaxart	Viral vectored – Adenovirus	Phase II
Altimmune	Viral vectored – Adenovirus	Phase II
AlphaVax	Viral vectored – Alphavirus	Phase II
Moderna	mRNA	Phase II
Inovio	DNA	Phase I
Mymetics	Virosomes	Phase I

Source: clinicaltrials.gov, Accessed on: September 2021 [52].

requiring specific quantification methods. Antibody-based techniques, including the SRID assay—the only assay formally approved by the WHO and other regulatory bodies for influenza quantification for vaccine formulation – measure the biologically active trimeric form of the HA protein. Since these techniques are time consuming and depend on the generation of specific antibodies, they are mainly employed for vaccine formulation [53]. While faster methods for influenza quantification are described, such as flow-virometry [54], RP-HPLC [55], and ion exchange HPLC [56], various challenges still hamper their use for routine influenza quantification [53]. The hemagglutination assay, that measures HA content based on the reticulation of red blood cells, is one of the most common assays for influenza quantification. Although it is a fast and inexpensive alternative for routine quantification during the production stage, it presents high variations based on the quality of cells and expertise of the operator, resulting in increased errors that restrict its use for process development. A decrease in HA assay errors was demonstrated by Kalbfuss, et al. [57] when an automated read-out for plates and a shift in dilution factor were introduced. The neuraminidase assay, based on the measure of the NA activity, has shown to be a fast and accurate assay for influenza quantification that could be applied during the production process of some types of vaccines (before virus inactivation). However, increased cost of reagents and high background noise depending on sample composition have shown to be major problems for assay application [57]. Finally, infectivity measures such as the TCID50 assay, quantify total infectious particles based on the ability of the viral suspension to kill host cells [58]. While extremely time consuming and labor intensive, these methods are crucial for viral stock preparation. The development of fast and cost-effective methods for influenza quantification could have a great impact over the development of upstream processes. New methods, such as the ddPCR for total viral particles assessments [59,60] and the surface plasmon resonance for HA quantification [61–64] are being developed, the latter being especially interesting given the possibility of its application for online measurements during production [64]. Nilsson, et al. [62] developed an antibody-dependent SPR assay for the quantification of HA in process samples. The assay, developed in an inhibition format, showed quantification results and coefficients of variation similar to those obtained by SRID, and a quantification range between 1–10 μgHA/mL. Durous, et al. [64] demonstrated the use of an antibody-independent SPR assay for HA quantification, employing fetuin containing α-2,3 and α-2,6-linked terminal sialic acid as ligands. The assay, shown to be specific only to active trimeric HA, displayed a large dynamic range (between 0.03–20 μgHA/mL) and negligible non-specific interactions with different culture medias or MDCK by-products.

9.6 DOWNSTREAM PROCESSING OF INFLUENZA VACCINES

The development of new processes for influenza vaccine production raises new challenges for downstream processing. Depending on the vaccine type (IIV, LAIV, RV) and production substrate (eggs, cell culture), the required steps for product purification, as well as the techniques to be employed, may differ [65]. The purification process starts with the harvest of virus-containing media, which is done by extracting the allantoic fluid in egg-based systems, or by collecting supernatant/cell pellet after centrifugation for cell substrates. An additional step of extraction with detergent, performed on the collected cell pellet, is required for RVs produced using the baculovirus system in insect cells. Clarification is then performed either through a centrifugation or a filtration step, depending on the production platform. A virus inactivation step is required for cell and egg-based IIVs, which is usually done chemically by the addition of formaldehyde or β-propiolactone [65]. After that, purification steps vary largely for different types of vaccines.

For IIVs produced in eggs, following steps include concentration and purification by zonal centrifugation, virus disruption performed by centrifugation in the presence of cetyltrimethylammonium bromide CTAB, followed by polishing, and vaccine formulation. Flucelvax (Seqirus), a sub-unit IIV produced in MDCK cells, follows a different process: concentration and initial purification is done through a chromatographic step. Host DNA is then removed by benzonase treatment and virus disruption is performed by centrifugation with CTAB, followed by an ultracentrifugation polishing step and vaccine formulation [66,67]. A different process is employed for Flublok (Sanofi-Pasteur), a RV composed of recombinant HA. After clarification, a capture step is performed using an ion-exchange chromatography resin, followed by another chromatographic step (hydrophobic resin) for further removal of contaminants. Host cell DNA is removed by Q membrane filtration followed by a final ultrafiltration step and vaccine formulation [68].

9.7 CONCLUSION

In the last decades, a robust system for influenza surveillance and vaccine production was developed, interconnecting the World Health Organization, national centers, regulatory agencies, and manufacturers around the globe. Other fast-mutating infectious agents, such as viruses from the coronavirus family and notably the SARS-CoV-2, responsible for the COVID-19 pandemic, could benefit from similar systems to reduce the severity of outbreaks and increase responsiveness to eventual pandemics. The development of new viral vaccine technologies, such as mRNA/DNA and viral-vectored vaccines, may revolutionize influenza vaccine manufacturing in the next decade and downstream processes will adapt to the new production platforms. However, influenza quantification remains a challenge and new techniques such as the SPR could play a critical role in the development of new methods.

REFERENCES

[1] World Health Organization. (2018, September 20, 2021). *Influenza – Seasonal*. Available: https://www.who.int/news-room/fact-sheets/detail/influenza-(seasonal)

[2] F. Krammer *et al.*, "Influenza," *Nat. Rev. Dis. Primers*, vol. 4, no. 1, p. 3, Jun. 2018.

[3] A. D. Iuliano *et al.*, "Estimates of global seasonal influenza-associated respiratory mortality: A modelling study," *Lancet*, vol. 391, no. 10127, pp. 1285–1300, 2018.

[4] J. K. Taubenberger and J. C. Kash, "Influenza virus evolution, host adaptation, and pandemic formation," *Cell Host Microbe*, vol. 7, no. 6, pp. 440–451, Jun. 2010.

[5] A. H. Reid and J. K. Taubenberger, "The origin of the 1918 pandemic influenza virus: a continuing enigma," *J. Gen. Virol.*, vol. 84, no. Pt 9, pp. 2285–2292, Sep. 2003.

[6] F. S. Dawood *et al.*, "Estimated global mortality associated with the first 12 months of 2009 pandemic influenza A H1N1 virus circulation: a modelling study," *Lancet Infect. Dis.*, vol. 12, no. 9, pp. 687–695, 2012.

[7] A. J. Hay, V. Gregory, A. R. Douglas, and Y. P. Lin, "The evolution of human influenza viruses," *Philos. Trans. R Soc. Lond B Biol. Sci.*, vol. 356, no. 1416, pp. 1861–1870, Dec. 2001.

[8] D. P. Nayak, R. A. Balogun, H. Yamada, Z. H. Zhou, and S. Barman, "Influenza virus morphogenesis and budding," *Virus Res.*, vol. 143, no. 2, pp. 147–161, Aug. 2009.

[9] M. C. Zambon, "Epidemiology and pathogenesis of influenza," *J. Antimicrob. Chemother.*, vol. 44 Suppl B, pp. 3–9, Nov. 1999.

[10] S. Su, X. Fu, G. Li, F. Kerlin, and M. Veit, "Novel Influenza D virus: Epidemiology, pathology, evolution and biological characteristics," *Virulence*, vol. 8, no. 8, pp. 1580–1591, Nov. 2017.

[11] B. E. Johansson and I. C. Brett, "Changing perspective on immunization against influenza," *Vaccine*, vol. 25, no. 16, pp. 3062–3065, Apr. 2007.

[12] M. Tafalla, M. Buijssen, R. Geets, and M. Vonk Noordegraaf-Schouten, "A comprehensive review of the epidemiology and disease burden of Influenza B in 9 European countries," *Hum. Vaccin. Immunother*, vol. 12, no. 4, pp. 993–1002, Apr. 2016.

[13] S. Boivin, S. Cusack, R. W. Ruigrok, and D. J. Hart, "Influenza A virus polymerase: structural insights into replication and host adaptation mechanisms," *J. Biol. Chem.*, vol. 285, no. 37, pp. 28411–28417, Sep. 2010.

[14] World Health Organization. (2021, September 20, 2021). *Global Influenza Surveillance and Response System – GISRS*. Available: https://www.who.int/initiatives/global-influenza-surveillance-and-response-system

[15] C. Gerdil, "The annual production cycle for influenza vaccine," *Vaccine*, vol. 21, no. 16, pp. 1776–1779, 2003.

[16] F. Krammer and P. Palese, "Advances in the development of influenza virus vaccines," *Nat. Rev. Drug. Discov.*, vol. 14, no. 3, pp. 167–182, Mar. 2015.

[17] E. Sparrow *et al.*, "Global production capacity of seasonal and pandemic influenza vaccines in 2019," *Vaccine*, vol. 39, no. 3, pp. 512–520, Jan. 2021.

[18] E. Milian and A. A. Kamen, "Current and emerging cell culture manufacturing technologies for influenza vaccines," *Biomed. Res. Int.*, vol. 2015, p. 504831, 2015.

[19] S. S. Wong and R. J. Webby, "Traditional and new influenza vaccines," *Clin. Microbiol. Rev.*, vol. 26, no. 3, pp. 476–492, Jul. 2013.

[20] S. Rockman and L. Brown, "Pre-pandemic and pandemic influenza vaccines," *Hum. Vaccin.*, vol. 6, no. 10, pp. 792–801, Oct. 2010.

[21] P. A. Scuffham and P. A. West, "Economic evaluation of strategies for the control and management of influenza in Europe," *Vaccine*, vol. 20, no. 19–20, pp. 2562–2578, 2002.

[22] M. Kiso *et al.*, "Resistant influenza A viruses in children treated with oseltamivir: descriptive study," *Lancet*, vol. 364, no. 9436, pp. 759–765, 2004.

[23] O. Kistner, P. N. Barrett, W. Mundt, M. Reiter, S. Schober-Bendixen, and F. Dorner, "Development of a mammalian cell (Vero) derived candidate influenza virus vaccine," *Vaccine*, vol. 16, no. 9–10, pp. 960–968, May–Jun. 1998.

[24] O. G. Engelhardt, "Many ways to make an influenza virus--review of influenza virus reverse genetics methods," *Influenza Other Respir. Viruses*, vol. 7, no. 3, pp. 249–256, May 2013.

[25] J. Medina *et al.*, "Vero/CHOK1, a novel mixture of cell lines that is optimal for the rescue of influenza A vaccine seeds," *J. Virol. Methods*, vol. 196, pp. 25–31, Feb. 2014.

[26] E. Hoffmann, "Eight-plasmid system for rapid generation of influenza virus vaccines," *Vaccine*, vol. 20, no. 25–26, pp. 3165–3170, 2002.

[27] E. Milian *et al.*, "Accelerated mass production of influenza virus seed stocks in HEK-293 suspension cell cultures by reverse genetics," *Vaccine*, vol. 35, no. 26, pp. 3423–3430, Jun. 2017.

[28] F. Krammer, "The human antibody response to influenza A virus infection and vaccination," *Nat. Rev. Immunol.*, vol. 19, no. 6, pp. 383–397, Jun. 2019.

[29] A. Sabbaghi, S. M. Miri, M. Keshavarz, M. Zargar, and A. Ghaemi, "Inactivation methods for whole influenza vaccine production," *Rev. Med. Virol.*, vol. 29, no. 6, p. e2074, Nov. 2019.

[30] M. F. Clincke, C. Molleryd, Y. Zhang, E. Lindskog, K. Walsh, and V. Chotteau, "Very high density of CHO cells in perfusion by ATF or TFF in WAVE bioreactor. Part I. Effect of the cell density on the process," *Biotechnol. Prog.*, vol. 29, no. 3, pp. 754–767, May–Jun. 2013.

[31] B. Kelley, "Industrialization of mAb production technology: the bioprocessing industry at a crossroads," *MAbs*, vol. 1, no. 5, pp. 443–452, Sep–Oct. 2009.

[32] C. A. T. Silva, A. A. Kamen, and O. Henry, "Recent advances and current challenges in process intensification of cell culture-based influenza virus vaccine manufacturing," *Can. J. Chem. Eng.*, vol. 99, no. 11, pp. 2525–2535, 2021.

[33] J. Coronel, G. Granicher, V. Sandig, T. Noll, Y. Genzel, and U. Reichl, "Application of an Inclined Settler for Cell Culture-Based Influenza A Virus Production in Perfusion Mode," *Front. Bioeng. Biotechnol.*, vol. 8, p. 672, 2020.

[34] D. Vazquez-Ramirez, I. Jordan, V. Sandig, Y. Genzel, and U. Reichl, "High titer MVA and influenza A virus production using a hybrid fed-batch/perfusion strategy with an ATF system," *Appl. Microbiol. Biotechnol.*, vol. 103, no. 7, pp. 3025–3035, Apr. 2019.

[35] F. Tapia *et al.*, "Production of high-titer human influenza A virus with adherent and suspension MDCK cells cultured in a single-use hollow fiber bioreactor," *Vaccine*, vol. 32, no. 8, pp. 1003–1011, Feb. 2014.

[36] E. Petiot and A. Kamen, "Real-time monitoring of influenza virus production kinetics in HEK293 cell cultures," *Biotechnol. Prog.*, vol. 29, no. 1, pp. 275–284, Jan–Feb. 2013.

[37] E. Petiot, D. Jacob, S. Lanthier, V. Lohr, S. Ansorge, and A. A. Kamen, "Metabolic and kinetic analyses of influenza production in perfusion HEK293 cell culture," *BMC Biotechnol.*, vol. 11, p. 84, Sep. 2011.

[38] F. Feidl *et al.*, "Process-wide control and automation of an integrated continuous manufacturing platform for antibodies," *Biotechnol. Bioeng.*, vol. 117, no. 5, pp. 1367–1380, May 2020.

[39] C. T. Hardy, S. A. Young, R. G. Webster, C. W. Naeve, and R. J. Owens, "Egg fluids and cells of the chorioallantoic membrane of embryonated chicken eggs can select different variants of influenza A (H3N2) viruses," *Virology*, vol. 211, no. 1, pp. 302–306, Aug. 1995.

[40] S. J. Zost *et al.*, "Contemporary H3N2 influenza viruses have a glycosylation site that alters binding of antibodies elicited by egg-adapted vaccine strains," *Proc. Natl. Acad. Sci. U S A*, vol. 114, no. 47, pp. 12578–12583, Nov. 2017.

[41] P. D. Minor *et al.*, "Current challenges in implementing cell-derived influenza vaccines: implications for production and regulation, July 2007, NIBSC, Potters Bar, UK," *Vaccine*, vol. 27, no. 22, pp. 2907–2913, May 14 2009.

[42] Y. Tao, J. Shih, M. Sinacore, T. Ryll, and H. Yusuf-Makagiansar, "Development and implementation of a perfusion-based high cell density cell banking process," *Biotechnol. Prog.*, vol. 27, no. 3, pp. 824–829, May–Jun. 2011.

[43] P. R. Dormitzer, T. F. Tsai, and G. Del Giudice, "New technologies for influenza vaccines," *Hum. Vaccin. Immunother.*, vol. 8, no. 1, pp. 45–58, Jan. 2012.

[44] J. Liu, X. Shi, R. Schwartz, and G. Kemble, "Use of MDCK cells for production of live attenuated influenza vaccine," *Vaccine*, vol. 27, no. 46, pp. 6460–6463, Oct. 2009.

[45] Y. Z. Ghendon *et al.*, "Development of cell culture (MDCK) live cold-adapted (CA) attenuated influenza vaccine," *Vaccine*, vol. 23, no. 38, pp. 4678–4684, Sep. 2005.

[46] M. George, M. Farooq, T. Dang, B. Cortes, J. Liu, and L. Maranga, "Production of cell culture (MDCK) derived live attenuated influenza vaccine (LAIV) in a fully disposable platform process," *Biotechnol. Bioeng.*, vol. 106, no. 6, pp. 906–917, Aug. 2010.

[47] J. Romanova *et al.*, "Live cold-adapted influenza A vaccine produced in Vero cell line," *Virus Res.*, vol. 103, no. 1–2, pp. 187–193, Jul. 2004.

[48] M. M. Cox, "Recombinant protein vaccines produced in insect cells," *Vaccine*, vol. 30, no. 10, pp. 1759–1766, Feb. 2012.

[49] R. Baxter, P. A. Patriarca, K. Ensor, R. Izikson, K. L. Goldenthal, and M. M. Cox, "Evaluation of the safety, reactogenicity and immunogenicity of FluBlok(R) trivalent recombinant baculovirus-expressed hemagglutinin influenza vaccine administered intramuscularly to healthy adults 50–64 years of age," *Vaccine*, vol. 29, no. 12, pp. 2272–2278, Mar. 2011.

[50] J. C. King, Jr., M. M. Cox, K. Reisinger, J. Hedrick, I. Graham, and P. Patriarca, "Evaluation of the safety, reactogenicity and immunogenicity of FluBlok trivalent recombinant baculovirus-expressed hemagglutinin influenza vaccine administered intramuscularly to healthy children aged 6-59 months," *Vaccine*, vol. 27, no. 47, pp. 6589–6594, Nov. 2009.

[51] S. Rockman, K. L. Laurie, S. Parkes, A. Wheatley, and I. G. Barr, "New Technologies for Influenza Vaccines," *Microorganisms*, vol. 8, no. 11, Nov. 2020.

[52] US National Library of Medicine. (September 20, 2021). *ClinicalTrials.gov*. Available: https://clinicaltrials.gov/

[53] A. P. Manceur and A. A. Kamen, "Critical review of current and emerging quantification methods for the development of influenza vaccine candidates," *Vaccine*, vol. 33, no. 44, pp. 5913–5919, Nov. 2015.

[54] P. C. Stepp, K. A. Ranno, E. D. Dawson, K. L. Rowlen, and M. M. Ferris, "Comparing H1N1 virus quantification with a unique flow cytometer and quantitative PCR," *Bioprocess I*, vol. 9, pp. 50–56, 2011.

[55] B. Lorbetskie *et al.*, "Optimization and qualification of a quantitative reversed-phase HPLC method for hemagglutinin in influenza preparations and its comparative evaluation with biochemical assays," *Vaccine*, vol. 29, no. 18, pp. 3377–3389, Apr. 2011.
[56] J. Transfiguracion, A. P. Manceur, E. Petiot, C. M. Thompson, and A. A. Kamen, "Particle quantification of influenza viruses by high performance liquid chromatography," *Vaccine*, vol. 33, no. 1, pp. 78–84, Jan. 2015.
[57] B. Kalbfuss, A. Knochlein, T. Krober, and U. Reichl, "Monitoring influenza virus content in vaccine production: precise assays for the quantitation of hemagglutination and neuraminidase activity," *Biologicals*, vol. 36, no. 3, pp. 145–161, May 2008.
[58] D. D. LaBarre and R. J. Lowy, "Improvements in methods for calculating virus titer estimates from TCID50 and plaque assays," *J. Virol. Methods*, vol. 96, no. 2, pp. 107–126, 2001.
[59] S. L. Schwartz and A. C. Lowen, "Droplet digital PCR: A novel method for detection of influenza virus defective interfering particles," *J. Virol. Methods*, vol. 237, pp. 159–165, Nov. 2016.
[60] A. J. Veach, C. Beard, F. Porter, M. Wilson, and F. B. Scorza, "Digital droplet PCR for influenza vaccine development," *Procedia Vaccinol.*, vol. 9, pp. 96–103, 2015.
[61] C. F. Mandenius *et al.*, "Monitoring of influenza virus hemagglutinin in process samples using weak affinity ligands and surface plasmon resonance," *Anal. Chim. Acta*, vol. 623, no. 1, pp. 66–75, Aug. 2008.
[62] C. E. Nilsson, S. Abbas, M. Bennemo, A. Larsson, M. D. Hamalainen, and A. Frostell-Karlsson, "A novel assay for influenza virus quantification using surface plasmon resonance," *Vaccine*, vol. 28, no. 3, pp. 759–766, Jan. 2010.
[63] S. Khurana, L. R. King, J. Manischewitz, E. M. Coyle, and H. Golding, "Novel antibody-independent receptor-binding SPR-based assay for rapid measurement of influenza vaccine potency," *Vaccine*, vol. 32, no. 19, pp. 2188–2197, Apr. 2014.
[64] L. Durous *et al.*, "SPRi-based hemagglutinin quantitative assay for influenza vaccine production monitoring," *Vaccine*, vol. 37, no. 12, pp. 1614–1621, Mar. 2019.
[65] S. B. Carvalho, C. Peixoto, M. J. T. Carrondo, and R. J. S. Silva, "Downstream processing for influenza vaccines and candidates: An update," *Biotechnol Bioeng*, vol. 118, no. 8, pp. 2845–2869, Aug. 2021.
[66] I. Manini *et al.*, "Flucelvax (Optaflu) for seasonal influenza," *Expert Rev. Vacc.*, vol. 14, no. 6, pp. 789–804, Jun. 2015.
[67] European Medicines Agency, "Flucelvax Tetra: EPAR – Public assessment report," 2018, Available: https://www.ema.europa.eu/en/medicines/human/EPAR/flucelvax-tetra
[68] K. Wang, K. M. Holtz, K. Anderson, R. Chubet, W. Mahmoud, and M. M. Cox, "Expression and purification of an influenza hemagglutinin--one step closer to a recombinant protein-based influenza vaccine," *Vaccine*, vol. 24, no. 12, pp. 2176–2185, Mar. 2006.

10 Recombinant vaccines: Gag-based VLPs

Laura Cervera, Irene González-Domínguez, Jesús Lavado-García, and Francesc Gòdia
Grup d'Enginyeria Cel·lular i Bioprocés, Universitat Autònoma de Barcelona, Bellaterra, Barcelona, Spain

CONTENTS

10.1 VIRUS-LIKE PARTICLES

Virus-like particles (VLPs) resemble the viral native conformation by the recombinant expression of their structural proteins. Their highly organized and repetitive structure has proven that they generate a potent immunogenic response activating both cellular and humoral immunogenicity responses [1,2]. VLPs can be classified in non-enveloped and enveloped structures (Figure 10.1). Within this general classification, there is a large diversity of VLP configurations: from the simplest non-enveloped single-protein structure, such as in the Hepatitis B, to multilayered protein configurations [3]. Enveloped VLPs consist of one or more structural protein surrounded by the producer cell membrane that is acquired when the VLP buds from the cell. If there are antigens expressed at the producer cell membrane, when it buds, they are incorporated at the surface of the VLP. The ease of production, safety, and their efficient recognition and cellular uptake, has expanded the interest on the possible applications of these structures in the last

DOI: 10.1201/9781003229797-10

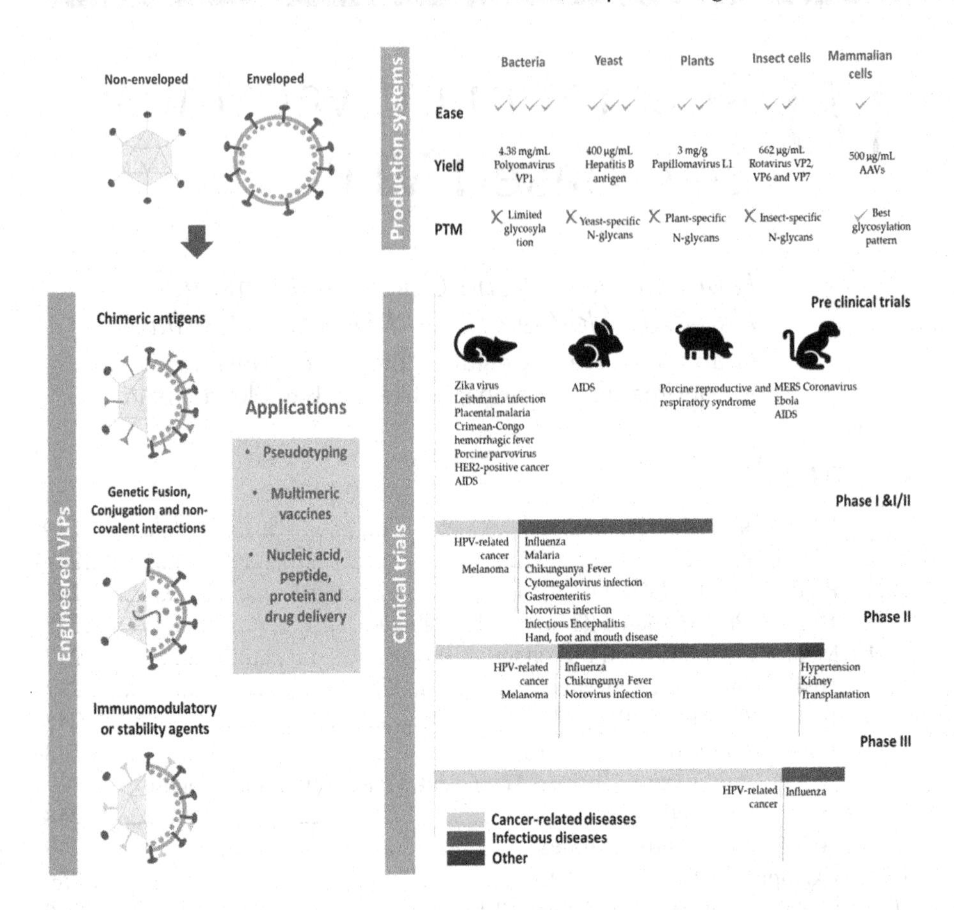

FIGURE 10.1 Virus-like particles state-of-the-art. VLPs can be classified in non-enveloped and enveloped structures depending on the nature of the wild-type virus. For both types single or multilayered protein can be found (adapted from [3]). Engineered VLPs: synthetic biology tools are applied to wide VLP applications. By the addition of chimeric antigens, multimeric vaccines or pseudotyping of different VLP scaffolds are proposed. Furthermore, surface modification and cell-specific targeting molecules are also described in the literature for the controlled release of nucleic acids or drugs, increased immune response, or improved stability of the VLP candidates (adapted from [3,4,6]). Production systems: bacteria, yeast, plants, baculovirus-infected insect cells and mammalian cells are used for the production of different VLP candidates (adapted from [3,13,15,16]). Clinical trials: VLPs are currently tested against several diseases in pre-clinical studies in several animal models (mouse, rabbit, pig, or rhesus macaque). Currently there are more than 125 clinical trials (on-going and completed) based on VLPs mainly targeting cancer and infectious diseases; data obtained from clinicaltrials.gov (Accessed January 2022) and [5,17]. AAV: adeno-associated virus; AIDS: acquired immunodeficiency syndrome; HPV: human papilloma virus; MERS: Middle East respiratory syndrome.

decade. VLPs can be further improved by encapsulation, chemical conjugation, and genetic manipulation (Figure 10.1). Bioengineering has been applied to strengthen their stability and immunostimulatory properties and to generate novel engineered VLPs, as well as vectors for DNA and drug delivery strategies [4–8].

Depending on the complexity of the final structure, there are several production platforms available (Figure 10.1). The simpler VLP types could be produced in prokaryotes and assembled in cell-free environments [9,10]. The bacteria *Escherichia coli* or the yeast *Pichia pastoris* have been described as the most productive platforms with bulk concentrations up to 4.38 mg/mL [11] and 400 mg/mL [12], respectively, although post-translational modification (PTMs) may limit their application to the production of complex VLPs [3,7,13]. Transgenic plants such as potato or tobacco [3,7,14] and baculovirus-vector expression system (BEVS) with High Five or Sf9 insect cell lines are also used to produce VLPs. Despite their different PTM characteristics [14,15], several products have been licensed and numerous phase III clinical trials are ongoing with VLPs produced in these systems [5,10]. Finally, the mammalian CHO and HEK 293 cell lines, which present better glycosylation patterns, are preferred for the expression of highly complex VLP candidates.

Up to date, more than 100 ongoing or completed clinical trials have evaluated VLPs as vaccine candidates [5]. From these studies, two main applications can be highlighted: infectious viral diseases and cancer [4], influenza and HPV-cancer-related vaccine candidates being the ones in more advanced phases. Importantly, there are already two licenced VLP vaccines against HPV and widely distributed (Gardasil and Cervarix). Of note, many diseases such as malaria or Chikungunya fever are being tested in Phase I, whereas new VLP vaccine candidates against Ebola, Zika, MERS (Middle East Respiratory Syndrome), coronavirus, or AIDS (acquired immune deficiency syndrome) have been investigated in pre-clinical trials, among others [5]. Also, during the COVID-19 pandemic, several candidates tested for vaccination were based on VLPs.

10.2 HIV-1 GAG VLPs

HIV-1 contains two copies of single-stranded RNA (ssRNA) genome composed of nine open reading frames (ORF), with three main structural genes: *gag, pol,* and *env*, two regulatory proteins (*tat* and *rev*) and four accessory genes (*nef, vif, vpu,* and *vpr*) (Figure 10.2). From the three main genes, *gag* codes for the structural polyprotein, encompassing the nucleocapsid (NC), capsid (CA), and matrix (MA), whereas *pol* and *env* encode for the main enzymes (PR, RT, and IN) and the receptor binding proteins (gp120 and gp41), respectively (Figure 10.2). At the end of 20th century, Göttlinger described the capacity of the Gag polyprotein to generate non-infectious viral particles on its own [18]. Once produced in the cytoplasm of a host cell, myristoylation of the N-terminal is produced, increasing its affinity for the cell membrane and with the aid of the ESCRT (endosomal sorting complexes required for transport) machinery, immature particles are secreted to the extracellular space taking part of the cell membrane through a budding process [19]. The first application of VLPs consisted of the generation of HIV-1 vaccine candidates by expressing the native Env glycoprotein on its surface. However, this strategy was shown to be very inefficient and the successive approaches for HIV vaccine development moved toward using recombinant-truncated and conserved Env fragments [17–20]. Although increased knowledge has been generated on HIV-1 VLPs,

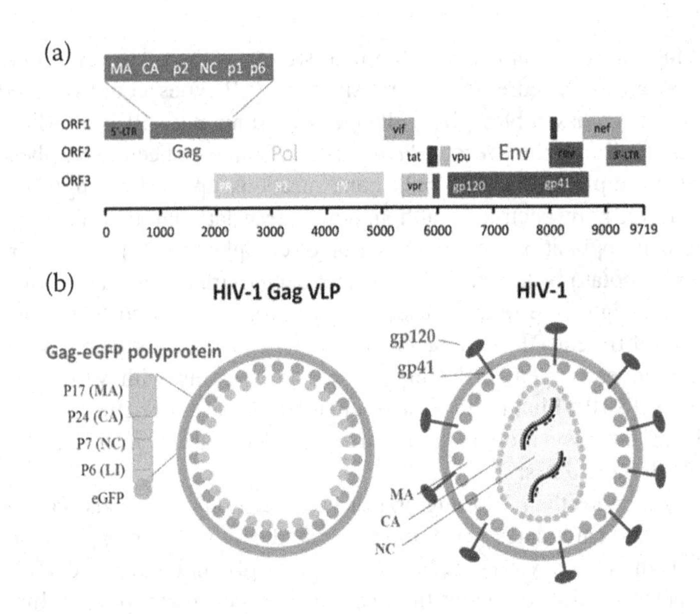

FIGURE 10.2 Structure of HIV-1 Gag VLP and mature HIV-1. (a) Schematic representation of HIV-1 genome with its 9 coding genes. The genome is composed of three structural genes (*gag, pol* and *env*), four accessory genes in grey, and two regulatory genes located throughout the 3 ORFs. Gag is further processed into six protein domains known as matrix (MA or p17), capsid (CA or p24), spacer peptide 1 (SP1 or p2), nucleocapsid (NC or p7), spacer peptide 2 (SP2 or p1), and p6 (Adapted from "Landmarks of the HIV genome"). www.hiv.lanl.gov/content/sequence/HIV/MAP/landmark.html (b) Morphological structure of Gag VLP and wildtype HIV-1. Wild-type HIV-1 is composed of the cleavage products of three major viral polyproteins: Gag, Pol and Env. The sole expression of Gag polyprotein gives rise to the generation of Gag virus-like particles, which are basically immature HIV-1 particles carrying uncleaved Gag capsids surrounded by a host cell lipid layer (adapted from [17] with editorial permission).

no vaccine candidate has reached clinical trials to date. Advances in prime boost immunization regimens, mucosal immunizations or the addition of novel adjuvants in the formulation have resulted in positive outcomes in preclinical studies in mice, rabbit, or rhesus macaque models, and great promises for the development of future vaccine candidates (Table 10.1) [17,21,22].

Interest in Gag VLPs has not been limited to HIV-1 vaccines; the development of chimeric VLPs against different diseases or for delivery strategies has been also described in the literature. In Table 10.1, different applications entailing the use of Gag VLPs are listed. Several authors have exploited the Gag polyprotein for antigen presentation of influenza, Dengue, West Nile Virus, HPV, equine herpes virus and pseudorabies immunogens (Table 10.1). Furthermore, genetic modifications into the Gag sequence have been explored for the delivery of nucleic acids, enzymes, or drugs. Voráckova and co-workers used M-PMV (Mason-Pfizer monkey virus) Gag VLPs, as nanocages, producing the recombinant Gag subunits in *E. coli* and performing their assembly *ex vivo* loading small interference RNA (siRNA) inside the

TABLE 10.1
Summary of different applications of Gag VLPs

Application	Target	Gag sequence	Antigen/cargo	Strategy	Cell line	Animal Model	Reference
HIV-1 Vaccines		HIV-1 pr55, SIVmac239, dGag	Env Variants gp120, gp140, gp41, gp145	BEVS and TGE	Sf9, HEK 293F, HEK 293 T, COS, S2 cells	BALB/c mice, C57BL/6 J, rabbit, guinea pig, Rhesus macaque, Chacma baboon	Cervera, et.al. 2019 (review) [17]
Pseudotyped Vaccines	Influenza	HIV-1 pr55	HA and NA (H1N1)	TGE +SGE	HEK 293 SF cells	BALB/c mice	Venereo-Sanchez, et. al 2016 [26]
	FMDV	HIV-1 pr55	RVG carring G-H loop	TGE	HEK 293 SF cells	BALB/c mice	Fontana, et al. 2021 [27]
	Influenza	HIV-1	HA and NA (H1N1 & H5N1)	TGE	HEK 293T cells	BALB/c mice	Giles, et al. 2011[28] and Carter, et al. 2016 [29]
	Dengue West Nile Virus	HIV-1 pr55	DIII-DENV^{1}_RigE DIII-WNV^{Kun}–RigE	BEVS	Sf9 cells	BALB/c mice	Chua, et al. 2013 [30]
	HPV	HIV-1	HPV-16 E7 and VSV-G	TGE+ Packaging cell	gag-pol 293 GPR packaging cells	C57BL/6 mice	Di Bonito, et al. 2009 [31]
	Equine Herpes Virus	HIV-1 b-p55	Truncated gp14	BEVS	High Five	BALB/c mice	Osterrieder, et al. 1995 [32]
	Pseudorabies	HIV-1 pr55	gD	BEVS	Sf9 cells		Garnier, et al. 1995 [33]

(Continued)

TABLE 10.1 (Continued)
Summary of different applications of Gag VLPs

Application	Target	Gag sequence	Antigen/cargo	Strategy	Cell line	Animal Model	Reference
Delivery Strategies	Nucleic Acid delivery	M-PMV ΔProCa-NC	AKR-targeted siRNA	Transformation	*e.Coli* BL21 (DE3)		Vorácková, et al. 2014 [23]
	Protein Delivery	RSV Pr76gag	Gag-cre recombinase, Gag-Fcy::Fur, and Gag-human caspase-8, VSV-G, truncated NA and HA, NA-IFN-γ	TGE	HEK 293T cells		Kaczmarczyk, et al. 2011 [24]
	Protein Delivery	HIV-1 Pr55	Vpr1 and Vpx-2 in fusion with SN, SN* and CAT	TGE	HeLa cells		Wu, et al. 1995 [34]
	Drug Delivery	RSV gag-577	calcein-AM and doxorubicin	Bacmid injection	Silkworm	BALB/c mice	Deo, et al. 2015 [35]
Therapeutic Vaccines	Pancreatic Cancer	SIVmac239 gag	Trop2	BEVS	Sf9	C57BL/6 mice	Cubas, et al. 2011 [36]
		SIVmac239 gag	murine mesothelin	BEVS	Sf9	C57BL/6 mice	Zhang, et al. 2013 [37]

AKR: aldoketoreductase; BEVS: baculovirus vector expression system; CAT: chloramphenicol acetyltransferase; Gag-Fcy::Fur: Gag-Cytosine Deaminase and Uracil Phosphory-bosyltransferase; HPV: human papilloma virus; M-PMV: Mason-Pfizer monkey virus; RSV: Rous Sarcoma Virus; SGE: stable gene expression; SIV: Simian immunodeficiency virus; SN: staphylococcal nuclease; SN*: inactive mutant SN; TGE: transient gene expression.

nanoparticles [23]. Kaczmarczyk et al. generated fusion Gag proteins with several prodrugs and enzymes and demonstrated its directed delivery *in vitro* [24]. The immunogenicity of VLPs has been also studied in cancer research [25].

10.3 PRODUCTION OF HIV-1 GAG VLPs

HIV-1 Gag VLP production has been achieved in several cellular platforms [17], but most of the research is performed with animal cell cultures, as shown in Table 10.1. Three main strategies can be distinguished in the production of viral-based products in animal cell cultures: viral infection, transient gene expression (TGE), and stable gene expression (SGE). From the three mentioned strategies, infection with alphavirus, vaccinia virus, adenoviral vectors, or baculovirus has been used for the rapid production of bioproducts [38]. Among them, the BEVS is the most used system in the production of Gag VLPs with insect cells (Table 10.1) [39]. BEVS has been widely used with the insect cell lines, Sf9 [40] and High Five [41], and the new TNMS-42 [42]. BEVS is a very productive system achieving productions of milligrams per liter [13]. However, the obtention of the recombinant viral stock is not always straightforward and the lytic cycle caused by the infection may affect product quality. Furthermore, its main disadvantage is the baculovirus interference in the purification process, due to their similarities with enveloped viral structures, as with Gag VLPs.

TGE on the other hand, uses non-viral vectors to introduce a DNA plasmid coding for the protein of interest. In this regard, mammalian cell lines, especially HEK 293, is the workhorse in TGE strategies for viral-based products [39]. Still, shorter production times of 2–4 weeks are required to obtain up to grams of a protein of interest [43]. Several physical and chemical methods could be found in literature to this end, as reviewed in [44]. However, in industrial biotechnology, the use of transfection reagents such as calcium phosphate and in a more extent the cationic polymer, polyethylenimine (PEI), are the most extensively used [43]. Alternative transfection reagents, such as cationic lipids, have been also used for DNA delivery in a highly efficient way. Nonetheless, their high cost has relegated their use to SGE strategies [45]. HEK 293 cell cultures [26,46], but also the mammalian cell lines HeLa [34], CHO [47], or CAP-T [48] have been used for the production of HIV-1 Gag VLPs by means of TGE and SGE. Alternatively, the development of SGE and TGE strategies in insect cell lines devoid of BEVS has been proven in recent years [49–52].

10.3.1 PEI-Mediated Transient Transfection

The use of PEI for gene delivery *in vitro* and *in vivo* has been on the spotlight for more than three decades [43,53]. PEI is simple to use, efficient with suspension cells, compatible with serum free media and cost-effective. Since its first use as transfection reagent described by Boussif et al. in 1995 [54], several polymer lengths and structures, namely linear and branched PEI, have been tested in DNA delivery approaches [55]. Improved PEI formulations are nowadays commercially available like JetPEI, FectoPro, PeiPro, ExGene500, or the clinical-grade PEIPro®-HQ from Polyplus transfection (Illkirch, France) [43] and Transporter 5®, MAXgene®, PEI MAX® from PolySciences.

PEI amine groups interact with the negatively charged phosphate groups of nucleic acids generating positively charged complexes (polyplexes). By doing so, the DNA sequences are condensed and protected from nuclease degradation (Figure 10.6). Once formed, polyplexes are attracted by the negatively charged cell membrane, where cellular uptake mechanisms have been described through endocytic and non-endocytic pathways [53]. Indeed, when a given pathway is inhibited, complexes may enter by alternative ones [56]. Besides, the main uptake mechanism described is the endocytic, where complexes are thought to be trapped into endosomes that lately fuse to lysosomes. At that point, the so-called "proton sponge effect" of PEI might cause an influx of chloride ions and increase the osmotic pressure into the lysosome, which eventually would burst and release the complexes to the cytoplasm. However, this theory is still questioned [55].

DNA/PEI complexes must then reach the nucleus, where the DNA is transcribed (Figure 10.3). Despite this last step is neither fully understood yet, and controversy on a possible passive transport during cell division [57,58], or an active transport through the nuclear pores [59,60] is still unsolved. Besides, a lack of consensus on the intracellular dynamics of the DNA delivery process is still a matter of debate [53,55].

One of the reasons that might contribute to this controversy is the large number of variables that affect PEI-based processes [44,61]. Several chemical factors such as pH, salt concentration, temperature or incubation time have been described to highly

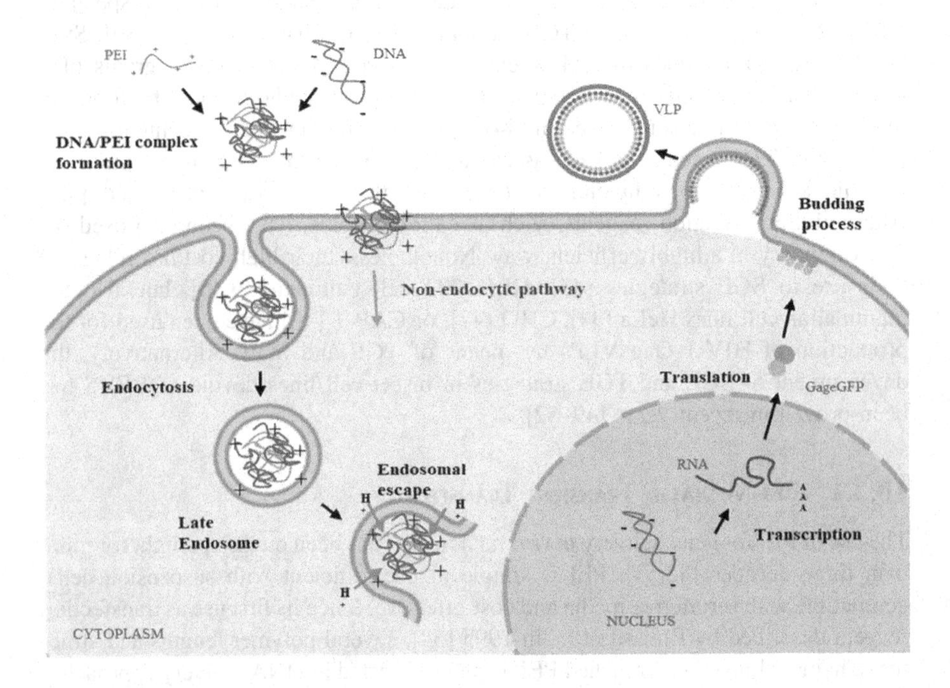

FIGURE 10.3 PEI-mediated TGE for the production of HIV-1 Gag VLPs (figure kindly provided by Dr. Puente-Massaguer, adapted from [74]).

affect the transfection efficiency [59,61–64]. DNA and PEI chain length, PEI charge density, structure and chemical modifications, the nitrogen/phosphate (N/P) ratio, the DNA and PEI concentrations, or the complex preparation and addition process have also a major impact on the TGE efficiency [43,44,55,64,65]. Furthermore, the cell line, cell culture medium composition, cell concentration, time of contact of DNA/PEI complexes with cells, or the addition or replacement of medium in the transfection protocol have also a remarkable influence on the *in vitro* results [43,61,66]. Design of experiments (DoE) have been implemented to optimize some of these multiple variables in a faster and rational way [41,48,67,68]. On top of that, the idea of DNA/PEI complexes as nanoparticles has brought new parameters, such as size, particle concentration, or morphology, that might also play a role in the delivery process [69–73]. Apart from the myriad of variables, a large heterogeneity in the DNA/PEI polyplex population itself has been described [70].

10.4 METHODS TO IMPROVE THE PRODUCTION PROCESS

In the Gag-VLPs production process different elements can be optimized. One efficient way to optimize the relevant aspects in the process can be design of experiments (DoE). By using this methodology, the different variables can be optimized at the same time studying their interactions and the overall contributions to the result. The classical approach where several variables need to be optimized performs a primary screening experiment to determine which variables have a positive effect on the result. These screening can be performed using the Plackett-Burmann design of experiments. After identifying the variables with a positive effect, the next step is to find the optimal levels for each one of them using surface response designs. This can be done using central composite design (CCD) and, when the number of variables to optimize increases, the Box-Behnken approach. The result of these experiments are statistically significant data that are used to find a mathematical model and an optimal combination of the different variables. Finally, the determined optimal conditions must be validated experimentally.

10.4.1 Serum-Free Media

Culture media has a high impact on the maximum cell density reached, duration of the production phase and final product concentration.

The need of serum removal from bioprocess was recognized decades ago. First attempts in serum-free media (SFM) development included components of animal origin mimicking the components supplied by the serum, such as insulin, transferrin and lipids of animal origin as well as other poorly defined mixtures (extracts, hydrolysates). Currently, the overall trend in SFM formulation is to suppress completely animal-derived components to avoid any potential contact with new viruses or prion strains. The increasing number of animal-derived component-free (ADCF) and chemically defined (CD) media formulations available in the market and the relatively recent commercial availability of recombinant versions of key serum proteins produced in *Escherichia coli* or yeast (e.g., albumin and transferrin), as well as supplements of plant origin or synthetic nature and facilitates switching to efficient animal-derived component-free production processes.

Importantly, not all culture media can be used for PEI-mediated transient transfection [46,75–77] because of the presence of anti-clumping agents such as heparin or dextran sulfate polyanions (frequently added to suspension culture media formulations to prevent cells from aggregating) in some formulations.

10.4.2 Cell Lines/Plasmids

A key point in the recombinant protein expression is the stability of the plasmid within the cell. To increase plasmid persistence, a successful strategy developed has been the constitutive expression of the large T antigen of simian virus 40 (SV40) in 293-T, CHO-T or CAP-T cell lines, increasing therefore the replication of vectors containing the SV40 origin of replication [78–82].

Another genetic element engineered for the optimization in mammalian cells, is the addition of EBNA-1 gene of the Epstein –Barr virus in conjunction with its replication origin, oriP, when provided in trans are claimed to boost protein expression plasmids. Furthermore, EBNA-1 appears to act as a transcriptional enhancer in human as well as rodent cells [78,81–84].

10.4.3 Optimization of TGE

Transient transfection needs to be optimized for each cell line and product. Once the culture media is chosen, the variables to optimize are cell concentration at the time of transfection, concentration of plasmids, and transfection reagent. The current optimized protocol for VLP production in batch involves transfecting the culture at 2 million cells/mL with a DNA concentration of 1 μg/mL, a DNA:PEI ration of 1:2 [46].

10.4.4 Additives to Increase Transient Transfection and Protein Production

Several additives have been tested to enhance transient transfection efficiency. Lithium acetate and DMSO are used to increase cell membrane porosity to increase the capacity of the polyplexes to enter the cell [85]. One of them is Nocodazole, that acts in terms of cell cycle arrest in the G2/M phase, which is thought to enhance nuclear uptake of the DNA/PEI complexes when the nuclear membrane dissolves during mitosis [57].

Once the DNA is in the nucleus, another group of additives used to increase production has also proven to be efficient in several cases. Sodium butyrate [86–93], Valproic acid [90,94–97], and Trichostatin A [90,97,98] are used to inhibit histone (HDAC), resulting in hyperacetylation of histones and, consequently, alterations in DNA transcription [86].

Hydroxyurea is used to block cell cycle in the G1 phase of the cell cycle, which also leads to an increase in production [57,99,100].

Caffeine is a well-established inhibitor of several kinases, including ATM (ataxia telangiectasia mutated), ATR (ataxia telangiectasia and Rad3-related protein), and DNA-PKcs (DNA-dependent protein kinase catalytic subunit), which are important signaling proteins involved in the repair of DNA double-stranded breaks [101–103]. This feature is able to increase lentivirus titer in HEK 293 cells [104].

Several compounds were described to enhance Gag-based VLPs. [105]. Two main groups of transfection enhancers were tested. One group was selected on the basis that they can either facilitate the entry of PEI/DNA transfection complexes into the cell or cell nucleus. Another group was selected according to their capacity to increase the levels of gene expression. Among the eight reagents tested (trichostatin A, valproic acid, sodium butyrate, DMSO, lithium acetate, caffeine, hydroxiurea, and nocodazole), an optimal combination of compounds exhibiting the greatest effect on gene expression was identified. The addition of 20 mM lithium acetate, 3.36 mM of valproic acid, and 5.04 mM of caffeine increased production levels by fourfold, while maintaining cell culture viability at 94%.

10.4.5 Cell Culture Modes

The system used for production of VLPs by transient transfection plays a central role to increase the yield of the protein of interest.

One of the main characteristics of transient transfection is that the plasmid is maintained in episomal form in the nucleus and as the cell divides there is a dilution of the plasmid. For this reason, transient transfection processes last a reduced number of days. To overcome this limitation a new process was developed, based on the findings made by an extensive process characterization using techniques such as flow cytometry, confocal microscopy, and fluorometry [106].

HEK293 cells were transfected using PEI as a transfection reagent with half of the DNA labeled with Cy3 before transfection, allowing it to follow the complexes during endocytosis and entrance to the nucleus. Cell membrane was dyed with cell mask (red), the cell nucleus with Hoescht (blue), and the VLPs were observed in green as the GAG gene is fused to the GFP. The number of cells expressing the protein increased during the first 60 min of contact between the cells and polyplexes. No additional improvement in the number of cells expressing Gag-GFP (up to 60%) or VLP production (up to 1×10^{10} VLPs/mL) was observed with additional contact time between the cells and polyplexes (Figure 10.4) [74].

When the producer cells are observed by flow cytometry, it can be observed that at 0hpt all the cells are interacting with a polyplex (Figure 10.4), and after 4hpt it appears a cell population that is both Cy3 positive and expressing the Gag-GFP protein. This population increases until 48hpt, when a new population of cells not transfected appears derived from the dilution of the plasmid over cell division.

When these cells are observed under confocal microscopy, at 24 hours, a homogeneous green signal can be observed in the cytoplasm of the cells. After 48 hours post-transfection, there is an accumulation of the Gag VLPs at the vicinity of the cell membrane that can be observed by the colocalization of the red signal of the cell membrane and the green signal of the Gag-GFP polyprotein, which translates in

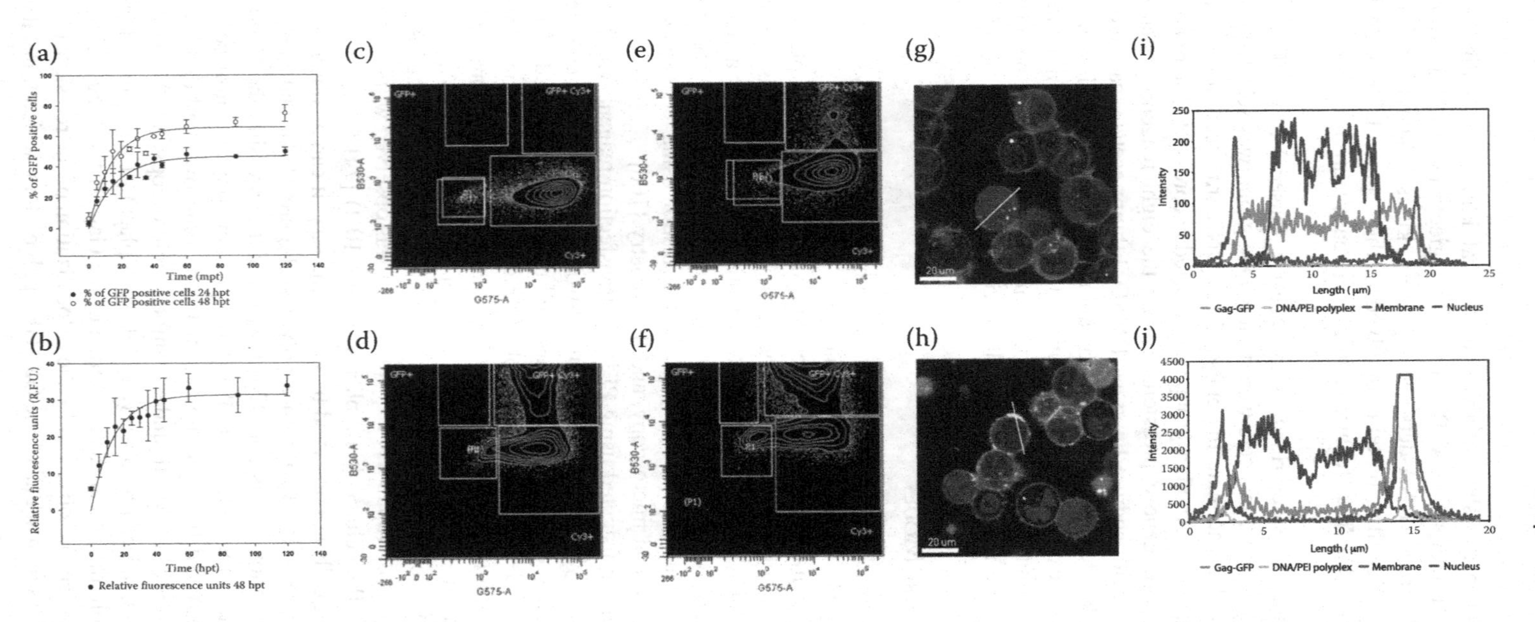

FIGURE 10.4 (a) Percentage of GFP positive cells depending on the contact time allowed between cells and complexes (b) production of VLPs (in relative fluorescence units) attained. Percentage of cells attached to a Cy3 complex, expressing GFP or Cy3 positive and expressing GFP protein at 0hpt (c), 4hpt (e), 10hpt (10hpt), and 48hpt (f). Confocal image showing the cell membrane in red (dyed with cell mask), the nucleus in blue (dyed with Hoescht) and the expression of the Gag-GFP VLPs (observed in green) in the cytoplasm of the cells at 24hpt (g) and colocalizing at the cell membrane (observed in yellow) at 48 hpy (h). The colocalization can also be observed by representing the intensity of the different channels (i, j).

yellow. By representing the intensity of the different channels, it can be observed that the two wavelengths are at the same position (Figure 10.4).

With the knowledge gained with this characterization, a new bioprocess called extended gene expression was designed, in order to extend the production time while maintaining the percentage of transfected cells and therefore increase the production of VLPs (Figure 10.5) [106]. In the extended gene expression (EGE), a medium exchange is performed in combination with a re-transfecting of the culture every 48 hours.

As a control, in the batch standard process in transient gene expression (standard TGE), a medium exchange before transfection was performed prior transfection, and then VLPs were recovered at 72 or 96 hours post-transfection. Additionally, the standard TGE was compared to another protocol called medium exchange (ME) in which after the standard transfection a complete medium exchange was performed every 48 hours, without any retransfection.

Since it is known that the DNA/PEI complexes are toxic to the cells, two other protocols to reduce the amount of DNA/PEI added to the culture were studied. In the 0.5 EGE protocol after the first transfection, retransfections are performed with half of the DNA. In the last protocol (96h EGE), more time between transfections is given; re-transfections are performed only at 96 hours and 192 hours post-transfection.

The results of these experiments show that the viability of the batch culture (standard TGE) starts to drop after 48 hours. For the other experiments, the viability is higher, and the cell growth does not change significantly between the different protocols.

The percentage of GFP-positive cells drops only for the ME protocol, the condition with, only medium exchange. For the other conditions in which additional transfections were performed, the percentage of GFP-positive cells was maintained until the end of the culture.

Regarding the production of VLPs, an increase can be observed in any of the conditions where retransfections are performed. The best and similar results are obtained in the EGE and in the 0.5 EGE protocols in which the production was increased 12-fold compared with the standard batch production.

To further optimize the protocol, each of the re-transfections that in the normal protocol are performed using the Gag-GFP plasmid was replaced with a cherry plasmid to know the effect of a specific re-transfection on the overall production. Five different experiments were performed evaluating the effect of each transfection.

The results show that the most important transfection is the first one. The second and third transfections also have a significant effect on transfection efficiency and production. When the fourth and the fifth transfections are not performed, there is not a decrease in transfection efficiency nor production, implying that these two transfections do not have a significant effect on the performance of the EGE protocol.

These results were validated in an experiment in which the EGE protocol using half of the DNA (0.5 EGE) and the same protocol without the last two transfections (0.5 EGE (−2)) were compared. The differences in both transfection efficiency and production of VLPs were not statistically significant. This allowed defining an optimized protocol in which the first transfection is performed using 1 ug/mL of DNA after a medium exchange, then a medium exchange is performed every 48 hours post-transfection, and the culture is re-transfected with half of the DNA at 48 and 96 hours post-transfection.

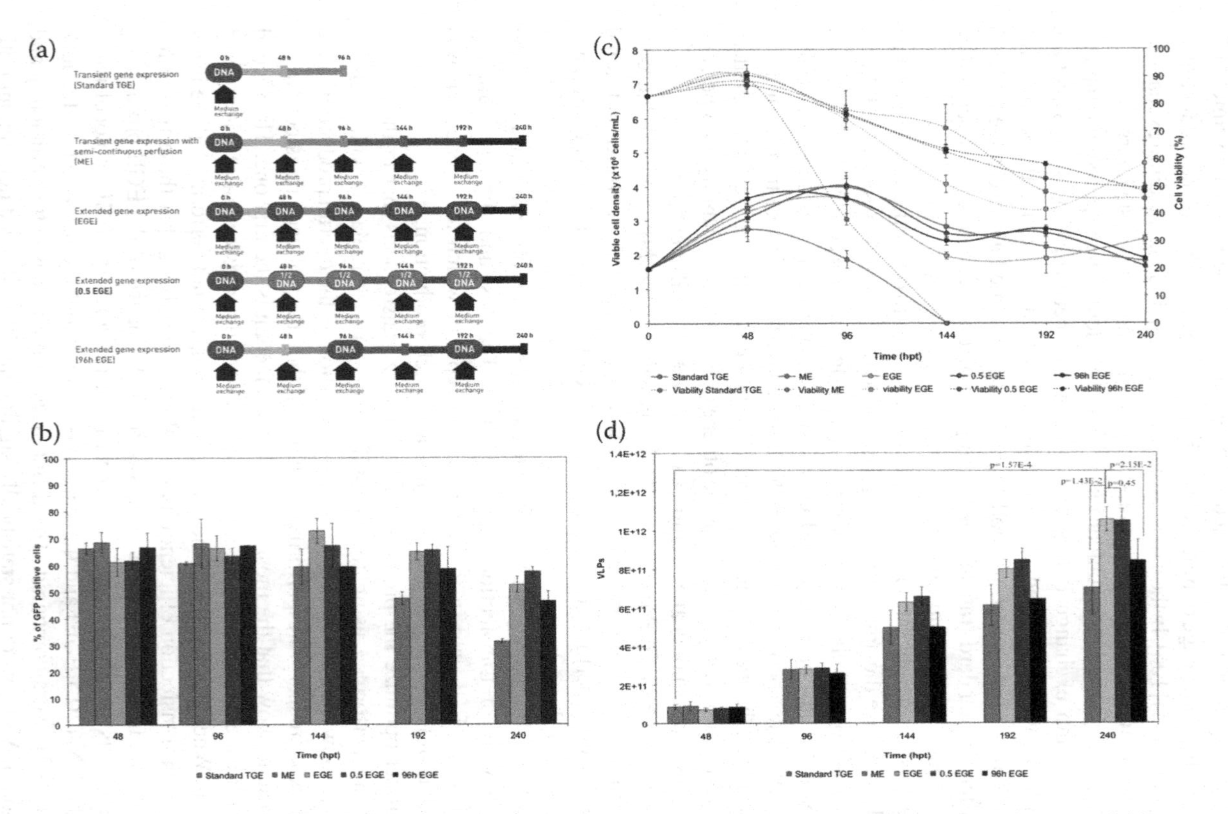

FIGURE 10.5 Schematic description of the different protocols tested (a), growth curves when the different protocols are used (c), percentage of GFP positive cels (b) and VLP production (d).

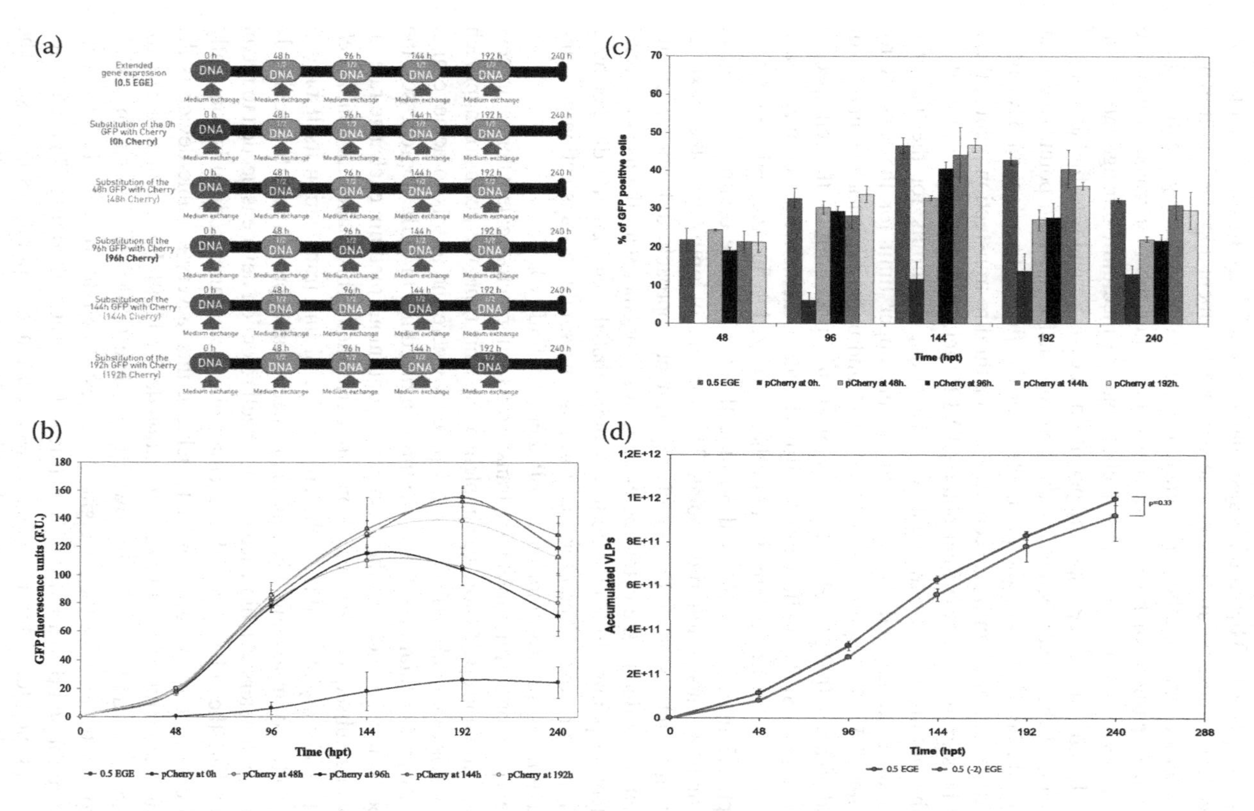

FIGURE 10.6 Schematic description of the different protocols tested (a), percentage of GFP positive cels (c), GFP production (b), and comparison of the production when 0.5 EGE and 0.5 EGE (−2) are used (d).

As previously explained, modifying the number of retransfections, the amount of DNA used in each re-transfection and the combination of different medium exchange time points resulted in variations of VLP titers and overall yield and efficiency of the process. Therefore, a systematic methodology for optimization was used to combine these parameters in a design of experiment approach. Following a Box-Behnken design, a mathematical model could be rendered using the time of re-transfection (measured in hours post-transfection after the first transfection), the amount of DNA used for this retransfection and the medium exchange rate (measured in pL·cell^{-1}·day^{-1}, cell-specific perfusion rate) as variables to find an optimal combination of these parameters for the first re-transfection time point. A similar method could be used, once the parameters for the first re-transfection have been optimized, to further optimize a second or even a third re-transfection point if these prove to be significant to the final obtained VLP titer.

For the optimization of the first re-transfection, the limits for the analyzed variables were set 24 and 72 hpt as lower and upper limit for the time of re-transfection, 0.5 and 2 μg/mL of DNA as lower and upper limit for the amount of DNA and 30 and 1,000 pL·cell^{-1}·day^{-1} as lower and upper limit for the cell-specific perfusion rate (CSPR). Following the Box-Behnken method for optimization, a mathematical model was obtained and further optimized to find the optimal solution of 24 hpt, 1.7 μg/mL of DNA and 30 pL·cell^{-1}·day^{-1} as the optimal parameters for the first re-transfection time point. The method of optimizing the next re-transfection point using a design of experiment approach did not offer any significant result due to the cytotoxic effect of PEI together with the use of several re-transfection points close in time to the point that the cell culture did not have enough time to recover. Therefore, performing one single re-transfection with the given parameters already improved VLP titers 7.5-fold [107].

These operational conditions were transferred to bioreactor scale where a process based on continuous perfusion was implemented. This time, the perfusion approach was not a conventional perfusion-based process where the aim is to achieve high cell density and maintain a steady-state through bleeding. In TGE bioprocesses, the aim of perfusion is to achieve the appropriate continuous medium exchange rate, or CSPR, while subsequent retransfections are performed. This makes the viable cell density to enter a plateau due to the effects of PEI and transfection itself on cell growth [108]. To implement this continuous perfusion approach and the previously optimized parameters at a bioreactor scale, a cell retention device is needed to retain and recirculate the cells back to bioreactor while the culture medium is constantly being circulated through it. In order to achieve this, a very suitable device is the alternating tangential flow (ATF) cell retention device [109]. For the final bioreactor setup, the vessel and the ATF hollow fiber module (HFM) are placed on a scale which constantly monitor their weight. A harvest flow rate from the HFM is controlled using a pump that constantly extracts the filtrated medium. This mass displacement is monitored by the scale, to which a mass setpoint has been set. The scale is connected to another pump triggering the addition of fresh culture medium to balance the mass displacement driven by the harvest flow. Like this, the weight is maintained constant in the bioreactor and the culture medium is renew at a given CSPR.

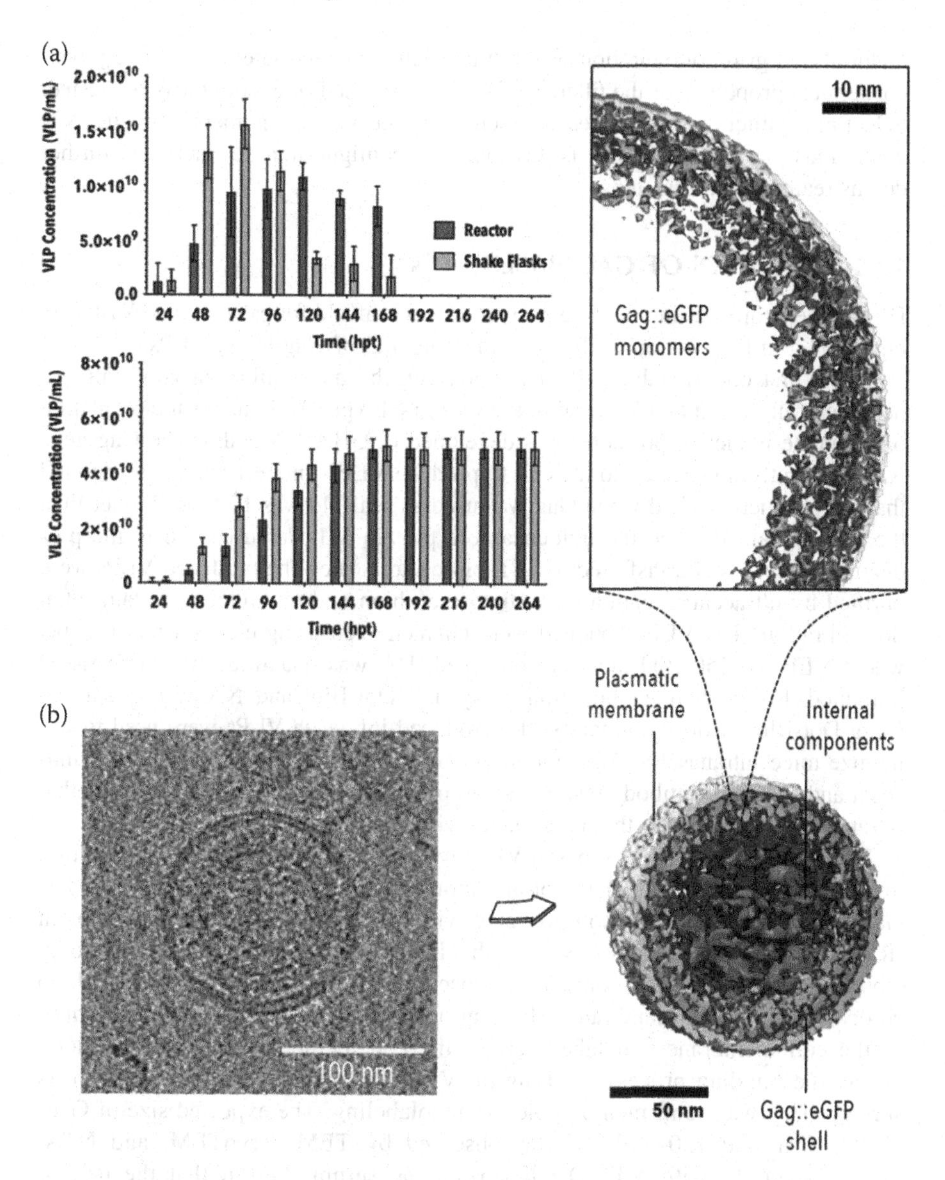

FIGURE 10.7 (a) Daily (top) and cumulative (bottom) production of VLPs using a perfusion-based mode of operation at bioreactor scale. (b) Adapted from [110]. CryoTEM micrograph of a VLP showing the characteristic ordered ring-shaped below the cell membrane. (c) Adapted from [110]. 3D model of a VLP rendered using electron cryotomography. The different segmented parts of the VLP can be observed in detail, such as the individual Gag monomers attached below the cell membrane.

With this bioreactor setup, VLP production can be successfully transferred from shake flasks to a bioreactor scale (Figure 10.7). These VLPs can be either retained in the bioreactor or continuously harvested in the harvest fraction. This depends on the HMF and the nature of the filter used. Depending on the pore size, filter material, the grid configuration of the filter wall, or even the electrical and physicochemical properties of the filter, VLPs can be retained or continuously harvested. Selecting a filter that separates unassembled free Gag monomers and allows a continuous harvest of VLPs is the optimal configuration to facilitate further downstream processing steps.

10.5 EXAMPLES OF GAG-BASED VLPs

Different examples using the Gag polyprotein as scaffold for pseudotyped VLPs are highlighted in this section to illustrate the potential of Gag-based VLPs.

In the first one, a stable cell line expressing the main influenza viral antigens hemagglutinin (HA) and neuraminidase (NA) (subtypeH1N1) under the regulation of a cumate inducible promoter was developed (293HA-NA cells). The Gag gene was transiently transfected to the cell to produce VLPs. Interestingly, it was found that the production yield when Gag was used as scaffold was 10 times higher than the yield obtained when the influenza core protein M1 was used. Then, the production process was transferred to a 3L bioreactor scale. The produced VLPs were purified by ultracentrifugation on a sucrose cushion and ultrafiltered by tangential flow filtration. The VLPs obtained were characterized using electron microscope, where VLPs of 150–200 nm were observed, HA was quantified by single radial immunodiffusion, hemagglutination assay and Dot-Blot and NA was quantified using Dot-Blot. More importantly, the produced influenza VLPs were used to immunize mice intranasally. This immunization induced strong antigen-specific mucosal and systemic antibody responses and provided full protection against a lethal intranasal challenge with the homologous virus strain.

In a second example, Gag-based VLPs were produced by transient transfection to develop a vaccine for the foot-and-mouth disease virus (FMDV) [27]. The Gag-GFP polyprotein was cotransfected with a novel fusion rabies glycoprotein (RVG), which carries in its N-term the FMDV main antigen: the G-H loop. Observing the cells under confocal microscopy it is confirmed that the G-H loop colocalize at the cell membrane. The Gag polyprotein accumulates at the vicinity of the cell membrane and takes part of the cell membrane and the G-H loop during the budding process resulting on VLPs that display the GH-RVG on its surface. This was confirmed by gold immunolabeling. The expected size of Gag-VLPs of around 130–140 nm was observed by TEM, CryoTEM, and NTA. Immunostainings with a FMDV hyperimmune serum showed that the heterologous antigenic site, genetically fused to RVG, is recognized by specific G-H loop antibodies. Additionally, the cVLPs produced expose the G-H loop to the liquid surrounding (analyzed by specific ELISA). Finally, these FMD cVLPs can induce a specific humoral immune response, based on antibodies directed to the G-H loop in experimental animals.

10.6 SCALABLE DSP FOR HIV-1 GAG VLPS

The first step in VLP purification is clarification. In the case of HIV-1 Gag VLPs, they benefit from being an extracellular product; thus, the cell lysis step is not required [17]. More than one unit operation is sometimes used to achieve a desired clarified product. The first clarification aims to remove larger specimens, like cells, while a second one will reduce colloids and other sub-micron particles [111]. Conventional unit operations such as sedimentation or flocculation are still being used in vaccine development. These methods are simple to design and operate, but do not remove cellular debris or sub-micron particles [111,112]. Normal flow filtration (NFF, also known as dead-end filtration) or tangential flow filtration (TFF) are also used in large-scale production processes. Nonetheless, process optimization and low flow rates are sometimes reported due to mechanical stress or titer loss caused by virus adsorption. Depth filters, which contain filter aid, enhance retention of cell debris and contribute to improve NFF results, while the use of hollow fibers or membrane cassettes with high cutoffs, reduces the shear stress of viral particles improving TFF. In addition, membrane-based approaches might benefit from the use of inert materials such as regenerated cellulose (RC), polysulfone (PS), polyethersulfone (PES), and polyvinylidene fluoride (PVDF) [113,114].

One or more concentration steps are typically performed after clarification. Affinity, ion-exchange (IEX), or hydrophobic interaction chromatography (HIC) with different matrices, such as polymer-grafted beads, monolith, membrane adsorbers, or gigapore resins have been successfully applied to VLPs and viral vectors [115–121]. Alternative methods such as two-phase extraction or flocculation have been also described [121,122]. TFF was successfully reported for influenza VLPs, where Carvalho et al. described a total recovery of 76% in a complete membrane-based purification process [123,124]. In addition, a nuclease treatment is often added to reduce dsDNA, especially when anion IEX chromatography is applied [17].

In order to polish VLPs from process-related impurities and prepare them for formulation, diafiltration or size exclusion chromatography (SEC) are mainly used [121]. Polishing steps aim to remove dsDNA, HCP and other contaminant particles that might co-elute with VLPs, such as EVs, adventitious viruses or BEVS. While the quantification and removal of the first two has been widely described, the separation of VLPs from other contaminant particles is highly difficult due to their similar physicochemical properties. Moleirinho et al. recently reported the development of an affinity chromatography method to separate VLPs from BV [125], whereas Reiter and co-workers achieved a reduction on BEVS content by polymer-grafted chromatography [126]. Regarding EVs, no separation method as such has been described. Limitations at analytical level for the specific quantification of VLPs and EVs present in the same sample may strongly contribute to this fact. Concretely, the available analytical tools do not allow the direct differentiation between VLPs and EVs due to their same origin, composition and physicochemical properties. On the other hand, the enrichment of HIV-1 Gag VLPs over total particles has been reported by TFF, IEX chromatography and heparin affinity chromatography [119,124,127].

Removal of adventitious agents must be demonstrated according to the procedures established by regulatory authorities. Common methods to remove or inactivate viruses are UV or gamma-irradiation, sterile filtration, detergent treatments, or high-temperature incubation, which may compromise the candidate integrity or biological activity. Mitigation of such risk should start at the selection of raw material level. The use of animal-component free, chemically defined cell culture media and GMP-compliant cell lines is encouraged to prevent the introduction of adventitious agents [17]. HIV-1 Gag VLPs, as well as lentivirus and other viruses, have a mean particle diameter between 100 and 200 nm. Because of that, when sterile filtration is applied (<0.2 um), a substantial titer reduction might occur [128]. Therefore, this step could be omitted when the whole process is demonstrated to be performed under aseptic conditions (FDA 1997) [17].

10.7 CHARACTERIZATION AND QUANTIFICATION OF VLPs

The quantification and characterization of VLPs is a challenge. As VLPs are not infective viruses, the use of classical techniques like infectivity assays (TCID50) or the quantification of the viral genomes by qPCR is not possible.

Compared to simple protein-based bioproducts, the characterization of VLPs becomes more difficult, since not only protein composition but also their three-dimensional structure should be assessed. Biochemical, biological and biophysical methods have been used in VLP characterization [3]. Biochemical protein gels and biological enzyme-linked immunosorbent assay (ELISA) or immunoblot are normally used to assess product purity and VLP quantification, respectively [26,47,129,130]. However, these assays cannot distinguish assembled from non-assembled structures [131] and may require biophysical methods to study the structural integrity of the obtained VLPs.

Analytical ultracentrifugation, dynamic light scattering (DLS), and transmission electron microscopy (TEM) were primarily used to assess VLP physical properties [3]. Recently, technical progress in the field of microscopy, as well as the application of nanotechnology to virology, have given rise to several single-particle analytical technologies. Nanoparticle tracking analysis (NTA), tunable-resistive pulse sensor (TRPS), flow virometry, cryogenic electron microscopy (cryo-EM), or atomic force microscopy (AFM) represent very advanced methods to evaluate VLP size, polydispersity, purity, and even nanoparticle composition [131].

The specific detection and quantification of VLPs entails several difficulties, especially for enveloped VLPs, which are composed of a protein capsid surrounded by the host-cell lipid membrane. These structures must be distinguished from other similar nanovesicle structures like adventitious viruses, baculoviruses, [132] and extracellular vesicles (EVs), including exosomes and microvesicles [133].

When using transmission electron microscopy, the classical method is to use negative staining sample preparation, in which VLPs are mixed with a very dense salt concentration, normally uranyl acetate. The stain interacts with the particles that are present in the sample, so the presence of Gag polyprotein inside the particles and the difference in density of the lipid membrane can be observed. The drawback of this method is that different unwanted backgrounds can appear depending on the production platform, which can alter the final visualization of the samples. Also, the interaction with the staining reagents can deform the membrane, hence the visualization of the native structure of the particle in the sample is no longer possible.

The use of Cryo-TEM allows the visualization of the samples without any sample pre-treatment, allowing the observation of the native structure of the particles

(Figure 10.7). The VLPs' liquid suspension is rapidly vitrified at −180°C in liquid ethane, and then this frozen preparation is placed in the chamber of the electron microscope, which is maintained at −173°C with liquid nitrogen. Cryo-TEM gives more representative images of the particles as there is no interaction with the stain.

A further step is the use of electron cryotomography. Here, several cryo images are taken at different angles collecting tilt series data sets so they can be later processed to obtain a 3D model of the structure (Figure 10.7). Segmentation of the different parts composing the VLP can be used to render a detailed 3D model defining each of the Gag monomers that form the Gag shell [110].

Regarding quantification, there are some classical methods that have typically been used for quantifying VLPs, such as the counting of particles in TEM with the presence of a standard, ELISA to quantify the amount of an antigen, or HPLC methods. There are also tag-based methods in which a fluorescence-based method is used to facilitate the monitoring of these particles in a simple way [134].

In the field of nanotechnology and nanoscience, different techniques have emerged, such as nanoparticle tracking analysis (NTA) and flow virometry methods based in light scattering, that provide further insights in the quantification of VLPs. They could be further combined with other fluoresce tags, not only characterizing the total amount of nanoparticles that are present in the samples but if the protein conforming the VLP is fused to a tag protein, the populations of VLPs and microvesicles can be differentiated within the samples. Flow virometry has the same concept as flow cytometry but is used to quantify particles that are 1,000-fold smaller than cells achieving the limit of detection of this equipment.

The advances in all these different technologies to characterize VLPs pave the way for further development of this platform, contributing to overcome the current challenges. In the last few years, the capacity to enhance VLP production, purification, and characterization has significantly increased. However, as a complex cellular product, this platform still faces the challenges of understanding the molecular pathways governing the loading of internal components and custom modification of the bounding cellular membrane. These next steps will certainly send forth this promising platform beyond vaccine development towards the use of VLPs for specific delivery strategies and specific protein-receptor interactions in future therapies.

REFERENCES

[1] M. Braun *et al.*, "Virus-like particles induce robust human T-helper cell responses," *Eur. J. Immunol.*, vol. 42, no. 2, pp. 330–340, 2012, doi: 10.1002/eji.201142064

[2] F. Zabel *et al.*, "Viral particles drive rapid differentiation of memory B cells into secondary plasma cells producing increased levels of antibodies," *J. Immunol.*, vol. 192, no. 12, pp. 5499–5508, 2014, doi: 10.4049/jimmunol.1400065

[3] L. H. Lua, N. K. Connors, F. Sainsbury, Y. P. Chuan, N. Wibowo, and A. P. Middelberg, "Bioengineering virus-like particles as vaccines," *Biotechnol. Bioeng.*, vol. 111, no. 3, pp. 425–440, 2014, doi: 10.1002/bit.25159

[4] H. K. Charlton Hume, J. Vidigal, M. J. T. Carrondo, A. P. J. Middelberg, A. Roldão, and L. H. L. Lua, "Synthetic biology for bioengineering virus-like particle vaccines," *Biotechnol. Bioeng.*, vol. 116, no. 4, pp. 919–935, Apr. 2019, doi: 10.1002/bit.26890

[5] B. Donaldson, Z. Lateef, G. F. Walker, S. L. Young, and V. K. Ward, "Virus-like particle vaccines: immunology and formulation for clinical translation," *Expert Rev. Vaccines*, vol. 17, no. 9, pp. 833–849, Sep. 2018, doi: 10.1080/14760584.2018.1516552
[6] B. D. Hill, A. Zak, E. Khera, and F. Wen, "Engineering virus-like particles for antigen and drug delivery," *Curr. Protein Pept. Sci.*, vol. 19, no. 1, pp. 112–127, Nov. 2017, doi: 10.2174/1389203718666161122113041
[7] A. Roldao, M. C. Mellado, L. R. Castilho, M. J. Carrondo, and P. M. Alves, "Virus-like particles in vaccine development," *Expert Rev. Vaccines*, vol. 9, no. 10, pp. 1149–1176, 2010, doi: 10.1586/erv.10.115
[8] D. Yan, Y.-Q. Wei, H.-C. Guo, and S.-Q. Sun, "The application of virus-like particles as vaccines and biological vehicles," *Appl. Microbiol. Biotechnol.*, vol. 99, no. 24, pp. 10415–10432, Dec. 2015.
[9] B. C. Bundy, M. J. Franciszkowicz, and J. R. Swartz, "Escherichia coli-based cell-free synthesis of virus-like particles," *Biotechnol. Bioeng.*, vol. 100, no. 1, pp. 28–37, May 2008, doi: 10.1002/bit.21716
[10] N. Kushnir, S. J. Streatfield, and V. Yusibov, "Virus-like particles as a highly efficient vaccine platform: Diversity of targets and production systems and advances in clinical development," *Vaccine*, vol. 31. pp. 58–83, 2012, doi: 10.1016/j.vaccine.2012.10.083
[11] M. W. O. Liew, A. Rajendran, and A. P. J. Middelberg, "Microbial production of virus-like particle vaccine protein at gram-per-litre levels," *J. Biotechnol.*, vol. 150, no. 2, pp. 224–231, 2010, doi: 10.1016/j.jbiotec.2010.08.010
[12] J. M. Cregg *et al.*, "High–level expression and efficient assembly of Hepatitis B surface antigen in the methylotrophic Yeast, Pichia Pastoris," *Nat. Biotechnol.*, vol. 5, no. 5, pp. 479–485, May 1987, doi: 10.1038/nbt0587-479
[13] J. Fuenmayor, F. Gòdia, and L. Cervera, "Production of virus-like particles for vaccines," *Nat. Biotechnol.*, 2017, doi: 10.1016/j.nbt.2017.07.010
[14] L. Durous, M. Rosa-Calatrava, and E. Petiot, "Advances in influenza virus-like particles bioprocesses," *Expert Rev. Vaccines*, vol. 18, no. 12, pp. 1285–1300, Dec. 2019, doi: 10.1080/14760584.2019.1704262
[15] M. J. Betenbaugh, N. Tomiya, S. Narang, J. T. A. Hsu, and Y. C. Lee, "Biosynthesis of human-type N-glycans in heterologous systems," *Curr. Opin. Struct. Biol.*, vol. 14, no. 5, pp. 601–606, 2004, doi: 10.1016/j.sbi.2004.09.001
[16] J. L. Hye, K. K. Yeon, S. H. Dong, and J. C. Hyung, "Expression of functional human transferrin in stably transfected Drosophila S2 cells," *Biotechnol. Prog.*, vol. 20, no. 4, pp. 1192–1197, Aug. 2004, doi: 10.1021/bp034375a
[17] L. Cervera, F. Gòdia, F. Tarrés-freixas, C. Aguilar-gurrieri, and J. Carrillo, "Production of HIV-1-based virus-like particles for vaccination: achievements and limits,"*Appl. Microbiol. Biotechnol.* 103, pp. 7367–7384, 2019. https://doi.org/10.1007/s00253-019-10038-3
[18] H. G. Göttlinger, *HIV-1 Gag: a Molecular Machine Driving Viral Particle Assembly and Release*. Los Alamos, New Mexico: Los Alamos National Laboratory, Theoretical Biology and Biophysics, 2001.
[19] J. Votteler and W. I. Sundquist, "Virus budding and the ESCRT pathway," *Cell Host Microbe*, vol. 14, no. 3, pp. 232–241, Sep. 2013, doi: 10.1016/j.chom.2013.08.012
[20] L. X. Doan, M. Li, C. Chen, and Q. Yao, "Virus-like particles as HIV-1 vaccines," *Rev. Med. Virol.*, vol. 15. pp. 75–88, 2005, doi: 10.1002/rmv.449
[21] C. Zhao, Z. Ao, and X. Yao, "Current advances in virus-like particles as a vaccination approach against HIV infection," *Vaccines*, vol. 4, no. 1, pp. 2, 2016, doi: 10.3390/vaccines4010002

[22] M. P. Girard and W. C. Koff, "Human Immunodeficiency Virus Vaccines," in *Plotkin's Vaccines Seventh Edition*, 7th Ed., vol. 29, S. A. Plokin, O. Walter A. P. A. Offit, and K. M. Edwards, Eds. Philadelphia: Elsevier, 2018, pp. 400–429.
[23] I. Voráčková, P. Ulbrich, W. E. Diehl, and T. Ruml, "Engineered retroviral virus-like particles for receptor targeting," *Arch. Virol.*, vol. 159, no. 4, pp. 677–688, 2014, doi: 10.1007/s00705-013-1873-6
[24] S. J. Kaczmarczyk, K. Sitaraman, H. A. Young, S. H. Hughes, and D. K. Chatterjee, "Protein delivery using engineered virus-like particles," *Proc. Natl. Acad. Sci.*, vol. 108, no. 41, pp. 16998–17003, Oct. 2011, doi: 10.1073/pnas.1101874108
[25] H. K. Ong, W. S. Tan, and K. L. Ho, "Virus like particles as a platform for cancer vaccine development," *Peer J.*, vol. 2017, no. 11, pp. 1–31, 2017, doi: 10.7717/peerj.4053
[26] A. Venereo-Sanchez *et al.*, "Hemagglutinin and neuraminidase containing virus-like particles produced in HEK-293 suspension culture: An effective influenza vaccine candidate," *Vaccine*, vol. 34, no. 29, pp. 3371–3380, 2016, doi: 10.1016/j.vaccine.2016.04.089
[27] D. Fontana, E. Garay, L. Cervera, R. Kratje, C. Prieto, and F. Gòdia, "Chimeric VLPs based on HIV-1 gag and a fusion rabies glycoprotein induce specific antibodies against Rabies and foot-and-mouth disease Virus," *Vaccines*, vol. 9, no. 3, p. 251, Mar. 2021, doi: 10.3390/VACCINES9030251
[28] B. M. Giles and T. M. Ross, "A computationally optimized broadly reactive antigen (COBRA) based H5N1 VLP vaccine elicits broadly reactive antibodies in mice and ferrets," *Vaccine*, vol. 29, no. 16, pp. 3043–3054, 2011, doi: 10.1016/j.vaccine.2011.01.100
[29] D. M. Carter *et al.*, "Design and Characterization of a Computationally Optimized Broadly Reactive Hemagglutinin Vaccine for H1N1 Influenza Viruses," *J. Virol.*, vol. 90, no. 9, pp. 4720–4734, 2016, doi: 10.1128/JVI.03152-15
[30] A. J. Chua *et al.*, "A novel platform for virus-like particle-display of flaviviral envelope domain III: Induction of Dengue and West Nile virus neutralizing antibodies," *Virol. J.*, vol. 10, no. 1, p. 129, 2013, doi: 10.1186/1743-422X-10-129
[31] P. Di Bonito *et al.*, "Anti-tumor CD8 + T cell immunity elicited by HIV-1-based virus-like particles incorporating HPV-16 E7 protein," *Virology*, vol. 395, no. 1, pp. 45–55, 2009, doi: 10.1016/j.virol.2009.09.012
[32] N. Osterrieder, R. Wagner, C. Brandmüller, P. Schmidt, H. Wolf, and O.-R. Kaaden, "Protection against EHV-1 challenge infection in the murine model after vaccination with various formulations of recombinant glycoprotein gp14 (gB)," *Virology*, vol. 208, no. 2, pp. 500–510, Apr. 1995, doi: 10.1006/viro.1995.1181
[33] L. Garnier *et al.*, "Incorporation of pseudorabies virus gD into human immunodeficiency virus type 1 Gag particles produced in baculovirus-infected cells.," *J. Virol.*, vol. 69, no. 7, pp. 4060–4068, 1995, doi: 10.1128/jvi.69.7.4060-4068.1995
[34] X. Wu *et al.*, "Targeting foreign proteins to human immunodeficiency virus particles via fusion with Vpr and Vpx, "*J Virol.,* vol. 69, no. 6, pp. 3389–3398, 1995.
[35] V. K. Deo, T. Kato, and E. Y. Park, "Chimeric virus-like particles made using GAG and M1 capsid proteins providing dual drug delivery and vaccination platform," *Mol. Pharm.*, vol. 12, no. 3, pp. 839–845, 2015, doi: 10.1021/mp500860x
[36] R. Cubas, S. Zhang, M. Li, C. Chen, and Q. Yao, "Chimeric Trop2 virus-like particles: A potential immunotherapeutic approach against pancreatic cancer," *J. Immunother.*, vol. 34, no. 3, pp. 251–263, Apr. 2011, doi: 10.1097/CJI.0b013e318209ee72
[37] S. Zhang, L. Yong, D. Li, R. Cubas, C. Chen, and Q. Yao, "Mesothelin virus-like particle immunization controls pancreatic cancer growth through CD8 + T cell induction and reduction in the frequency of CD4 + foxp3 + ICOS 2 regulatory T cells," *PLOS ONE,* vol. 8, no. 7, 2013, doi: 10.1371/journal.pone.0068303

[38] V. Jäger, K. Büssow, and T. Schirrmann, "Transient Recombinant Protein Expression in Mammalian Cells," in *Animal Cell Culture*, vol. 9, M. Al-Rubeai, Ed. Basel: Springer, Cham, 2015, pp. 27–64.
[39] Y. Genzel, "Designing cell lines for viral vaccine production: Where do we stand?" *Biotechnol. J.*, vol. 10, no. 5, pp. 728–740, May 2015.
[40] E. Puente-Massaguer, M. Lecina, and F. Gòdia, "Application of advanced quantification techniques in nanoparticle-based vaccine development with the Sf9 cell baculovirus expression system," *Vaccine*, vol. 38, no. 7, pp. 1849–1859, Feb. 2020, doi: 10.1016/j.vaccine.2019.11.087
[41] E. Puente-Massaguer, M. Lecina, and F. Gòdia, "Integrating nanoparticle quantification and statistical design of experiments for efficient HIV-1 virus-like particle production in High Five cells," *Appl. Microbiol. Biotechnol.*, vol. 104, no. 4, pp. 1569–1582, Feb. 2020, doi: 10.1007/s00253-019-10319-x
[42] F. Strobl, S. M. Ghorbanpour, D. Palmberger, and G. Striedner, "Evaluation of screening platforms for virus-like particle production with the baculovirus expression vector system in insect cells," *Sci. Rep.*, vol. 10, no. 1, pp. 1–9, 2020, doi: 10.1038/s41598-020-57761-w
[43] S. Gutiérrez-Granados, L. Cervera, A. A. Kamen, and F. Gòdia, "Advancements in mammalian cell transient gene expression (TGE) technology for accelerated production of biologics," *Crit. Rev. Biotechnol.*, pp. 1–23, Jan. 2018, doi: 10.1080/07388551.2017.1419459
[44] N. Bono, F. Ponti, D. Mantovani, and G. Candiani, "Non-viral in vitro gene delivery: It is now time to set the bar!," *Pharmaceutics*, vol. 12, no. 2, pp. 1–23, Feb. 2020, doi: 10.3390/pharmaceutics12020183
[45] L. R. Castilho, Â. M. Moraes, E. F. P. Augusto, and M. Butler, *Animal Cell Technology: From Biopharmaceuticals to Gene Therapy*. New York: Taylor & Francis, 2008.
[46] L. Cervera, S. Gutiérrez-Granados, M. Martínez, J. Blanco, F. Gòdia, and M. M. Segura, "Generation of HIV-1 Gag VLPs by transient transfection of HEK 293 suspension cell cultures using an optimized animal-derived component free medium," *J. Biotechnol.*, vol. 166, no. 4, pp. 152–165, 2013, doi: 10.1016/j.jbiotec.2013.05.001
[47] P. Steppert *et al.*, "Purification of HIV-1 gag virus-like particles and separation of other extracellular particles," *J. Chromatogr. A*, vol. 1455, pp. 93–101, 2016, doi: 10.1016/j.chroma.2016.05.053
[48] S. Gutiérrez-Granados, L. Cervera, M. M. Segura, J. Wölfel, and F. Gòdia, "Optimized production of HIV-1 virus-like particles by transient transfection in CAP-T cells," *Appl. Microbiol. Biotechnol.*, vol. 100, no. 9, 3935–3947, 2016, doi: 10.1007/s00253-015-7213-x
[49] X. Shen, P. O. Michel, Q. Xie, D. L. Hacker, and F. M. Wurm, "Transient transfection of insect Sf-9 cells in TubeSpin® bioreactor 50 tubes," *BMC Proc.*, vol. 5, no. Suppl 8, p. P37, 2011, doi: 10.1186/1753-6561-5-s8-p37
[50] J. Vidigal *et al.*, "RMCE-based insect cell platform to produce membrane proteins captured on HIV-1 Gag virus-like particles," *Appl. Microbiol. Biotechnol.*, vol. 102, no. 2, pp. 655–666, Jan. 2018.
[51] M. Tagliamonte, M. L. Visciano, M. L. Tornesello, A. De Stradis, F. M. Buonaguro, and L. Buonaguro, "Constitutive expression of HIV-VLPs in stably transfected insect cell line for efficient delivery system," *Vaccine*, vol. 28, no. 39, pp. 6417–6424, Sep. 2010.
[52] E. Puente-Massaguer, M. Lecina, and F. Gòdia, "Nanoscale characterization coupled to multi-parametric optimization of Hi5 cell transient gene expression," *Appl. Microbiol. Biotechnol.*, vol. 102, no. 24, pp. 10495–10510, Dec. 2018, doi: 10.1007/s00253-018-9423-5

[53] B. Shi *et al.*, "Challenges in DNA delivery and recent advances in multifunctional polymeric DNA delivery systems," *Biomacromolecules*, vol. 18, no. 8, pp. 2231–2246, Aug. 2017, doi: 10.1021/acs.biomac.7b00803
[54] O. Boussif *et al.*, "A versatile vector for gene and oligonucleotide transfer into cells in culture and in vivo: Polyethylenimine," *Proc. Natl. Acad. Sci.*, vol. 92, no. 16, pp. 7297–7301, Aug. 1995, doi: 10.1073/pnas.92.16.7297
[55] Y. Yue and C. Wu, "Progress and perspectives in developing polymeric vectors for in vitro gene delivery," *Biomater. Sci.*, vol. 1, no. 2, pp. 152–170, 2013, doi: 10.1039/c2bm00030j
[56] M. E. Hwang, R. K. Keswani, and D. W. Pack, "Dependence of PEI and PAMAM gene delivery on Clathrin- and Caveolin-dependent trafficking pathways," *Pharm. Res.*, vol. 32, no. 6, pp. 2051–2059, Jun. 2015, doi: 10.1007/s11095-014-1598-6
[57] A. S. Tait *et al.*, "Transient production of recombinant proteins by Chinese hamster ovary cells using polyethyleneimine/DNA complexes in combination with microtubule disrupting anti-mitotic agents," *Biotechnol. Bioeng.*, vol. 88, pp. 707–721, 2004, doi: 10.1002/bit.20265
[58] S. Grosse, G. Thévenot, M. Monsigny, and I. Fajac, "Which mechanism for nuclear import of plasmid DNA complexed with polyethylenimine derivatives?" *J. Gene. Med.*, vol. 8, no. 7, pp. 845–851, Jul. 2006, doi: 10.1002/jgm.915
[59] X. Han *et al.*, "The heterogeneous nature of polyethylenimine-DNA complex formation affects transient gene expression," *Cytotechnology*, vol. 60, pp. 63–75, Aug. 2009, doi: 10.1007/s10616-009-9215-y
[60] M. Gillard *et al.*, "Intracellular trafficking pathways for nuclear delivery of plasmid DNA complexed with highly efficient endosome escape polymers," *Biomacromolecules*, vol. 15, no. 10, pp. 3569–3576, 2014, doi: 10.1021/bm5008376
[61] E. V. B. van Gaal *et al.*, "How to screen non-viral gene delivery systems in vitro?" *J. Control. Release*, vol. 154, no. 3, pp. 218–232, Sep. 2011, doi: 10.1016/j.jconrel.2011.05.001
[62] Y. Fukumoto *et al.*, "Cost-effective gene transfection by DNA compaction at pH 4.0 using acidified, long shelf-life polyethylenimine," *Cytotechnology*, vol. 62, no. 1, pp. 73–82, Jan. 2010, doi: 10.1007/s10616-010-9259-z
[63] Y. Sang *et al.*, "Salt ions and related parameters affect PEI-DNA particle size and transfection efficiency in Chinese hamster ovary cells," *Cytotechnology*, vol. 67, no. 1, pp. 67–74, Jan. 2015, doi: 10.1007/s10616-013-9658-z
[64] A. Raup *et al.*, "Compaction and transmembrane delivery of pDNA: Differences between l-PEI and two types of amphiphilic block Copolymers," *Biomacromolecules*, vol. 18, no. 3, pp. 808–818, Mar. 2017, doi: 10.1021/acs.biomac.6b01678
[65] A. K. Blakney, G. Yilmaz, P. F. McKay, C. R. Becer, and R. J. Shattock, "One size does not fit all: The effect of chain length and charge density of poly(ethylene imine) based copolymers on delivery of pDNA, mRNA, and RepRNA polyplexes," *Biomacromolecules*, vol. 19, no. 7, pp. 2870–2879, Jul. 2018, doi: 10.1021/acs.biomac.8b00429
[66] A. V. Ulasov, Y. V. Khramtsov, G. A. Trusov, A. A. Rosenkranz, E. D. Sverdlov, and A. S. Sobolev, "Properties of PEI-based polyplex nanoparticles that correlate with their transfection efficacy," *Mol. Ther.*, vol. 19, no. 1, pp. 103–112, Jan. 2011, doi: 10.1038/mt.2010.233
[67] J. Fuenmayor, L. Cervera, S. Gutierrez-Granados, and F. Godia, "Transient gene expression optimization and expression vector comparison to improve HIV-1 VLP production in HEK293 cell lines," *Appl. Microbiol. Biotechnol.*, vol. 102, no. 1, pp. 165–174, 2018, doi: 10.1007/s00253-017-8605-x

[68] F. Bollin, V. Dechavanne, and L. Chevalet, "Design of Experiment in CHO and HEK transient transfection condition optimization," *Protein Expr. Purif.*, vol. 78, no. 1, pp. 61–68, 2011, doi: 10.1016/j.pep.2011.02.008

[69] N. R. Visaveliya and J. M. Köhler, "Single-step in situ assembling routes for the shape control of polymer nanoparticles," *Biomacromolecules*, vol. 19, no. 3, pp. 1047–1064, Mar. 2018, doi: 10.1021/acs.biomac.8b00034

[70] S. Choosakoonkriang, B. A. Lobo, G. S. Koe, J. G. Koe, and C. R. R. Middaugh, "Biophysical characterization of PEI/DNA complexes," *J. Pharm. Sci.*, vol. 92, no. 8, pp. 1710–1722, Aug. 2003, doi: 10.1002/jps.10437

[71] E. J. Cho, H. Holback, K. C. Liu, S. A. Abouelmagd, J. Park, and Y. Yeo, "Nanoparticle characterization: State of the art, challenges, and emerging technologies," *Mol. Pharm.*, vol. 10, no. 6, pp. 2093–2110, Jun. 2013, doi: 10.1021/mp300697h

[72] W. Zhang *et al.*, "Nano-structural effects on gene transfection: Large, botryoid-shaped nanoparticles enhance DNA delivery via macropinocytosis and effective dissociation," *Theranostics*, vol. 9, no. 6, pp. 1580–1598, 2019, doi: 10.7150/thno.30302

[73] D. Pezzoli, E. Giupponi, D. Mantovani, and G. Candiani, "Size matters for in vitro gene delivery: Investigating the relationships among complexation protocol, transfection medium, size and sedimentation," *Sci. Rep.*, vol. 7, no. 1, p. 44134, Apr. 2017, doi: 10.1038/srep44134

[74] L. Cervera, I. Gonzalez-Dominguez, M. M. Segura, and F. Godia, "Intracellular characterization of Gag VLP production by transient transfection of HEK 293 cells," *Biotechnol. Bioeng.*, vol. 114, no. 11, pp. 2507–2517, 2017, doi: 10.1002/bit.26367

[75] S. Geisse and M. Henke, "Large-scale transient transfection of mammalian cells: A newly emerging attractive option for recombinant protein production," *J. Struct. Funct. Genomics*, vol. 6, pp. 165–170, 2005, doi: 10.1007/s10969-005-2826-4

[76] P. L. Pham *et al.*, "Large-scale transient transfection of serum-free suspension-growing HEK293 EBNA1 cells: Peptone additives improve cell growth and transfection efficiency," *Biotechnol. Bioeng.*, vol. 84, pp. 332–342, 2003, doi: 10.1002/bit.10774

[77] R. Tom, L. Bisson, and Y. Durocher, "Transfection of HEK293-EBNA1 cells in suspension with linear PEI for production of recombinant proteins," *CSH Protoc.*, vol. 2008, no. 3, Jan. 2008.

[78] S. Geisse, "Reflections on more than 10 years of TGE approaches," *Protein Expr Purif*, vol. 64, no. 2, pp. 99–107, 2009, doi: 10.1016/j.pep.2008.10.017

[79] L. Baldi, D. L. Hacker, M. Adam, and F. M. Wurm, "Recombinant protein production by large-scale transient gene expression in mammalian cells: State of the art and future perspectives," *Biotechnol. Lett.*, vol. 29, no. 5, pp. 677–684, 2007, doi: 10.1007/s10529-006-9297-y

[80] G. Berntzen, E. Lunde, M. Flobakk, J. T. Andersen, V. Lauvrak, and I. Sandlie, "Prolonged and increased expression of soluble Fc receptors, IgG and a TCR-Ig fusion protein by transiently transfected adherent 293E cells," *J. Immunol. Methods*, vol. 298, no. 1–2, pp. 93–104, Mar. 2005, doi: 10.1016/j.jim.2005.01.002

[81] Y. Durocher, S. Perret, and A. Kamen, "High-level and high-throughput recombinant protein production by transient transfection of suspension-growing human 293-EBNA1 cells," *Nucleic Acids Res.*, vol. 30, no. 2, p. E9, Jan. 2002, Accessed: May 31, 2014. [Online]. Available: http://www.pubmedcentral.nih.gov/articlerender.fcgi?artid=99848&tool=pmcentrez&rendertype=abstract

[82] K. Van Craenenbroeck, P. Vanhoenacker, and G. Haegeman, "Episomal vectors for gene expression in mammalian cells," *Eur. J. Biochem.*, vol. 267, pp. 5665–5678, 2000.

[83] P. Meissner, H. Pick, A. Kulangara, P. Chatellard, K. Friedrich, and F. M. Wurm, "Transient gene expression: recombinant protein production with suspension-adapted HEK293-EBNA cells," *Biotechnol. Bioeng.*, vol. 75, no. 2, pp. 197–203,

Oct. 2001, Accessed: Jul. 30, 2014. [Online]. Available: http://www.ncbi.nlm.nih.gov/pubmed/11536142.
[84] J. M. Young, C. Cheadle, J. S. Foulke, W. N. Drohan, and N. Sarver, "Utilization of an Epstein-Barr virus replicon as a eukaryotic expression vector," *Gene*, vol. 62, pp. 171–185, 1988, doi: 10.1016/0378-1119(88)90556-2
[85] J. Ye, V. Kober, M. Tellers, Z. Naji, P. Salmon, and J. F. Markusen, "High-level protein expression in scalable CHO transient transfection," *Biotechnol. Bioeng.*, vol. 103, pp. 542–551, 2009, doi: 10.1002/bit.22265
[86] H. Rodrigues Goulart, F. Dos, S. Arthuso, M. V. N. Capone, T. L. de Oliveira, P. Bartolini, and C. R. J. Soares, "Enhancement of human prolactin synthesis by sodium butyrate addition to serum-free CHO cell culture," *J. Biomed. Biotechnol.*, vol. 2010, p. 405872, Jan. 2010, doi: 10.1155/2010/405872
[87] R. Damiani, B. E. Almeida, J. E. Oliveira, P. Bartolini, and M. T. C. P. Ribela, "Enhancement of human thyrotropin synthesis by sodium butyrate addition to serum-free CHO cell culture," *Appl. Biochem. Biotechnol.*, vol. 171, no. 7, pp. 1658–1672, Dec. 2013, doi: 10.1007/s12010-013-0467-9
[88] Y. H. Sung, Y. J. Song, S. W. Lim, J. Y. Chung, and G. M. Lee, "Effect of sodium butyrate on the production, heterogeneity and biological activity of human thrombopoietin by recombinant Chinese hamster ovary cells," *J. Biotechnol.*, vol. 112, pp. 323–335, 2004, doi: 10.1016/j.jbiotec.2004.05.003
[89] Y. Mimura *et al.*, "Butyrate increases production of human chimeric IgG in CHO-K1 cells whilst maintaining function and glycoform profile," *J. Immunol. Methods*, vol. 247, pp. 205–216, 2001, doi: 10.1016/S0022-1759(00)00308-2
[90] G. Backliwal, M. Hildinger, I. Kuettel, F. Delegrange, D. L. Hacker, and F. M. Wurm, "Valproic acid: A viable alternative to sodium butyrate for enhancing protein expression in Mammalian cell cultures," *Biotechnology*, vol. 101, pp. 182–189, 2008, doi: 10.1002/bit.21882
[91] D. P. Palermo, M. E. DeGraaf, K. R. Marotti, E. Rehberg, and L. E. Post, "Production of analytical quantities of recombinant proteins in Chinese hamster ovary cells using sodium butyrate to elevate gene expression," *J. Biotechnol.*, vol. 19, pp. 35–47, 1991, doi: 10.1016/0168-1656(91)90073-5
[92] S. Ansorge, S. Lanthier, J. Transfiguracion, Y. Durocher, O. Henry, and A. Kamen, "Development of a scalable process for high-yield lentiviral vector production by transient transfection of HEK293 suspension cultures," *J. Gene Med.*, vol. 11, pp. 868–876, 2009, doi: 10.1002/jgm.1370
[93] Z. Jiang and S. T. Sharfstein, "Sodium butyrate stimulates monoclonal antibody over-expression in CHO cells by improving gene accessibility," *Biotechnol. Bioeng.*, vol. 100, pp. 189–194, 2008, doi: 10.1002/bit.21726
[94] S. Wulhfard, L. Baldi, D. L. Hacker, and F. Wurm, "Valproic acid enhances recombinant mRNA and protein levels in transiently transfected Chinese hamster ovary cells," *J. Biotechnol.*, vol. 148, pp. 128–132, 2010, doi: 10.1016/j.jbiotec.2010.05.003
[95] G. Backliwal, M. Hildinger, S. Chenuet, M. DeJesus, and F. M. Wurm, "Coexpression of acidic fibroblast growth factor enhances specific productivity and antibody titers in transiently transfected HEK293 cells," *N. Biotechnol.*, vol. 25, pp. 162–166, 2008, doi: 10.1016/j.nbt.2008.08.007
[96] G. Backliwal, M. Hildinger, S. Chenuet, S. Wulhfard, M. De Jesus, and F. M. Wurm, "Rational vector design and multi-pathway modulation of HEK 293E cells yield recombinant antibody titers exceeding 1 g/l by transient transfection under serum-free conditions," *Nucleic Acids Res.*, vol. 36, no. 15, p. e96, Sep. 2008, doi: 10.1093/nar/gkn423
[97] S. Fan *et al.*, "Valproic acid enhances gene expression from viral gene transfer vectors," *J. Virol. Methods*, vol. 125, pp. 23–33, 2005, doi: 10.1016/j.jviromet.2004.11.023

[98] A. Spenger, W. Ernst, J. P. Condreay, T. A. Kost, and R. Grabherr, "Influence of promoter choice and trichostatin a treatment on expression of baculovirus delivered genes in mammalian cells," *Protein Expr. Purif.*, vol. 38, pp. 17–23, 2004, doi: 10.1016/j.pep.2004.08.001

[99] M. Fussenegger, J. E. Bailey, H. Hauser, and P. P. Mueller, "Genetic optimization of recombinant glycoprotein production by mammalian cells," *Trends Biotechnol.*, vol. 17. pp. 35–42, 1999, doi: 10.1016/S0167-7799(98)01248-7

[100] E. Suzuki and D. F. Ollis, "Enhanced antibody production at slowed growth rates: experimental demonstration and a simple structured model," *Biotechnol. Prog.*, vol. 6, no. 3, pp. 231–236, Jan. 1990, doi: 10.1021/bp00003a013

[101] C. A. Hall-Jackson, D. A. Cross, N. Morrice, and C. Smythe, "ATR is a caffeine-sensitive, DNA-activated protein kinase with a substrate specificity distinct from DNA-PK," *Oncogene*, vol. 18, no. 48, pp. 6707–6713, Nov. 1999, doi: 10.1038/sj.onc.1203077

[102] J. N. Sarkaria *et al.*, "Inhibition of ATM and ATR kinase activities by the radiosensitizing agent, caffeine," *Cancer Res.*, vol. 59, no. 17, pp. 4375–4382, Sep. 1999, Accessed: Feb. 11, 2015. [Online]. Available: http://www.ncbi.nlm.nih.gov/pubmed/10485486

[103] W. D. Block, D. Merkle, K. Meek, and S. P. Lees-Miller, "Selective inhibition of the DNA-dependent protein kinase (DNA-PK) by the radiosensitizing agent caffeine," *Nucleic Acids Res.*, vol. 32, no. 6, pp. 1967–1972, Jan. 2004, doi: 10.1093/nar/gkh508

[104] B. L. Ellis, P. R. Potts, and M. H. Porteus, "Creating higher titer lentivirus with caffeine," *Hum. Gene Ther.*, vol. 22, no. 1, pp. 93–100, Jan. 2011, doi: 10.1089/hum.2010.068

[105] L. Cervera, J. Fuenmayor, I. González-Domínguez, S. Gutiérrez-Granados, M. M. Segura, and F. Gòdia, "Selection and optimization of transfection enhancer additives for increased virus-like particle production in HEK293 suspension cell cultures," *Appl. Microbiol. Biotechnol.*, vol. 99, no. 23, pp. 9935–9949, 2015, doi: 10.1007/s00253-015-6842-4

[106] L. Cervera, S. Gutiérrez-Granados, N. S. Berrow, M. M. Segura, and F. Gòdia, "Extended gene expression by medium exchange and repeated transient transfection for recombinant protein production enhancement," *Biotechnol. Bioeng.*, vol. 112, no. 5, pp. 934–946, 2015, doi: 10.1002/bit.25503

[107] J. Lavado-García, L. Cervera, and F. Gòdia, "An alternative perfusion approach for the intensification of virus-like particle Production in HEK293 Cultures," *Front. Bioeng. Biotechnol.*, vol. 8, Article number 617, 2020, doi: 10.3389/fbioe.2020.00617

[108] J. Lavado-García, I. Jorge, L. Cervera, J. Vázquez, and F. Gòdia, "Multiplexed quantitative proteomic analysis of HEK293 provides insights into molecular changes associated with the cell density effect, transient transfection, and virus-like particle production," *J. Proteome Res.*, vol. 19, no. 3, pp. 1085–1099, Mar. 2020, doi: 10.1021/acs.jproteome.9b00601

[109] D. J. Karst, E. Serra, T. K. Villiger, M. Soos, and M. Morbidelli, "Characterization and comparison of ATF and TFF in stirred bioreactors for continuous mammalian cell culture processes," *Biochem. Eng. J.*, vol. 110, pp. 17–26, 2016, doi: 10.1016/j.bej.2016.02.003

[110] J. Lavado-García, I. Jorge, A. Boix-Besora, J. Vázquez, F. Gòdia, and L. Cervera, "Characterization of HIV-1 virus-like particles and determination of Gag stoichiometry for different production platforms," *Biotechnol. Bioeng.*, vol. 118, pp. 2660–2675, Apr. 2021, doi: 10.1002/bit.27786

[111] L. Besnard *et al.*, "Clarification of vaccines: An overview of filter based technology trends and best practices," *Biotechnol. Adv.*, vol. 34, no. 1, pp. 1–13, Jan. 2016, doi: 10.1016/j.biotechadv.2015.11.005

[112] M. Westoby, J. Chrostowski, P. De Vilmorin, J. P. Smelko, J. K. Romero, and N. Carolina, "Effects of solution environment on Mammalian cell fermentation broth properties: Enhanced impurity removal and clarification performance," vol. 108, no. 1, pp. 50–58, 2011, doi: 10.1002/bit.22923

[113] S. B. Carvalho *et al.*, "Efficient filtration strategies for the clarification of influenza virus-like particles derived from insect cells," *Sep. Purif. Technol.*, vol. 218, pp. 81–88, Jul. 2019, doi: 10.1016/j.seppur.2019.02.040

[114] M. G. Moleirinho, R. J. S. Silva, P. M. Alves, M. J. T. Carrondo, and C. Peixoto, "Current challenges in biotherapeutic particles manufacturing," *Expert Opin. Biol. Ther.*, vol. 20, no. 5, pp. 451–465, May 2020, doi: 10.1080/14712598.2020.1693541

[115] P. Pereira Aguilar, I. González-Domínguez, T. A. Schneider, F. Gòdia, L. Cervera, and A. Jungbauer, "At-line multi-angle light scattering detector for faster process development in enveloped virus-like particle purification," *J. Sep. Sci.*, vol. 42, pp. 2640–2648, Jun. 2019, doi: 10.1002/jssc.201900441

[116] P. Pereira Aguilar *et al.*, "Polymer-grafted chromatography media for the purification of enveloped virus-like particles, exemplified with HIV-1 gag VLP," *Vaccine*, vol. 37, no. 47, pp. 7070–7080, 2019, doi: 10.1016/j.vaccine.2019.07.001

[117] T. P. Pato *et al.*, "Development of a membrane adsorber based capture step for the purification of yellow fever virus," *Vaccine*, vol. 32, no. 24, pp. 2789–2793, May 2014, doi: 10.1016/j.vaccine.2014.02.036

[118] P. Steppert *et al.*, "Purification of HIV-1 gag virus-like particles and separation of other extracellular particles," *J. Chromatogr. A*, vol. 1455, pp. 93–101, 2016, doi: 10.1016/j.chroma.2016.05.053

[119] P. Steppert *et al.*, "Separation of HIV-1 gag virus-like particles from vesicular particles impurities by hydroxyl-functionalized monoliths," *J. Sep. Sci.*, vol. 40, no. 4, pp. 979–990, 2017, doi: 10.1002/jssc.201600765

[120] C. S. M. Fernandes *et al.*, "Retroviral particles are effectively purified on an affinity matrix containing peptides selected by phage-display," *Biotechnol. J.*, vol. 11, no. 12, pp. 1513–1524, 2016, doi: 10.1002/biot.201600025

[121] C. Ladd Effio, L. Wenger, O. Ötes, S. A. Oelmeier, R. Kneusel, and J. Hubbuch, "Downstream processing of virus-like particles: Single-stage and multi-stage aqueous two-phase extraction," *J. Chromatogr. A*, vol. 1383, pp. 35–46, Feb. 2015, doi: 10.1016/j.chroma.2015.01.007

[122] M. F. Gencoglu and C. L. Heldt, "Enveloped virus flocculation and removal in osmolyte solutions," *J. Biotechnol.*, vol. 206, pp. 8–11, Jul. 2015, doi: 10.1016/j.jbiotec.2015.03.030

[123] S. B. Carvalho *et al.*, "Membrane-based approach for the downstream processing of influenza virus-like particles," *Biotechnol. J.*, vol. 14, no. 1800570, pp. 1–12, Aug. 2019, doi: 10.1002/biot.201800570

[124] A. Venereo-Sanchez *et al.*, "Process intensification for high yield production of influenza H1N1 Gag virus-like particles using an inducible HEK-293 stable cell line," *Vaccine*, vol. 35, no. 33, pp. 4220–4228, 2017, doi: 10.1016/j.vaccine.2017.06.024

[125] M. G. Moleirinho *et al.*, "Baculovirus affinity removal in viral-based bioprocesses," *Sep. Purif. Technol.*, vol. 241, no. 116693, pp. 1–9, Jun. 2020, doi: 10.1016/j.seppur.2020.116693

[126] K. Reiter, P. P. Aguilar, D. Grammelhofer, J. Joseph, P. Steppert, and A. Jungbauer, "Separation of influenza virus-like particles from baculovirus by polymer grafted anion-exchanger," *J. Sep. Sci.*, 43, 12, pp. 1–21, Mar. 2020, doi: 10.1002/jssc.201901215

[127] K. Reiter, P. P. Aguilar, V. Wetter, P. Steppert, A. Tover, and A. Jungbauer, "Separation of virus-like particles and extracellular vesicles by flow-through and heparin affinity chromatography," *J. Chromatogr. A*, vol. 1588, pp. 77–84, Mar. 2019, doi: 10.1016/J.CHROMA.2018.12.035

[128] O.-W. Merten, M. Hebben, and C. Bovolenta, "Production of lentiviral vectors," *Mol. Ther. Methods Clin. Dev.*, vol. 3, no. September 2015, p. 16017, 2016.

[129] J. Transfiguracion, A. P. Manceur, E. Petiot, C. M. Thompson, and A. A. Kamen, "Particle quantification of influenza viruses by high performance liquid chromatography," *Vaccine*, vol. 33, no. 1, pp. 78–84, Jan. 2015, doi: 10.1016/j.vaccine.2014.11.027

[130] D. Fontana, R. Kratje, M. Etcheverrigaray, and C. Prieto, "Immunogenic virus-like particles continuously expressed in mammalian cells as a veterinary rabies vaccine candidate," *Vaccine*, vol. 33, no. 35, pp. 4238–4246, 2015, doi: 10.1016/j.vaccine.2015.03.088

[131] S. Heider and C. Metzner, "Quantitative real-time single particle analysis of virions," *Virology*, vol. 462–463, no. 100, pp. 199–206, Aug. 2014, doi: 10.1016/j.virol.2014.06.005

[132] Q. Wang, B.-J. Bosch, J. M. Vlak, M. M. van Oers, P. J. Rottier, and J. W. M. van Lent, "Budded baculovirus particle structure revisited," *J. Invertebr. Pathol.*, vol. 134, pp. 15–22, Feb. 2016, doi: 10.1016/j.jip.2015.12.001

[133] J. C. Akers, D. Gonda, R. Kim, B. S. Carter, and C. C. Chen, "Biogenesis of extracellular vesicles (EV): exosomes, microvesicles, retrovirus-like vesicles, and apoptotic bodies," *J. Neurooncol.*, vol. 113, no. 1, pp. 1–11, May 2013, doi: 10.1007/s11060-013-1084-8

[134] S. Gutiérrez-Granados, L. Cervera, F. Gòdia, J. Carrillo, and M. M. Segura, "Development and validation of a quantitation assay for fluorescently tagged HIV-1 virus-like particles," *J Virol. Methods*, vol. 193, no. 1, pp. 85–95, 2013, doi: 10.1016/j.jviromet.2013.05.010

11 Vectored vaccines

Zeyu Yang, Kumar Subramaniam, and Amine Kamen
Viral Vectors and Vaccines Bioprocessing Group, Department of Bioengineering, McGill University, Montréal, QC, Canada

CONTENTS

11.1 INTRODUCTION

This section will cover the design of viral vectors for vaccination and will describe in some detail the cell-culture production and processing of adenoviral (AdV) and vesicular stomatitis virus (VSV) vectored vaccines. A broad definition of recombinant viral

DOI: 10.1201/9781003229797-11

vectors covers the use of low-pathogenicity viruses genetically modified with the purpose of vectoring the expression or display of the dominant antigens related to the targeted infectious diseases to induce a humoral and cellular response.

Both AdV- and VSV-based vectors led to the development of approved vaccines for human use against emerging and re-emerging infectious diseases, especially different serotypes of AdV-vectored vaccines in the case of the COVID-19 pandemic and VSV-vectored vaccine in the case of the Ebola epidemic [1,2]. These two viral vectors, representative of the classes of non-enveloped (AdV) and enveloped viruses (VSV) as well as DNA (AdV) and RNA (VSV) viruses, involve a broad range of biological and structural properties that would determine their modes of production, purification, and formulation [3,4].

Apart from AdV- and VSV-vectored vaccines exemplified in this chapter, many other viral vectors have been developed for vaccination including yellow fever virus, poxvirus, paramixovirus, and alphavirus and they have been used for control of infectious diseases [5–8]. Chimeric vaccines, such as yellow fever virus-based vaccines including Japanese encephalitis and Dengue vaccines from Sanofi-Pasteur would fall under the general definition of vectored vaccines [9,10]. The yellow fever virus backbone is modified to express antigens related to the Japanese encephalitis virus infection. Similarly, the tetravalent dengue vaccine uses the yellow fever virus backbone to express proteins from different serotypes of the dengue virus. These vectored vaccines have been evaluated for safety and efficacy in numerous clinical trials at different phases, as shown in Table 11.1.

From the perspective of design and processing, some of the vectors and particularly adenovirus-vectored vaccines benefited from extensive background knowledge and expertise originally developed within the field of gene therapy and virotherapy extending the vaccinology application field to oncotherapy and therapeutic vaccines. Replication-defective vectors are mostly used in oncotherapy; however, replication-competent viruses such as reovirus and NDV are currently used in many clinical trials in virotherapy targeting specifically tumor cells [11,12].

11.2 ADENOVIRUS-VECTORED VACCINES

AdV vectors derive from *Adenoviridae*, a diverse family of non-enveloped, icosahedral, double-stranded DNA viruses [3]. The first use of AdV as a vaccine vector can be traced back to the 1980s [13]. Since then, AdV vectors have shown great promise and consequently have been extensively explored. AdV vectors hold attractive advantages as vaccine vectors including safety, stability, and efficacy. As a replication-defective vector, there is no serious risk of horizontal transmission when using AdV. The genome of AdV vectors does not integrate into the chromosome of the host cells, which avoids the risk of insertional mutagenesis [14]. AdV vectors possess a robust transgene expression ability that can be strengthened by strong heterologous promoters. AdV vectors can be efficiently produced in large-scale cell cultures. AdV can infect a wide range of vertebrates, including humans and non-human primates. AdV vectors can activate strong immunoreaction when administered via intramuscular or mucosal routes.

Additionally, AdVs were the first viral gene transfer vectors developed for use in humans due to their high transduction efficiency and relatively low virulence. The

TABLE 11.1
Clinical trials involving several viral vectors[a]

Vector	Disease	Number of trials in phase 1/2 and 2, 3, & 4	NCT number
Yellow fever virus	Dengue fever	3 (phase 2)	NCT00993447; NCT01187433; NCT00788151
	West Nile fever	2 (phase 2)	NCT00442169; NCT00746798
	Japanese encephalitis	4 (phase 3)	NCT01001988; NCT01190228; NCT01188343; NCT01092507
Poxvirus	HIV	1 (phase 1)	NCT00001136
	Influenza	3 (phase 1)	NCT01818362; NCT00942071; NCT01465035
		2 (phase 2)	NCT03883113; NCT00993083
Paramixovirus	HIV	1 (phase 1)	NCT01320176
	Chikungunya	1 (phase 2)	NCT02861586
Arenavirus	Cytomegalovirus	1 (phase 2)	NCT03629080

Note
[a] Data from https://clinicaltrials.gov.

therapeutic use of AdV vectors is currently focused on medical applications responding to transient transgene expression and selective immunogenicity, such as the expression of vaccine antigens and oncolytic therapy [15]. Several recombinant AdV vectors using different serotypes have emerged as antigen or gene delivery systems for vaccination and therapeutic interventions. Numerous clinical trials have been initiated. As shown in Table 11.2, clinical studies involving Adv vectors were performed to control the diseases in situations of pandemics or epidemics such as COVID-19, Ebola, and HIV and chronic diseases, such as different types of cancers.

Since AdV was first discovered in 1953, over 50 human serotypes of AdV have been described [16]. Among these, Human Adenovirus type 5 (HAd5) has been conventionally utilized in laboratories and clinical trials as the backbone for AdV vectors [17]. Currently, extensive preclinical and clinical trials are in progress using HAd5 as vectored vaccines. Noteworthy among them are: (i) an HAd5 expressing Zika virus pre-membrane and envelope protein (HAd5-ZKV) [18]; (ii) an HAd5 vector vaccine as veterinary vaccine that induced immunity to rabies [19]; (iii) several HAd5-based COVID-19 vectored vaccines have been tested in phase 3 clinical, such as Ad5-nCOV and Gam-COVID-Vac, which were developed by CanSino Biologics

TABLE 11.2
Clinical trials involving adenovirus vector[a]

Vector	Target	Disease	Number of trials in phase 1/2 and 2, 3, & 4	NCT number
Adenovirus	Infectious diseases	COVID-19	5 (phase 1/2)	NCT05043259; NCT04436276; NCT04889209; NCT04840992; NCT04398147;
			8 (phase 2)	NCT05005156; NCT04765384; NCT04341389; NCT04998240; NCT04566770; NCT05000216; NCT04587219; NCT05016622
			11 (phase 3)	NCT04505722; NCT04614948; NCT04540419; NCT04838795; NCT04526990; NCT04510493; NCT04848467; NCT04642339; NCT04540393; NCT04908722; NCT04656613
			5 (phase 4)	NCT04952727; NCT04833101; NCT04892459; NCT05037266; NCT05030974;
		HIV	6 (phase 1/2)	NCT02919306; NCT04983030; NCT02788045; NCT02099994; NCT02315703; NCT02935686
			7 (phase 2)	NCT00080106; NCT03060629; NCT00350623; NCT00095576; NCT00183261; NCT00976404

TABLE 11.2 (Continued)
Clinical trials involving adenovirus vector[a]

Vector	Target	Disease	Number of trials in phase 1/2 and 2, 3, & 4	NCT number
			1 (phase 3)	NCT03964415
		Malaria	3 (phase 1/2)	NCT01397227; NCT00392015; NCT03203421
			1 (phase 2)	NCT01366534;
		Hepatits C virus	1 (phase 1/2)	NCT02309086
		Ebola virus	1 (phase 1/2)	NCT02661464;
			9 (phase 2)	NCT04028349; NCT04186000; NCT02598388; NCT02564523; NCT04711356; NCT03929757; NCT02416453; NCT02876328; NCT02575456
			6 (phase 3)	NCT02661464; NCT04152486; NCT02509494; NCT04228783; NCT04556526; NCT02543567
		Tuberculosis	1 (phase 1/2)	NCT01017536
			1 (phase 2)	NCT01198366
		Rotavirus	1 (phase 3)	NCT01305109
	Therapeutic cancer vaccines	Colon/breast/ head/neck/ renal	7 (phase 1/2)	NCT01147965; NCT03127098; NCT02423902; NCT01042535; NCT03754933; NCT02842125; NCT02879760
			8 (phase 2)	NCT04111172; NCT03050814; NCT01703754; NCT00044993; NCT00017173; NCT00617409; NCT00776295; NCT00704938

(Continued)

TABLE 11.2 (Continued)
Clinical trials involving adenovirus vector[a]

Vector	Target	Disease	Number of trials in phase 1/2 and 2, 3, & 4	NCT number
		Prostate	2 (phase 1/2)	NCT01931046; NCT04097002
			1 (phase 2)	NCT00583492
		Lymphoma	3 (phase 2)	NCT00849524; NCT00942409; NCT03544723
		Solid tumor/ tissue sarcoma	2 (phase 1/2)	NCT02842125; NCT01042535
			3 (phase 2)	NCT03544723; NCT04111172; NCT00704938
		Leukemia/ glioblastoma	3 (phase 2)	NCT00849524; NCT00942409; NCT04006119
		Melanoma	2 (phase 1/2)	NCT01082887; NCT01397708
			2 (phase 2)	NCT03190824; NCT00704938

Note
[a] Data from https://clinicaltrials.gov.

Inc. and Gamaleya Research Institute, respectively. hAd5-S-Fusion+NETSD, VXA-COV2-1 and Ad-COVID are under phase 1 clinical evaluations [1].

Despite numerous advantages, the use of HAd5 in clinic is severely hindered by the widespread pre-exposure of the majority of the population to HAd5. The gene transfer efficiency of vectors will be reduced by impairing the B- and T-cell responses when the HAd5 neutralizing antibodies are prevalent in the human population. A survey shows that the seroprevalence rates of HAd5 are 40–60% in America, while in China, 74.2% of the population carries the neutralizing antibodies against HAd5 [20,21]. The situation may be even worse in other regions of the world such as Africa and Asian countries.

To overcome this issue, researchers developed alternative human AdV serotypes which possess low prevalence among the population such as HAd35, HAd11, HAd26, etc [22–24]. Though less immunogenic, these rare serotypes are used as an alternative to HAd5, as vaccine vectors (Figure 11.1). Another alternative strategy is to isolate AdVs from other species such as cattle, chimpanzees, pigs, etc. and

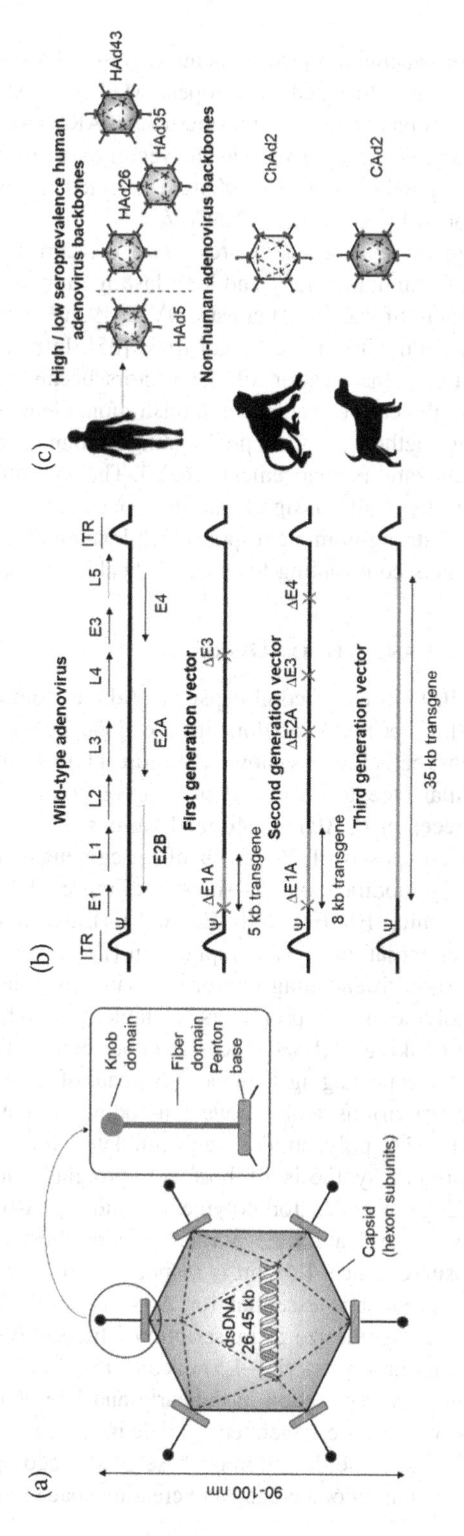

FIGURE 11.1 (a) Structure of adenovirus showing size, spike protein, dsDNA etc. (b) Construction of different generations of AdV vectors. (c) Human and non-human AdV backbones.

evaluate their efficacy for vaccination [15]. A heterologous Ad26-vectored vaccine against Zaire Ebola virus was authorized by European Medicines Agency on July 1, 2020 (https://www.ema.europa.eu/en/medicines/human/EPAR/zabdeno). In the race of COVID-19 vaccines, Janssen used HAd26 as the vector to express stabilized pre-fusion spike protein of SARS-CoV-2. AstraZeneca selected a chimpanzee AdV (ChAdY25) as the vector to build ChAdOX1-nCoV.

Despite successful application and safety results of the current AdV vectored vaccines, enhancement of immunogenicity and long-lasting immunity are still to be assessed. In the development of vaccines against COVID-19, an intranasal route has been used for sterilizing immunity in preclinical trials [25]. Intranasal administration of vaccine can mitigate inactivation of the vectors because of pre-existing immunity, additionally to the ease of vaccine administration. Genetic modifications proved to be a promising method to evade pre-existing immunity, and create cell-specific vectors for immunization enhancement [26,27]. The continuing research on AdV vectors will hopefully lead to significant improvement of AdV vectored vaccines by stimulation of strong immune response, while maintaining a high safety profile with rare side effects contributing to mankind health's protection.

11.2.1 AdV Structure and Vector Design

The non-enveloped 90–100 nm icosahedral capsid of AdV is composed of penton and hexon sub-units [28]. Fiber and knob domains are associated with pentons that mediate attachment to host cells [29], as shown in Figure 11.1. Affinity of the knob domain for various cellular receptors varies depending on serotype, with the coxsackievirus adenovirus receptor (CAR), CD46, and various integrins used [30].

The viral genome is composed of 26–45 kb of unsegmented, linear, dsDNA, which is amenable to easy modification. As shown in Figure 11.1, the genome is organized into five early units (E1, E2a, E2b, E3, and E4), two intermediate units which are expressed after initiation of viral replication (IVA2, IX), and late units (L1, L2, L3, L4, and L5), each encoding one or two virus-associated (VA) RNA controlled by internal polymerase III promoters. Notable non-coding elements of the genome include the flanking ITR sequences at either end of the genome that prime DNA replication, the ψ packaging sequence upstream of E1 necessary for the efficient assembly of mature virions, and various non-coding viral associated RNAs (VAI, VAII) [17,29,31]. E1A polypeptides can stimulate viral DNA synthesis, while E1B stops the protein synthesis of host cell proteins and contribute to transport viral RNA. E2 gene codes for polymerase and DNA-binding proteins which are essential for viral replication. E3 unit encodes proteins which can guide virus to escape immunosurveillance. E4 unit is responsible for viral RNA nuclear export. The IVA2 gene product is essential for AdV assembly and viral DNA packaging. The IX unit can synthesize a transcriptional transactivator, which can stabilize the virion. The late unit genes (L1–L5) encode 45 species of RNA, which play an important role in AdV replication in the early and late phases [3].

First-generation AdV vectors are generated by deleting the E1, and sometimes also E3, region of the viral genome. Deletion of these genes accomplishes the dual purpose of ensuring replication incompetence and creating space to accommodate a transgene cassette. AdV vector passages in a cell culture might result in *de novo* acquisition of E1 region and replication competent AdV (RCA) [32]. AdV-based COVID-19 vaccines such as Ad26.COV2-S and Ad5-nCoV use the first-generation adenovirus with the E1 and E3 genes deleted [1]. Detailed information of AdV-based COVID-19 vaccines will be further discussed in 11.2.3.3.

Third-generation AdV vectors, or gutless AdV vectors where all viral proteins have been removed which can significantly reduce the immune response of host cells upon AdV vector transduction [28] have been mainly used in gene therapy interventions.

11.2.2 Production Process of AdV Vectors

11.2.2.1 Cell Line Selection

Though many cell lines such as HER-911, HeLa, and HEL-299 are permissive to AdV infection, the most common cell line used for AdV replication is the Human Embryo Kidney 293 (HEK-293) cell [33]. The HEK-293 was transformed with sheared AdV DNA resulting in the incorporation of the gene expressing the HAd5 E1 protein [34,35]. Since the first-generation AdV vectors are replication-defective due to the loss of E1 region, the HEK-293 cell is used as a complementing cell line expressing the HAd5 E1 protein, making it possible to propagate the defective vector in HEK-293 cells. However, HEK-293 cell line incorporates not only the HAd5 E1 gene, but also viral sequences flanking the E1 region, and thus RCAs can then be generated through homologous recombination [36]. To overcome this issue, a PER.C6 cell line, derived from human embryonic retinal cells, was generated, which contains only E1 elements, such that the effect of homologous recombination is reduced [37]. For the COVID-19 vaccine developed by CanSino, the recombinant HAd5 vectored vaccine was amplified by serial passage on HEK293 cells [38].

Though HAd5-based vectors have efficient replication ability in HEK-293 and PER.C6 cells, some rare serotypes AdV vectors replicate at significantly lower yields [39]. Research work from Havenga et al. showed that the production of HAd35 can be significantly enhanced by expressing HAd5 E4-ORF6 protein in producer cell lines [40]. Comparable yields were observed for other AdV serotypes such as HAd26, HAd48, and HAd49, using a similar strategy. The E1/E3-deleted Ad26-vectors adopted by Janssen/Johnson & Johnson (Ad26.COV2-S) were generated in PER.C6.TetR cells using a plasmid containing a transgene expression cassette and the genome of Ad26 vector [41]. The production of non-human serotypes AdV vectors is more challenging because the species-specific producer cell lines may lack the scalability of HEK-293 or PER.C6 cells. The Madin-Darby Canine Kidney Epithelial (MDCK) cells were engineered to express the CAd2 E1 protein to produce Cad2 [42]. In general, HEK-293 is still an extensively used cell line for manufacturing the AdVs with the restriction in terms of RCA generation. The COVID-19 vaccine ChAdOX1-nCoV from Oxford/AstraZeneca used a ChAdY25 vector. The virus was propagated in T-Rex HEK293 cells [43].

11.2.2.2 Limitations in the Production of AdV

To increase the viral production, the first objective is to maximize the production yield on a per cell basis. However, one of the critical limitations in the production of AdV is the so-called cell density effect, which means the viral production is limited beyond a certain cell density and cannot increase in proportion to the total cell number. This observation is cell culture medium dependent, for example, with NSFM13 medium, the optimum productivity of viral particles expressing GFP-Q happened when infect the cells at 5.0E+05 cells/ml. If the cell density is doubled (infect at 1.0E+06 cells/ml), the viral production is reduced [44]. This "cell density effect" has been observed with a number of viruses including VSV [45].

Generally, cell-culture media were designed to maximize cell growth not necessarily viral production. Consequently, a better understanding of the metabolic limitations in the viral infection/production phase is needed to achieve high viral productivity. Currently, two approaches have been proposed by researchers [46]. One approach is to developing feeding strategies to add critical substrates and reduce/remove toxic metabolites during the production phase. The second approach is to design novel media that support not only the cell-culture growth but also the viral production effectively which requires extensive metabolic analyses over the growth and the production phases.

Osmolarity is also considered as a factor that significantly affects the viral yield [19,47]. Not only the osmolarity during production but also the osmolarity during cell growth affect the virus yield, which indicates that the history of the cells might determine their capacity for viral production. A productive system can be achieved by disassociating the growth phase from the production phase in terms of metabolic and cell physiology requirements. These observations might be generalized to other virus production systems.

11.2.2.3 Productivity of AdV in Different Operation Modes

Cultivation of mammalian cells has been realized using various technologies including roller bottles, microcarriers, bioreactor suspension cultures, etc. Suspension cell culture remains the most effective method for the production of AdV vectors at large scale especially when compared to processes using adherent cells. Additionally, with a homogeneous concentration of nutrients, metabolites, and cellular environment, the suspension culture is easy to monitor and scale-up to control process robustness and critical quality attributes of the AdV vectors. As the main cell line for production of AdV, HEK-293 cells have been adapted to suspension culture (293S). They were further adapted to serum-free medium 293SF, enhancing the scalability, batch-to-batch reproducibility, and regulatory approval [48,49]. Suspension cultures of HEK-293 in stirred tank bioreactors are projected to scale up to 10,000L, with yields for unpurified culture in the range of 10^9–10^{10} VP/ml [44,50]. Another important cell line PER.C6 has been successfully used in GMP manufacturing processes, growing to high cell densities in serum-free suspension culture, and can be used to produce AdV vectors in a similar fashion [50].

However, in viral production phase, the productivity of AdV vectors was limited by cell density effect in batch mode due to the depletion of key nutrients or the accumulation of inhibitors as previously discussed. The batch process, where cells are exposed to a constantly changing environment, limits the growth and production potential of the cells. Alternative modes of operation including fed-batch and perfusion might be operated to overcome these limitations [22]. These modes of operation are of incremental difficulty in implementation and operation. The fed-batch mode adopts a special feeding strategy to extend the culture lifetime. The addition of glucose, glutamine, and amino acids in a fed-batch bioreactor allows the infections at higher cell densities. However, the development of an efficient feeding strategy is still needed to enable virus replication at high cell densities. A rational design of a feeding strategy relies on an extensive understanding of the metabolism in cell culture and viral production phases. The perfusion mode usually results in high product titers. In this mode, cells are retained inside the bioreactor using different retention devices including acoustic filters, membrane units, centrifuges, etc. For more details, refer to chapter 6 on process intensification. The nutrients are fed and toxic by-products are removed continuously. After optimizing the perfusion conditions like feed rate, infection time, and harvest time, a higher specific productivity can be achieved when infecting the cells at a higher cell density [51,52].

11.2.2.4 Quantitation Methods of AdV Vector

The primary characterization of AdV are viral particle units (VPs) and infectious viral particle units (IVPs). The most common method to physically determine the total viral particles relies on the absorbance reading at 260 nm. A more precise method to quantify the AdV is the anion-exchange–high-performance liquid chromatography (AE–HPLC), in which the VPs were detected by photodiode array detector at 260 nm [53]. For example, the AdV serotype 5 particles can be measured by AE-HPLC method using a UNO Q column. The virus peak was eluted at 450 mM NaCl in about 8 min. Another quantification method for detecting the VPs uses quantitative polymerase chain reaction (qPCR), which serves as a gold standard method. Compared with qPCR, the digital-droplet PCR (ddPCR) is more robust and has higher throughput since it provides absolute quantification with higher sensitivity, omits the use of a standard, and requires less sample volume [54]. For the quantification of IVP, the titers were usually determined by fifty-percent tissue culture infective dose ($TCID_{50}$) assay. The $TCID_{50}$ quantifies the amount of virus which can kill 50% of host cells or produce a cytopathic effect in 50% of cultured cells [36,55]. For replication-competent AdV (RCA), determinations are generally performed on pre-clinical and clinical studies. Results can be observed by cytopathic effect on permissive cells such as A-549 and Hela cells after multiple dilutions. In addition, using PCR can enhance the sensitivity of the method. To standardize the results of different quantitation methods and facilitate transfer of pre-clinical and clinical data, a reference material for AdV vector has been generated [56]. For more details, refer to chapter 9.

11.2.3 Examples of AdV-Based Vaccines

11.2.3.1 Veterinary Vaccines

11.2.3.1.1 AdV-Based Rabies Vaccine

Control of rabies in wildlife remains an important challenge. The well-documented rabies reservoir is bats. Raccoons, skunks, and foxes can be contaminated with rabies, while domestic animals such as dogs can also be infected. In Canada, the rabies transmission is largely through national parks. To control the wildlife rabies, an AdV serotype 5 vectored vaccine was developed. The vaccine that expresses the rabies glycoprotein (AdRG1.3) is produced using HEK-293 cells. Then, AdRG1.3 AdV is formulated and packaged into baits by Artemis Technologies Inc. using proprietary technology [19]. To meet the requirement of field trails, AdRG1.3 AdV production was successfully scaled up from 1 to 500 L. The results of field trials demonstrated advantages of the vaccine in immunizing wild animals that were previously difficult to vaccinate [19]. The vaccine is currently licensed for use in Canada.

11.2.3.1.2 AdV-Based Foot and Mouth Disease (FMD) Vaccine

Foot-and-mouth disease virus (FMDV) is a *Picornaviridae* RNA virus. FMD afflicts cloven-hooved animals, including sheep, goats, cattle, pigs, and buffalo with pedal and oronasal vesicular lesions. The virus can be transmitted through aerosol droplets, direct contact, and ingestion with infected animals [57]. There are seven FMDV serotypes, and numerous strains. In most countries throughout Africa and Asia, inactivated vaccines are widely used to control outbreaks. Compared with inactivated vaccines, the next-generation recombinant FMD vaccines that are produced without virulent FMDV strains have more advantages, especially for a rapid response against newly emerging FMDV. In 2012, Adt.A24 FMD vaccine was licensed by the U.S. Department of Agriculture for conditional use to protect cattle [58]. The vaccine used a replication-defective human AdV containing the capsid- and 3C protease-coding regions of the A24 FMDV [59]. Previous studies demonstrated that this vaccine can protect both swine and cattle with 64% efficacy [60].

11.2.3.2 AdV-Based Ebola Vaccine

First appearing in 1976 in Africa, Ebola is one of the most severe hemorrhagic fever viruses. Causing a sudden high fever, followed by vomiting, diarrhea, and even bleeding, Ebola has fatality rates of 20–90%. There are five known species of Ebola including Zaire ebolavirus, Sudan ebolavirus, Taï Forest ebolavirus, Bundibugyo ebolavirus, and Reston ebolavirus [61]. Since there have been no treatments so far, it is vital to develop vaccines to control the disease. In 2017, China's food and drug authority approved a licence for the Ebola virus vaccine. An Ad5-EBOV vaccine was developed by CanSino Biologics Inc., based on the same principle. The product is stored as a stable lyophilized powder, which is convenient to transport and use in remote areas. GlaxoSmithKline (GSK) developed a recombinant chimpanzee adenovirus type-3 vectored vaccine against Zaire Ebola (ChAd3-EBO-Z). It has been registered with ClinicalTrials.gov (NCT02548078) for a phase two clinical trial. ChAd3-EBO-Z was proved to elicit immunity and good tolerance in children aged 1–17 years [2]. Since about 20% of cases were reported in children during the large Ebola virus outbreak in 2013–2016, the vaccine candidate developed by GSK shows a great value.

11.2.3.3 AdV-Based COVID-19 Vaccine

Since the first report of a patient infected with acute respiratory syndrome coronavirus 2 (SARS-CoV-2) in late 2019, over 240 million positive cases have been reported worldwide so far, causing over 4.8 fatalities. Currently, four AdV-based vaccines have been approved to be administered in parts of the world. The Ad5-nCOV developed by CanSino Biologics Inc. and Beijing Institute of Biotechnology used a HAd5 as the vector [38]. In Europe, Janssen Vaccines & Prevention B.V. (Johnson & Johnson) developed Ad26.COV2-S vaccine with a rare HAd26 vector [41]. Gamaleya Research Institute in Russia used a combination of HAd5 and HAd26 to develop Gam-COVID-Vac/SputnikV vaccine [25]. The University of Oxford collaborated with AstraZeneca to develop ChAdOX1-nCoV based on a chimpanzee (ChAdY25) vector [43].

11.3 VESICULAR STOMATITIS VIRUS (VSV) VECTORED VACCINES

11.3.1 Vesicular Stomatitis Virus

Vesicular stomatitis virus (VSV) is a member of the *Rhabdoviridae* family in the *Vesiculovirus* genus [62]. *Vesiculovirus* genus members are widely distributed in nature and mainly infect biting insects and livestock. According to the geographic distribution, at least 14 different phylogenetical and serological members in the *Vesiculovirus* genus are divided into two groups [63]. One includes the Indiana (IND) and New Jersey (NJ) serotypes of VSV, which were found in the Americas. The other was found in the Eastern Hemisphere including Chandipura, Yug Bogdanovac, and Isfahan [62]. The symptom of VSV infection of livestock is transient and accompanied by low level of viremia without a major virus spread. There are rare cases that VSV infects humans, but only when they are exposed to VSV in laboratories or when they come in close contact with the infected animals. Humans can be infected through skin and mucous tissues, while some cases are reported to be infected by insect bites. Human infection by VSV can be accompanied with disease symptoms including myalgia, fever, and headache, which resolve within days [64].

The standard VSV particle is a bullet-shaped, single-strand, negative-sense RNA virus with 65 × 180 nm. The viral genome is about 11 kilobases, which encodes five major viral proteins including nucleoprotein (N), phosphoprotein (P), matrix protein (M), glycoprotein (G), and the viral polymerase (L) [65]. The N protein associates to form viral nucleocapsid for genomic RNA, which serves as functional template for viral replication and transcription. This protein is also the most abundant protein expressed in infected cells. The M protein is the main protein in the VSV particle. The M protein has various functions in infected cells. This protein can regulate the viral transcription, inhibit the gene expression of host cells, and contribute to virus budding. The P and L genomic motifs associate to express the viral RNA polymerase with the functions of transcriptase and replicase. The G protein is a transmembrane glycoprotein on the virus surface with a trimeric spike-like structure. The G protein is responsible for virus attachment to the receptors of host cells.

11.3.2 Construction and Production Process of VSV Vectors

11.3.2.1 Pseudotype and Recombinant VSV

VSV is widely used as a viral vector in various fields including vaccines, gene therapy, and oncolytic virotherapy. Most virus-based vectors with envelopes are pseudotyped viruses. A pseudotyped virus is a viral particle bearing other viral envelopes or host cell proteins. Accordingly, generation of pseudotyped VSV with a heterologous viral envelope can be achieved (Figure 11.2). The rVSV gene encoding VSV.G is replaced with a reporter transgene; for example, enhanced green fluorescent protein (EGFP). The heterologous viral envelope protein is expressed in the host cells via an expression plasmid before infection with rVSV-ΔG. The rVSV-ΔG virions can finally encode EGFP with a heterologous viral envelope (Figure 11.3). Such pseudotyped viruses have been used in identifying novel viral receptors, understanding the

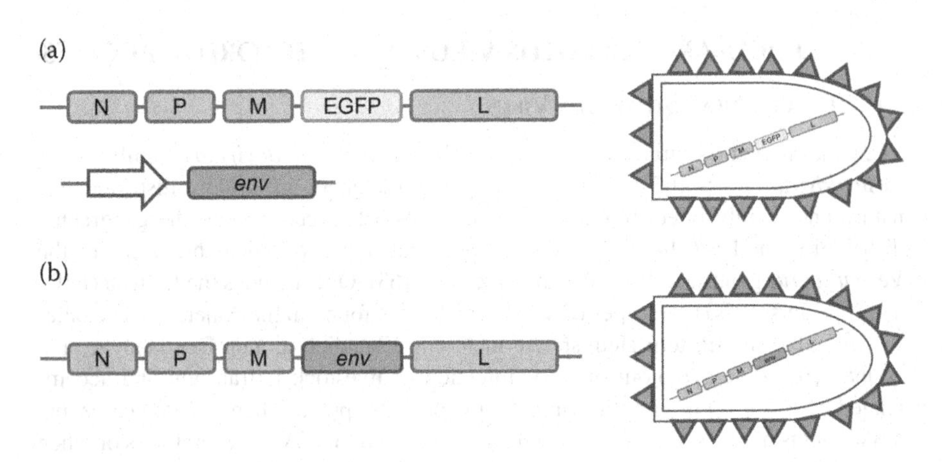

FIGURE 11.2 (a) Construction of rVSV-ΔG pseudotyped with a heterologous viral envelope. (b) Construction of replication-competent rVSV expressing a heterologous viral envelope.

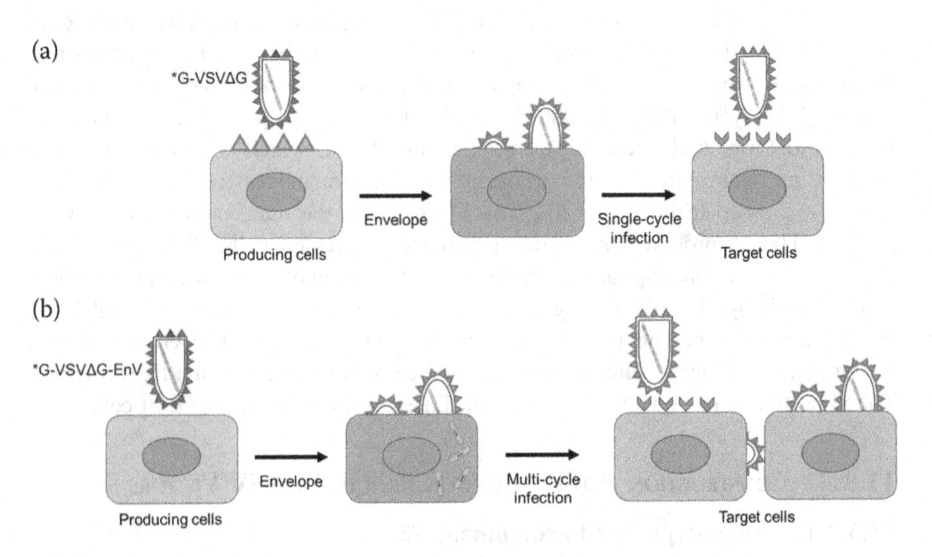

FIGURE 11.3 (a) Schematic diagram of the production of pseudotyped VSV. (b) Schematic diagram of the production of recombinant VSV.

entry mechanisms and vaccines. Additionally, pseudotyped viruses have some applications in the research on viruses with high risk [65].

Another recombinant mode of VSV is generated by replacing the gene encoding for VSV G by the gene encoding for another viral envelope protein (Figure 11.2). The rVSV is replication-competent both in vivo and in vitro. Unlike the pseudotyped virus which can realize single-step infection, the recombinant VSV can serve as an authentic tool for viral infection to produce a replication-competent vaccine or non-neurotropic oncolytic virus (Figure 11.3).

11.3.2.2 VSV As a Vaccine Vector

Many advantages have been demonstrated using vectored vaccines to tackle infectious diseases in recent years. The proteins expressed by the vectors aim to stimulate a specific and balanced immune response. Compared with most existing live-attenuated versions of vaccines, modified viral vectored vaccines serve as an alternative strategy when tackling high-risk diseases such as HIV and Ebola. VSV-based vaccines possess many advantages in stability, safety, and efficacy. VSV vector shows high capacity and stability in insertion of transgenes and relatively low toxicity since they do not integrate into the host cell genome when replicating. As a rare virus type in humans, pre-existing immunity for VSV is less an issue than it is for AdV. In addition, the results from preclinical and clinical studies demonstrate that VSV-based vaccines can induce strong cellular and humoral immune responses.

The recent VSV success story was the FDA-approved Ebola vaccine using a recombinant replication-competent VSV-based vector carrying the glycoprotein of a Zaire Ebolavirus (ZEBOV), the main antigen of the Ebolavirus [66]. The progress achieved with rVSV-ZEBOV indicates the potential to use this rVSV vector as a platform against other emerging viral infections. Recently, three novel rVSV constructions have been reported, which harbor distinct glycoproteins of the human immunodeficiency virus (HIV) [67]. Additionally, several rVSV-based vectored vaccines are in progress, including vaccines against measles, Middle East respiratory syndrome (MERS), Lassa fever, and Marburg virus disease [68–71]. Numerous clinical trials have been initiated with the VSV vector. As shown in Table 11.3, clinical studies involving VSV were performed to control the diseases in situations of pandemics or cancers.

Despite some successes, there are two main drawbacks using VSV-based vaccines. First, the natural VSV is a neurotropic virus. Therefore, the concern of neurotoxicity-related diseases such as encephalitis raises when using the live replication-competent virus. Second, when comes to the second dose, the pre-existing immunologic response issue occurs because of the administration of the same vaccine vector. However, research on heterologous prime-boost regimens has had some successes.

11.3.2.3 Cell-Line Selection

The Vero (African green monkey kidney) cells, recommended by the World Health Organization for human-use vaccine production, are the most widely used continuous cell line for vaccine production. The original Vero cells are anchorage-dependent cells. The current rVSV-ZEBOV vaccine is manufactured using adherent

TABLE 11.3
Clinical trials involving VSV vector[a]

Vector	Target	Disease	Number of trials in phase 1/2 and 2, 3, & 4	NCT number
VSV	Infectious diseases	Ebolaviruses	4 (phase 1)	NCT02280408; NCT04906629; NCT02933931; NCT02314923
		HIV	2 (phase 1)	NCT01438606; NCT01578889
		COVID-19	1 (phase 2/3)	NCT04990466
	Therapeutic cancer vaccines	Melanoma	1 (phase 1)	NCT03865212
		Solid tumor	1 (phase 1)	NCT02923466
			1 (phase 1/2)	NCT03647163
		Head/neck/lung	1 (phase 1/2)	NCT03647163
		Endometrial cancer	1 (phase 1)	NCT03456908

Note
[a] Data from https://clinicaltrials.gov.

Vero cells in roller bottles [72]. For cost-effective manufacturing, more scalable bioprocesses including fixed-bed bioreactors and microcarrier bioreactors have been developed [71]. However, the use of adherent cells still has scale-up limitations. For example, in cell expansion steps, the cells are required to detach and re-attach to surfaces involving enzymatic solutions such as trypsin, which complicate the process. For process scale-up, adaptation of anchorage-dependent cells to suspension cultures is valuable, since the transfer of cells to larger bioreactors is straightforward. In addition, serum is needed in most cell growth. However, the addition of serum presents critical drawbacks including the risk of hypersensitivity, serum variability, and contaminations. Therefore, the requirement of biological safety motivated industries to develop serum-free media.

Some successful results of adaptation have been reported using proprietary media. Shen et al. adapted the Vero cells to suspension culture and growth in serum-free media. They successfully use the system to produce rVSV-GFP in batch and perfusion bioreactors [73]. Kiesslich et al. further adapted the Vero cells to grow in a commercially available medium. In addition, they produced different constructions of rVSV including rVSV-ZEBOV, rVSV-HIV, and $rVSV_{Ind}$-msp-S_F-Gtc [71]. Additionally, Elahi et al. produced VSV-GFP in suspension cell culture and serum-free medium using, HEK293SF-3F6 derived from HEK293A cells and SF-BMAdR cells (a variant of A549) [45].

11.3.2.4 Production Process of VSV Vector

The manufacturing process of the rVSV vaccine starts with cell-line expansion in shake flasks. Then, the cells are seeded in a bioreactor. Fewer papers have focused on the bioreactor operation modes. Currently, single-use bioreactors are used more commonly in process development and large-scale production. Once harvested, the viral stock is subjected to enzyme treatment, essentially Benzonase. Then, the supernatant is transferred to purification processes and concentration. A critical step after purification is the formulation step, which has a significant effect on transport and storage (Figure 11.4). Two parameters, genomic viral particles and infectious viral titer, are usually used as the characterization of rVSV production. The genomic viral particles that can be measured by ddPCR are equivalent to the total particles. The infectious viral titer can be measured by TCID50, which can give us the infectivity or functionality of the viruses. Although people calculate the dose based on the infectious unit, the load that is injected to the patient in terms of total particles matters since too much virus particles can result in a cytokine storm.

11.3.3 Example of VSV-Based Vaccines

11.3.3.1 VSV-Based Ebola Vaccine

The Ebola virus (EBOV) outbreak in West Africa from 2013 to 2016 accelerated the development of several antivirals and vaccine candidates that have been evaluated in clinical trials. Among those, a replication-competent VSV-based vaccine which expresses the glycoprotein of Ebola was developed and tested in animal models and in phase 1–3 clinical trials in Europe, North America, and Africa [74–76]. Although some side effects were still observed in vaccination, the results of clinical trials for rVSV-ZEBOV demonstrated safety and strong immunogenicity in humans [61]. In 2019, rVSV-ZEBOV was approved by FDA for emergency use in individuals at risk for Ebola virus disease. In the rVSV-ZEBOV vaccine, the glycoprotein (GP) of the EBOV is inserted into a VSV vector in which the G open reading frame is deleted. The rVSV was generated as a replication-competent virus particle expressing EBOV GP on its surface.

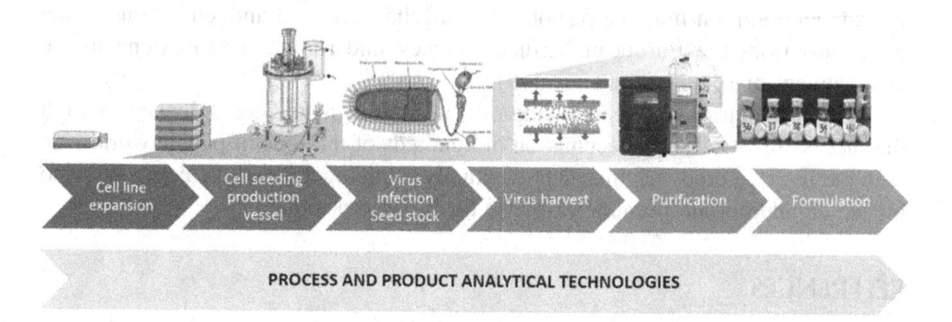

FIGURE 11.4 Schematic representation of the bioprocesses for VSV viral vector production.

11.3.3.2 VSV-Based COVID-19 Vaccine Candidates

In the race of COVID-19 vaccines, several rVSV-based vaccine candidates against COVID-19 expressing the spike protein of SARS-CoV-2 are evaluated in preclinical trials (University of Western Ontario, Canada; University of Manitoba, Canada; Aurobindo Pharma, India; FBRI SRC VB VECTOR, Russia; and Israel Institute for Biological Research/Weizmann Institute of Science, Israel) and clinical trials (Table 11.3). The construct of $rVSV_{Ind}$-*msp*-S_F-*Gtc* used in COVID-19 vaccine candidates is temperature-sensitive (VSV_{Ind}-GML mutant), which is avirulent in vivo and shows reduced cytopathic effect in vitro at 37°C, but replicates well at 31°C [77]. This characteristic can increase the safety of rVSV for human use. The $rVSV_{Ind}$-*msp*-S_F-*Gtc* expressed the SARS-CoV-2 spike protein, the honeybee melittin signal peptide, and the VSV-G transmembrane protein.

11.4 OVERALL CONCLUSION ON VECTORED VACCINES

In 2013, the WHO had informal consultations on the topic "Where are we with the development of viral vectored vaccine?" [78] The response to this question years later can be summarized in the following. In the case of AdV vectors, there are a number of human and non-human primates AdV serotypes that have been evaluated in preclinical and clinical trials, including HAd5, HAd26, HAd35, ChAd63, etc. Successful applications have been achieved during the COVID-19 pandemic with the licensing of vectored vaccines using HAd5, HAd26, and ChAdY25 serotypes. Vaccination against colds provoked by the HAd4 and HAd7 in the U.S. army is delivered orally in the form of tablets. There is also progress in developing orally delivered HAd4 with the H5N1 avian strain, as illustrated by advanced clinical trials.

The recommendation from the WHO working group is that platforms should be developed in the case of a pandemic situation. They called for the development of novel vectors as potential platforms, evaluation of viral vectors in heterologous prime-boost regimens, improvement on manufacturing technology toward continuous cell culture (e.g., poxvirus uses chicken fibroblasts that are generated each time), and closer collaboration between veterinary and human vaccine developers.

"Live viral vectors that express heterologous antigens are being extensively investigated in the development of novel vaccines and it is believed that these will provide an optimum immune response toward the expressed antigen" is one of the guidelines from the European Medical Agency underlining a solid trend in the development of vectored vaccines [79].

Years later, large-scale convincing demonstrations have been achieved through the successful COVID-19 vaccination of hundreds of millions of people worldwide with the live vectored vaccines that eventually bring together all the components to induce an appropriate immune response for broad protection.

REFERENCES

[1] S. A. Mendonca, R. Lorincz, P. Boucher, and D. T. Curiel, "Adenoviral vector vaccine platforms in the SARS-CoV-2 pandemic," *NPJ Vaccines*, vol. 6, no. 1, p. 97, Aug. 2021, doi: 10.1038/s41541-021-00356-x

[2] M. D. Tapia *et al.*, "Safety, reactogenicity, and immunogenicity of a chimpanzee adenovirus vectored Ebola vaccine in children in Africa: A randomised, observer-blind, placebo-controlled, phase 2 trial," *Lancet Infect. Dis.*, vol. 20, no. 6, pp. 719–730, 2020, doi: 10.1016/s1473-3099(20)30019-0
[3] N. Tatsis and H. C. Ertl, "Adenoviruses as vaccine vectors," *Mol. Ther.*, vol. 10, no. 4, pp. 616–629, Oct 2004, doi: 10.1016/j.ymthe.2004.07.013
[4] D. M. Knipe, D. Baltimore, and H. F. Lodish, "Separate pathways of maturation of the major structural proteins of vesicular stomatitis virus," *J. Virol.*, vol. 21, no. 3, pp. 1128–1139, 1977, doi: 10.1128/JVI.21.3.1128-1139.1977
[5] E. Sasso, A. M. D'Alise, N. Zambrano, E. Scarselli, A. Folgori, and A. Nicosia, "New viral vectors for infectious diseases and cancer," *Semin. Immunol.*, vol. 50, p. 101430, Aug 2020, doi: 10.1016/j.smim.2020.101430
[6] H. Y. Naim, "Measles virus A pathogen, vaccine, and a vector," *Hum. Vaccines Immunother*, Article; Proceedings Paper vol. 11, no. 1, pp. 21–26, Jan. 2015, doi: 10.4161/hv.34298
[7] K. Ljungberg and P. Liljestrom, "Self-replicating alphavirus RNA vaccines," *Expert Rev. Vaccines*, vol. 14, no. 2, pp. 177–194, Feb. 2015, doi: 10.1586/14760584.2015.965690
[8] Y. Kaname *et al.*, "Acquisition of Complement Resistance through Incorporation of CD55/Decay-Accelerating Factor into Viral Particles Bearing Baculovirus GP64," *J. Virol.*, vol. 84, no. 7, pp. 3210–3219, Apr 2010, doi: 10.1128/jvi.02519-09
[9] B. Guy, F. Guirakhoo, V. Barban, S. Higgs, T. P. Monath, and J. Lang, "Preclinical and clinical development of YFV 17D-based chimeric vaccines against dengue, West Nile and Japanese encephalitis viruses," *Vaccine*, vol. 28, no. 3, pp. 632–649, Jan. 2010, doi: 10.1016/j.vaccine.2009.09.098
[10] B. Guy, M. Saville, and J. Lang, "Development of sanofi pasteur tetravalent dengue vaccine," *Hum. Vaccines*, vol. 6, no. 9, pp. 696–705, 2014, doi: 10.4161/hv.6.9.12739
[11] R. A. Berkeley *et al.*, "Antibody-Neutralized Reovirus Is Effective in Oncolytic Virotherapy," *Cancer Immunol. Res.*, vol. 6, no. 10, pp. 1161–1173, Oct. 2018, doi: 10.1158/2326-6066.CIR-18-0309
[12] E. V. Shashkova, M. N. Kuppuswamy, W. S. Wold, and K. Doronin, "Anticancer activity of oncolytic adenovirus vector armed with IFN-alpha and ADP is enhanced by pharmacologically controlled expression of TRAIL," *Cancer Gene Ther.*, vol. 15, no. 2, pp. 61–72, Feb. 2008, doi: 10.1038/sj.cgt.7701107
[13] J. L. Imler, "Adenovirus vectors as recombinant viral vaccines," *Vaccine*, vol. 13, no. 13, pp. 1143–1151, Sep .1995, doi: 10.1016/0264-410x(95)00032-v
[14] D. S. Bangari and S. K. Mittal, "Development of nonhuman adenoviruses as vaccine vectors," *Vaccine*, vol. 24, no. 7, pp. 849–862, Feb. 2006, doi: 10.1016/j.vaccine.2005.08.101
[15] K. Ewer, S. Sebastian, A. J. Spencer, S. Gilbert, A. V. S. Hill, and T. Lambe, "Chimpanzee adenoviral vectors as vaccines for outbreak pathogens," *Hum. Vaccines Immunother.*, vol. 13, no. 12, pp. 3020–3032, 2017, doi: 10.1080/21645515.2017.1383575
[16] C. Zhang, Y. D. Chi, and D. M. Zhou, "Development of Novel Vaccines Against Infectious Diseases Based on Chimpanzee Adenoviral Vector," in *Recombinant Virus Vaccines: Methods and Protocols*, vol. 1581, M. C. Ferran and G. R. Skuse Eds., Humana New York, NY, Methods in Molecular Biology, 2017, pp. 3–13.
[17] A. J. Davison, M. Benko, and B. Harrach, "Genetic content and evolution of adenoviruses," *J. Gen. Virol.*, vol. 84, pp. 2895–2908, Nov. 2003, doi: 10.1099/vir.0.19497-0
[18] T. Steffen *et al.*, "Immunogenicity and efficacy of a recombinant human adenovirus type 5 vaccine against Zika virus," *Vaccines (Basel)*, vol. 8, no. 2, Apr. 2020, doi: 10.3390/vaccines8020170

[19] C. F. Shen *et al.*, "Process optimization and scale-up for production of rabies vaccine live adenovirus vector (AdRG1.3)," *Vaccine*, vol. 30, no. 2, pp. 300–306, Jan. 2012, doi: 10.1016/j.vaccine.2011.10.095
[20] T. C. Mast *et al.*, "International epidemiology of human pre-existing adenovirus (Ad) type-5, type-6, type-26 and type-36 neutralizing antibodies: Correlates of high Ad5 titers and implications for potential HIV vaccine trials," *Vaccine*, vol. 28, no. 4, pp. 950–957, Jan. 2010, doi: 10.1016/j.vaccine.2009.10.145
[21] X. Wang *et al.*, "Neutralizing antibody responses to enterovirus and adenovirus in healthy adults in China," *Emerg. Microbes Infect.*, vol. 3, May 2014, Art no. e30, doi: 10.1038/emi.2014.30
[22] L. Holterman *et al.*, "Novel replication-incompetent vector derived from adenovirus type 11 (Ad11) for vaccination and gene therapy: Low seroprevalence and non-cross-reactivity with Ad5," *J. Virol.*, vol. 78, no. 23, pp. 13207–13215, Dec. 2004, doi: 10.1128/jvi.78.23.13207-13215.2004
[23] R. Vogels *et al.*, "Replication-deficient human adenovirus type 35 vectors for gene transfer and vaccination: Efficient human cell infection and bypass of preexisting adenovirus immunity," *J. Virol.*, vol. 77, no. 15, pp. 8263–8271, Aug. 2003, doi: 10.1128/jvi.77.15.8263-8271.2003
[24] E. A. Weaver and M. A. Barry, "Low seroprevalent species D adenovirus vectors as influenza vaccines," *PloS one*, vol. 8, no. 8, Aug 2013, Art no. e73313, doi: 10.13 71/journal.pone.0073313
[25] N. C. Kyriakidis, A. Lopez-Cortes, E. V. Gonzalez, A. B. Grimaldos, and E. O. Prado, "SARS-CoV-2 vaccines strategies: a comprehensive review of phase 3 candidates," *NPJ Vaccines*, vol. 6, no. 1, p. 28, Feb. 2021, doi: 10.1038/s41541-021-00292-w
[26] N. Belousova *et al.*, "Genetically targeted adenovirus vector directed to CD40-expressing cells," *J. Virol.*, vol. 77, no. 21, pp. 11367–11377, Nov. 2003, doi: 10.1128/jvi.77.21.11367-11377.2003
[27] N. Belousova, V. Krendelchtchikova, D. T. Curiel, and V. Krasnykh, "Modulation of adenovirus vector tropism via incorporation of polypeptide ligands into the fiber protein," *J. Virol.*, vol. 76, no. 17, pp. 8621–8631, Sep. 2002, doi: 10.1128/jvi.76.1 7.8621-8631.2002
[28] D. Sharon and A. Kamen, "Advancements in the design and scalable production of viral gene transfer vectors," *Biotechnol. Bioeng.*, vol. 115, no. 1, pp. 25–40, Jan. 2018, doi: 10.1002/bit.26461
[29] G. R. Nemerow, P. L. Stewart, and V. S. Reddy, "Structure of human adenovirus," *Curr. Opin. Virol.*, vol. 2, no. 2, pp. 115–121, Apr. 2012, doi: 10.1016/j.coviro.2011.12.008
[30] N. Arnberg, "Adenovirus receptors: implications for targeting of viral vectors," *Trends in Pharmacological Sciences*, vol. 33, no. 8, pp. 442–448, Aug. 2012, doi: 10.1016/j.tips.2012.04.005
[31] D. Majhen and A. Ambriovic-Ristov, "Adenoviral vectors - How to use them in cancer gene therapy?," *Virus Res.*, vol. 119, no. 2, pp. 121–133, Aug. 2006, doi: 10.1016/j.virusres.2006.02.001
[32] X. Danthinne and M. J. Imperiale, "Production of first generation adenovirus vectors: a review," *Gene Therapy*, vol. 7, no. 20, pp. 1707–1714, Oct. 2000, doi: 10.103 8/sj.gt.3301301
[33] I. Kovesdi and S. J. Hedley, "Adenoviral producer cells," *Viruses-Basel*, vol. 2, no. 8, pp. 1681–1703, Aug. 2010, doi: 10.3390/v2081681
[34] Y. C. Lin *et al.*, "Genome dynamics of the human embryonic kidney 293 lineage in response to cell biology manipulations," *Nat. Commun.*, vol. 5, Sep. 2014, Art no. 4767, doi: 10.1038/ncomms5767

[35] G. N. Stacey and O. W. Merten, "Host Cells and Cell Banking," in *Viral Vectors for Gene Therapy: Methods and Protocols*, vol. 737, O. W. Merten and M. AlRubeai Eds. Methods in Molecular Biology, 2011, pp. 45–88.
[36] K. M. Hehir *et al.*, "Molecular characterization of replication-competent variants of adenovirus vectors and genome modifications to prevent their occurrence," *J. Virol.*, vol. 70, no. 12, pp. 8459–8467, Dec. 1996, doi: 10.1128/jvi.70.12.8459-8467.1996
[37] F. J. Fallaux *et al.*, "New helper cells and matched early region 1-deleted adenovirus vectors prevent generation of replication-competent adenoviruses," *Hum. Gene Therapy*, vol. 9, no. 13, pp. 1909–1917, Sep. 1998, doi: 10.1089/hum.1998.9.13-1909
[38] S. Wu *et al.*, "A single dose of an adenovirus-vectored vaccine provides protection against SARS-CoV-2 challenge," *Nat. Commun.*, vol. 11, no. 1, p. 4081, Aug. 2020, doi: 10.1038/s41467-020-17972-1
[39] J. Vellinga *et al.*, "Challenges in manufacturing adenoviral vectors for global vaccine product deployment," *Hum. Gene Therapy*, vol. 25, no. 4, pp. 318–327, Apr. 2014, doi: 10.1089/hum.2014.007
[40] M. Havenga *et al.*, "Novel replication-incompetent adenoviral B-group vectors: high vector stability and yield in PER.C6 cells," *J. Gen. Virol.*, vol. 87, pp. 2135–2143, Aug. 2006, doi: 10.1099/vir.0.81956-0
[41] N. B. Mercado *et al.*, "Single-shot Ad26 vaccine protects against SARS-CoV-2 in rhesus macaques," *Nature*, vol. 586, no. 7830, pp. 583–588, Oct. 2020, doi: 10.1038/s41586-020-2607-z
[42] M. Szelechowski, C. Bergeron, D. Gonzalez-Dunia, and B. Klonjkowski, "Production and purification of non replicative canine adenovirus type 2 derived vectors," *Jove-J. Visual. Exp.*, no. 82, Dec. 2013, Art no. e50833, doi: 10.3791/50833
[43] N. van Doremalen *et al.*, "ChAdOx1 nCoV-19 vaccine prevents SARS-CoV-2 pneumonia in rhesus macaques," *Nature*, vol. 586, no. 7830, pp. 578–582, Oct. 2020, doi: 10.1038/s41586-020-2608-y
[44] A. Kamen and O. Henry, "Development and optimization of an adenovirus production process," *J. Gene Med.*, vol. 6 Suppl 1, pp. S184–S192, Feb. 2004, doi: 10.1002/jgm.503
[45] S. M. Elahi, C. F. Shen, and R. Gilbert, "Optimization of production of vesicular stomatitis virus (VSV) in suspension serum-free culture medium at high cell density," *J. Biotechnol.*, vol. 289, pp. 144–149, Jan. 2019, doi: 10.1016/j.jbiotec.2018.11.023
[46] E. Petiot, M. Cuperlovic-Culf, C. F. Shen, and A. Kamen, "Influence of HEK293 metabolism on the production of viral vectors and vaccine," *Vaccine*, vol. 33, no. 44, pp. 5974–5981, Nov. 2015, doi: 10.1016/j.vaccine.2015.05.097
[47] H. Hovel, "INFLUENCE OF MEDIUM OSMOLARITY ON VIRUS INFECTED CELL CULTURES," *Arzneimittel-Forschung*, vol. 21, no. 6, pp. 899-&, 1971. doi://WOS:A1971J712200044.
[48] Y. S. Tsao, R. Condon, E. Schaefer, P. Lio, and Z. Liu, "Development and improvement of a serum-free suspension process for the production of recombinant adenoviral vectors using HEK293 cells," *Cytotechnology*, vol. 37, no. 3, pp. 189–198, 2001, doi: 10.1023/a:1020555310558
[49] H. Kallel and A. A. Kamen, "Large-scale adenovirus and poxvirus-vectored vaccine manufacturing to enable clinical trials," *J. Biotechnol.*, vol. 10, no. 5, pp. 741–U124, May 2015, doi: 10.1002/biot.201400390
[50] L. Z. Xie *et al.*, "Large-scale propagation of a replication-defective adenovirus vector in stirred-tank bioreactor PER.C6 (TM) cell culture under sparging conditions," *Biotechnol. Bioeng.*, vol. 83, no. 1, pp. 45–52, Jul. 2003, doi: 10.1002/bit.10644
[51] K. Yamada, N. Morishita, T. Katsuda, S. Kubo, A. Gotoh, and H. Yamaji, "Adenovirus vector production using low-multiplicity infection of 293 cells," *Cytotechnology*, vol. 59, no. 3, pp. 153–160, Apr. 2009, doi: 10.1007/s10616-009-9208-x

[52] A. Bavelloni, M. Piazzi, M. Raffini, I. Faenza, and W. L. Blalock, "Prohibitin 2: At a communications crossroads," *IUBMB Life*, vol. 67, no. 4, pp. 239–254, Apr. 2015, doi: 10.1002/iub.1366

[53] J. Transfiguracion, A. Bernier, N. Arcand, P. Chahal, and A. Kamen, "Validation of a high-performance liquid chromatographic assay for the quantification of adenovirus type 5 particles," *J. Chromatogr. B-Anal. Technol. Biomed. Life Sci.*, vol. 761, no. 2, pp. 187–194, Sep. 2001, doi: 10.1016/s0378-4347(01)00330-9

[54] J. Transfiguracion *et al.*, "Rapid in-process monitoring of lentiviral vector particles by high-performance liquid chromatography," *Mol. Ther. Methods Clin. Dev.*, vol. 18, pp. 803–810, Sep. 2020, doi: 10.1016/j.omtm.2020.08.005

[55] J. D. Zhu *et al.*, "Characterization of replication-competent adenovirus isolates from large-scale production of a recombinant adenoviral vector," *Hum. Gene Therapy*, vol. 10, no. 1, pp. 113–121, Jan. 1999, doi: 10.1089/10430349950019246

[56] B. A.-C. Hutchins, E. Simek, S. Bauer, S. Carson K.L., "Development of a Reference Material for Characterizing Adenovirus Vectors," *Molecular Therapy*, vol. 5, no. 5, Supplement, p. S62, 2002, doi: 10.1016/S1525-0016(16)43012-1

[57] V. Aida *et al.*, "Novel vaccine technologies in veterinary Medicine: A Herald to human medicine vaccines," *Front. Vet. Sci.*, vol. 8, Apr. 2021, Art no. 654289, doi: 10.3389/fvets.2021.654289

[58] M. J. Grubman *et al.*, "Adenovirus serotype 5-vectored foot-and-mouth disease subunit vaccines: the first decade," *Future Virol.*, vol. 5, no. 1, pp. 51–64, Jan. 2010, doi: 10.2217/fvl.09.68

[59] J. G. Neilan *et al.*, "Efficacy of an adenovirus-vectored foot-and-mouth disease virus serotype A sub-unit vaccine in cattle using a direct contact transmission model," *BMC Vet. Res.*, vol. 14, no. 1, p. 254, Aug. 2018, doi: 10.1186/s12917-01 8-1582-1

[60] J. Barrera *et al.*, "Safety profile of a replication-deficient human adenovirus-vectored foot-and-mouth disease virus serotype A24 sub-unit vaccine in cattle," *Transbound. Emerg. Dis.*, vol. 65, no. 2, pp. 447–455, Apr. 2018, doi: 10.1111/tbed.12724

[61] E. Suder, W. Furuyama, H. Feldmann, A. Marzi, and E. de Wit, "The vesicular stomatitis virus-based Ebola virus vaccine: From concept to clinical trials," *Hum. Vaccine Immunother*, vol. 14, no. 9, pp. 2107–2113, 2018, doi: 10.1080/21645515.2018.1473698

[62] D. K. Clarke, D. Cooper, M. A. Egan, R. M. Hendry, C. L. Parks, and S. A. Udem, "Recombinant vesicular stomatitis virus as an HIV-1 vaccine vector," *Springer Semin. Immunopathol.*, vol. 28, no. 3, pp. 239–253, Nov. 2006, doi: 10.1007/s002 81-006-0042-3

[63] R. B. Tesh, A. Darosa, and J. S. T. Darosa, "Antigenic relationship among rhabdoviruses infecting terrestrial vertebrates," *J. Gen. Virol.*, vol. 64, no. JAN, pp. 169–176, 1983, doi: 10.1099/0022-1317-64-1-169

[64] C. H. Calisher *et al.*, "A newly recognized vesiculovirus, calchaqui virus, and subtypes of melao and maguari viruses from argentina, with serologic evidence for infections of humans and horses," *Am. J. Trop. Med. Hyg.*, vol. 36, no. 1, pp. 114–119, Jan 1987, doi: 10.4269/ajtmh.1987.36.114

[65] H. Tani, S. Morikawa, and Y. Matsuura, "Development and applications of VSV vectors based on cell tropism," *Front. Microbiol.*, vol. 3, 2012, Art no. 272, doi: 10.3389/fmicb.2011.00272

[66] E. Suder, W. Furuyama, H. Feldmann, A. Marzi, and E. de Wit, "The vesicular stomatitis virus-based Ebola virus vaccine: From concept to clinical trials," *Hum. Vaccines Immunother.*, vol. 14, no. 9, pp. 2107–2113, 2018, doi: 10.1080/21645515.2018.1473698

[67] M. Mangion *et al.*, "Evaluation of novel HIV vaccine candidates using recombinant vesicular stomatitis virus vector produced in serum-free Vero cell cultures," *Vaccine*, vol. 38, no. 50, pp. 7949–7955, Nov. 2020, doi: 10.1016/j.vaccine.2020.10.058

[68] T. W. Geisbert and H. Feldmann, "Recombinant vesicular stomatitis virus–based vaccines against Ebola and Marburg virus infections," *J. Infect. Dis.*, vol. 204, no. suppl_3, pp. S1075–S1081, 2011, doi: 10.1093/infdis/jir349

[69] A. M. Henao-Restrepo *et al.*, "Efficacy and effectiveness of an rVSV-vectored vaccine in preventing Ebola virus disease: Final results from the Guinea ring vaccination, open-label, cluster-randomised trial (Ebola Ça Suffit!)," *Lancet*, vol. 389, no. 10068, pp. 505–518, 2017, doi: 10.1016/S0140-6736(16)32621-6

[70] A. M. Munis, E. M. Bentley, and Y. Takeuchi, "A tool with many applications: vesicular stomatitis virus in research and medicine," *Expert Opin. Biol. Ther.*, vol. 20, no. 10, pp. 1187–1201, Oct. 2020, doi: 10.1080/14712598.2020.1787981

[71] S. Kiesslich, G. N. Kim, C. F. Shen, C. Y. Kang, and A. A. Kamen, "Bioreactor production of rVSV-based vectors in Vero cell suspension cultures," *Biotechnol. Bioeng.*, vol. 118, no. 7, pp. 2649–2659, Jul. 2021, doi: 10.1002/bit.27785

[72] T. P. Monath *et al.*, "rVSV Delta G-ZEBOV-GP (also designated V920) recombinant vesicular stomatitis virus pseudotyped with Ebola Zaire Glycoprotein: Standardized template with key considerations for a risk/benefit assessment," *Vaccine: X*, vol. 1, Apr. 2019, Art no. 100009, doi: 10.1016/j.jvacx.2019.100009

[73] C. F. Shen *et al.*, "Development of suspension adapted Vero cell culture process technology for production of viral vaccines," *Vaccine*, vol. 37, no. 47, pp. 6996–7002, Nov. 2019, doi: 10.1016/j.vaccine.2019.07.003

[74] A. Huttner *et al.*, "The effect of dose on the safety and immunogenicity of the VSV Ebola candidate vaccine: a randomised double-blind, placebo-controlled phase 1/2 trial," *Lancet Infect. Dis.*, vol. 15, no. 10, pp. 1156–1166, Oct. 2015, doi: 10.1016/s1473-3099(15)00154-1

[75] B. A. G. Coller *et al.*, "Clinical development of a recombinant Ebola vaccine in the midst of an unprecedented epidemic," *Vaccine*, vol. 35, no. 35, pp. 4465–4469, Aug. 2017, doi: 10.1016/j.vaccine.2017.05.097

[76] J. A. Regules *et al.*, "A recombinant vesicular stomatitis virus Ebola vaccine," *N. Engl. J. Med.*, vol. 376, no. 4, pp. 330–341, Jan 2017, doi: 10.1056/NEJMoa1414216

[77] G. N. Kim, K. Wu, J. P. Hong, Z. Awamleh, and C. Y. Kang, "Creation of matrix protein Gene variants of two serotypes of vesicular stomatitis virus as prime-boost vaccine vectors," *J. Virol.*, vol. 89, no. 12, pp. 6338–6351, Jun. 2015, doi: 10.1128/jvi.00222-15

[78] "WHO informal consultation on characterization and quality aspect of vaccines based on live viral vectors,"Geneva: WHO HQ, Dec. 2003.

[79] "Guideline on quality, non-clinical and clinical aspects of live recombinant viral vectored vaccines," Committee for Medicinal Product for Human Use (CHMP), 24 Jun 2010.

12 Design and production of vaccines against COVID-19 using established vaccine platforms

Ryan Kligman
McGill, Faculty of medicine,Montréal, QC, Canada

Jesús Lavado-García
Grup d'Enginyeria Cel·lular i Bioprocés, Universitat Autònoma de Barcelona

Amine Kamen
Viral Vectors and Vaccines Bioprocessing Group, Department of Bioengineering, McGill University, Montréal, QC, Canada

CONTENTS

DOI: 10.1201/9781003229797-12

12.1 INTRODUCTION

In late December 2019, a new species of coronavirus was discovered in Wuhan, China, causing a severe acute respiratory syndrome and was aptly dubbed SARS-CoV-2. The virus quickly spread around the globe and the clinical disease it causes, coronavirus disease 2019 (or COVID-19), was declared a global emergency of very high risk on February 28th, 2020. Soon after, on March 12th, 2020, these COVID-19 outbreaks were declared a pandemic [1].

The clinical presentation of COVID-19 varies substantially among individuals ranging from asymptomatic infection to multi-organ failure, which accounts for the difficulty in controlling its transmission. It largely causes upper and lower respiratory pathology; however, the virus has tropism for virtually every tissue-type (respiratory tract, gastrointestinal tract, central nervous system, cardiovascular system, etc.), accounting for its diverse symptomology. Clinical features include fatigue, fever, myalgias, cough, anosmia, ageusia, headache, nausea, vomiting, diarrhea, and so on, which makes it nearly impossible to predict the clinical course in any given individual. Further, older adults and those with underlying co-morbidities are more likely to experience severe infection, and subsequent complications like acute respiratory distress syndrome (ARDS), cytokine release syndrome (CRS), systemic inflammatory response syndrome (SIRS), septic shock, acute kidney injury (AKI), multi-organ failure, and cardiovascular complications like acute coronary syndrome and stroke [2,3]. However, even in healthy individuals with mild infections, long-term consequences can be deleterious due to loss of tissue stem cells resulting in inhibited cellular repair and inflammatory fibrosis [2].

The global pandemic caused by this newly discovered betacoronavirus is not the first in recorded history. In fact, betacoronaviruses were the causative agents of two previous outbreaks in the last 20 years: namely, the SARS-CoV outbreak of 2003 and Middle Eastern respiratory syndrome (MERS-CoV) in 2012. SARS-CoV caused a similar respiratory syndrome and resulted in 8098 cases and 774 deaths, whereas the death rate for MERS-CoV was even higher at 35% (although this statistic omits asymptomatic transmission) [4]. As of the writing of this chapter, there have been approximately 235 million cases of COVID-19 and 4.8 million deaths, which is a death rate of ~2% [5].

It is interesting to note that the incidence of pandemics and epidemics has been steadily increasing over the last 200 years. This is most likely due to the increases in population density and most recently, globalization. From the Middle Ages to the 19th century there were really only two epidemics of note: the bubonic plague which began in 1347 and smallpox from the early 1500's. From the 19th century onwards, the world saw the following outbreaks: influenza "Great Pandemic" (1833), cholera (1881), Spanish influenza (1918), Asian influenza (1957), hepatitis C (1960s), Hong Kong influenza (1968), Russian influenza (1977), HIV (1981), SARS-CoV-1 (2003), H1N1 (2009), MERS-CoV (2012), Ebola virus (2013), chikungunya virus (2013), Zika virus (2015), and now COVID-19 [6]. While most authorities predicted that a new global outbreak was imminent, influenza was thought to be the most likely cause of humanity's next great pandemic and probabilistic modeling forecasted a 1% annual chance of an influenza pandemic that would result in 6 million deaths [6].

12.2 SARS-COV-2 BIOLOGY

Coronaviruses are single-stranded, positive-sense RNA viruses with the largest genome of any RNA virus at 26–32 kb [7]. They are enveloped with a diameter of 100–160 nm. They have 6–11 open reading frames (ORFs), the first of which codes for 16 different non-structural proteins involved in transcription and replication of the genome. The rest of the genome contains other ORFs that encode the four structural proteins including the envelope protein (E), the membrane glycoprotein (M), the nucleocapsid protein (N), and the spike glycoprotein (S) [1,8].

The S protein is composed of two sub-units, S1 and S2, and each S protein is a trimer of 3 S1/S2 sub-units [7]. It is the S protein that is responsible for host cell recognition, binding and cell entry. More specifically, it interacts with the angiotensin converting enzyme 2 (ACE2) receptor, which is expressed in nearly every tissue accounting for the virus's extensive tropism and diverse clinical presentation. The role of the S1 sub-unit is to bind to the ACE2 receptor via its receptor binding domain (RBD), while S2 promotes fusion of the viral envelope with the host's cell membrane [7].

Upon cell entry, the viral RNA genome is replicated by viral replicase and translated into viral proteins using host ribosomes. New viruses are then assembled causing cell lysis allowing them to spread to adjacent cells or to be expelled from the body via the respiratory tract in the form of droplets or aerosols [9].

There are seven different species of coronavirus known to infect humans–four of which cause mild infections and three which cause severe infections. The four milder sub-types are HCoV-NL63, HCoV-229E, HCoV-OC43, and HKU, which cause mild upper respiratory disease in immunocompetent hosts. Whereas the three more virulent subtypes are the aforementioned SARS-CoV, MERS-CoV, and, of course, SARS-CoV-2 [1].

There has been much debate as to the origins of SARS-CoV-2 with the most likely explanation being that it was derived via zoonotic transfer from a bat. It is unclear, however, whether it mutated in its animal host prior to transfer, or whether it underwent natural selection after crossing species [10].

As the pandemic has progressed, several variants of SARS-CoV-2 have emerged described as either variants of interest, variants of concern, or variants of high consequence. The criteria outlined by the WHO to define variants of concern are as follows: increased viral transmissibility, increased disease severity, and a decrease in the effectiveness of public health measures including among other things, vaccine effectiveness [11]. Three variants that rapidly gained global attention were B.1.1.7 (first identified in the UK), 501Y.V2 (first identified in South Africa), and P.1 (first identified in Brazil). All three of these variants acquired the N501Y mutation, which results in an amino acid substitution at position 501 of the RBD. Two of the variants have additional amino acid substitutions in the RBD which serve to increase the binding affinity of the RBD to ACE2 [12].

The B.1.1.7 variant, also known as the alpha variant, was first detected in England in November 2020 and until recently was the predominant strain in Europe and North America [11]. However, in December 2020, the B.1.617 strain (or delta) was discovered in India, which quickly spread around the globe and was designated as a variant of concern on May 6, 2021 [11]. In addition to being more transmissible than the alpha variant, its reinfection rate was also found to be higher. Studies are still underway to determine its effect on vaccine effectiveness, although there is evidence of a modest reduction and some variants have already demonstrated complete immune escape against certain vaccines [12,13]. Other variants of concern have been identified thereafter. Vaccines design and production are the subject of the remainder of this chapter. However, prior to diving into the various vaccines developed against SARS-CoV-2, we must first discuss the virus's immunology.

12.3 SARS-COV-2 IMMUNOLOGY AND VACCINE RATIONALE

As with any newly discovered pathogen, especially when it comes to designing vaccine candidates, understanding the way the body's immune system interacts with it is of vital importance. Studies thus far have demonstrated that SARS-CoV-2, as well as SARS-CoV and MERS-CoV, tend to suppress activation of the innate immune system. And since it is the effects of the immune system that result in clinical symptoms, this may help explain the long pre-symptomatic period (up to 14 days) that is seen in COVID-19. Furthermore, it has been suggested that suppression of the innate immune system may also contribute to the dysregulated inflammatory response seen in more severe cases [2].

However, vaccine design depends largely on the adaptive immune system. In general, there are two main components of the adaptive immune system: cellular immunity mediated by T-cells, and humoral immunity mediated by antibodies secreted by B-cells (as detailed in Chapter 3). It has been shown that upon natural infection, B-cells produce neutralizing antibodies against SARS-CoV-2 in two manners: firstly, by targeting the S protein and preventing its interaction with ACE2 and secondly, by binding to the virus cytoskeleton including the internal nucleoprotein and preventing release of the genome [4,7]. Early studies showed that in patients with COVID-19, antibodies were seen in their serum on average 8 days after exposure reaching a peak after 14 days [8]. Due to this natural response against the S protein, it is unsurprising that nearly all vaccines in development have chosen it as the target antigen for vaccine development. This was even seen in SARS-CoV where antibodies targeted to the S1 RBD blocked its interaction with ACE2 and antibodies targeted to other epitopes of the S1 sub-unit inhibited conformational changes of the S protein required for viral cell-entry [14].

Cellular immunity mediated by T-cells is equally as important in vaccine design. In a study looking at the immune response of COVID-19 patients, CD4+ and CD8+ T-cells were seen in 100% and 70% of patients, respectively. Furthermore, 27% of the CD4+ T-cell response was specific for the S protein [15]. Additionally, it has even been shown that patients with less severe COVID-19 infections have had a higher number of CD8+ T-cells, which further reinforces their role in the clinical outcome [16]. It can even be argued that the T-cell response is far more important than its B-cell counterpart, since, for example, contrary to B-cell epitopes, T-cell epitopes are located along the full length of the S protein. Therefore, since T-cells target multiple regions of the S protein, viral mutations have a lesser effect on cellular immunity [17].

Particularly earlier on in the pandemic, there was a lot of media attention relating to various experimental therapies for COVID-19 including antiviral agents, immunomodulators, anticoagulants, anti-inflammatories, etc. [2]. None of these pharmacotherapeutic modalities, as of the writing of this chapter, have proven to be effective, and besides systemic steroids in the case of SIRS and cytokine storm, COVID-19 continues to be treated symptomatically. Therefore, the current best strategy to ending this pandemic continues to be mass vaccination.

12.4 VACCINE DEVELOPMENT FOR SARS-COV-2

As previously detailed in Chapter 3, vaccines contain antigens in some form, either genetic or proteinaceous; and by exposing an individual to an inert antigen of interest, the immune system can be primed to recognize the pathogen in the event of a future infection. A good vaccine will create long-lasting immunity through both cellular and humoral memory in the form of T-cells and antibodies, respectively.

As previously discussed in Chapter 2, each virus is unique in terms of its route of infection, the types of cells it infects, and subsequently its clinical pathology. As such, the immune response required to fight off a virus will be unique to each type [18]. This understanding of the way the immune system controls a natural infection is crucial when designing a vaccine against a particular virus. Simply put, if the immune response elicited by a vaccine is not optimal for a given virus, it will not offer much protection against a natural infection.

Since the discovery of the first vaccine in the 18th century, the conventional platform for most vaccines have been either inactivated viruses (IVs) or live attenuated viruses (LAVs). These have generally been very successful techniques, as detailed in Chapter 9 and will be expanded upon below [19]. Despite their great success in controlling and even eradicating certain diseases, their use has always been limited in the control of pandemics and epidemics [2]. This is largely due to the labor-intensive production process, which imposes constraints on the amount of vaccine that could be produced, and the time required to produce it. Therefore, it is perhaps unsurprising that it was the new vaccine technologies that were the first available in the COVID-19 pandemic.

Unlike other therapeutics, vaccines are administered into healthy individuals. Therefore, the margins of safety must be extremely high for the public to willingly accept being injected with a foreign substance to gain protection against a pathogen that they may one day encounter [18]. The process of thorough examination that vaccines undergo for their development and approval normally takes 10–15 years. The process is as follows: [4,20,21]:

- *Exploratory and pre-clinical phase (2–3 years):* This stage begins with basic labwork and computational modeling to identify a vaccine candidate. Experiments are then performed on in vitro cell/tissue models to establish proof of concept and safety. If this is successful, the next step is experiments on animal models.
- *Phase I clinical trials (2–3 years):* The focus of this phase is safety, dosage, and immunogenicity. These are the first experiments performed in

humans on a small group of healthy volunteers who have not been previously exposed to the pathogen.
- *Phase 2 clinical trials (2–3 years):* The focus of this phase is expanded safety, immunogenicity, and, to a lesser extent, efficacy. These experiments are performed on a larger group of individuals and the effects of gender, age, and ethnicity on immune response will be assessed.
- *Phase 3 clinical trials (2–3 years):* The focus of this phase is vaccine efficacy and is conducted during an active outbreak. These experiments are performed in a large group of individuals across many sites to gain enough statistical power to properly determine efficacy in terms of reduction in cases or in severity of disease.
- *Review and Approval (1–2 years):* Regulatory bodies will review the clinical trial data and decide if the vaccine should be approved. Vaccines may also be approved for emergency use authorization in the case of a pandemic. Additionally, in the case of vaccine a post-licensure mandatory phase 4 is implemented to monitor any side effect related to vaccination campaigns.

One can clearly see how, when these phases are carried out in series, the vaccine development process can take a very long time. However, in the case of a pandemic, such as COVID-19, these phases can be carried out in parallel, thus drastically reducing the time required between pre-clinical and the end of phase 3. Furthermore, vaccines may be approved for emergency use rather than go through the full approval process to expedite their availability to the public [21].

When it comes to vaccine safety, there are many elements to consider. However, one that should be mentioned is a phenomenon called antibody dependent enhancement (ADE). ADE occurs when antibodies produced against a vaccine pathogen are unable to effectively neutralize the virus and end up exacerbating the natural infection. It is caused by the Fc antibody portion of the virus-antibody complex binding more efficiently to cells with Fc receptors like macrophages and dendritic cells, thus increasing viral cell-entry [6,22]. This is especially important for potential vaccines against SARS-CoV-2, because this phenomenon was previously seen in SARS-CoV, MERS-CoV, and other respiratory viruses such as RSV and measles [23]. Even though ADE is particularly a concern for inactivated vaccines, it must be kept in mind for all other vaccine platforms [21].

Ensuring that the structure of the vaccine antigen is identical or nearly identical to the natural antigen is also of vital importance. A poorly represented antigen may result in low quality antibodies and may also result in a skewed immune response towards CD4+ Th2 cells, which can serve to suppress the CD8+ T-cell response resulting in a more severe pathology [2,22].

The fact that there have already been over 6.2 billion vaccine doses administered worldwide only 1.5 years out from the beginning of the pandemic is truly an incredible feat [5]. Previous pandemics certainly did not see the level of resource mobilisation and global cooperation as the COVID-19 pandemic. That is not to say, though, that pharmacological interventions were not attempted during previous pandemics. For instance, in the 1918 Spanish Flu epidemic, passive immunization with immunoglobulins was implemented, certainly an innovative intervention for the time [6]. The 1957 influenza pandemic was the first in which a vaccine was available, but only 30 million doses were available globally and the vaccine had poor protection [6]. During the 2009 H1N1 pandemic, the vaccine did not become available until after the peak of the pandemic had passed. Therefore, it was clear that if a vaccine were to make any difference in the current pandemic, new methodologies would be required, rather than the conventional technologies of producing vaccines in embryonated chicken eggs. New platforms that could be quickly upscaled and that were not necessarily reliant on growing a virus in a cell culture were needed.

At the time of this writing, there are 121 vaccines in clinical development and 194 in pre-clinical development. Of those in the clinical phase, 40% are protein sub-unit or viral-like particle (VLP) vaccines, 21% are viral vector vaccines, 15% are IV or LAV vaccines, and 26% are nucleic acid vaccines. Of the eight candidates that have made it to phase 3 trials, three are protein sub-unit vaccines (38%), two are IV (25%), and three are RNA (38%). In Europe, licensed vaccines are 50% based on mRNA technologies and 50% based on viral vectors [24]. Most of these vaccines use a recombinant S glycoprotein as the vaccine antigen [17].

The general concepts, production methods, and the pros and cons of the various vaccine platforms including nucleic acid, viral vectors, protein sub-units/VLPs, and IV/LAVs are discussed below.

12.4.1 RNA Vaccines

Using DNA or RNA as a means of in vivo therapeutic treatment is not a new concept. DNA gene therapies, and work with mRNA for vaccines and cancer has been a field of research since the 1990s. Since there are not currently any DNA vaccines approved for the treatment of COVID-19, this section will primarily focus on mRNA technology.

The basis of nucleic acid vaccine technology is simple. These methods rely on the delivery of genetic sequences either in the form of RNA or DNA to host cells, and the host cells produce the antigens in vivo. The fully formed antigen is then recognized by the immune system and immune memory ensues [25,26]. Because this platform can induce an immune response against any protein of interest, it provides new opportunities to design vaccines and drugs for previously undruggable targets ranging from infectious pathogens to cancer and even heart disease [27,28].

These vaccines have been widely and rapidly developed due to their cost-effectiveness, safety profile, ease of design, and potential for rapid scale-up [26], in contrast to conventional vaccines such as IVs or LAVs, which are time consuming due to the cell culture involved and have safety concerns due to working with live, potentially very virulent viruses [2]. For example, the Moderna mRNA vaccine reached clinical trials 63 days after identification of the sequence for the S protein, a full month before conventional platforms using IVs and LAVs [18]. This is also in stark contrast to the SARS-CoV and MERS-CoV outbreaks where clinical trials did not begin for 25 and 22 months, respectively; or in an even more striking case, during the Dengue and Chikungunya outbreaks, where trials were not reached for

52 and 19 years, respectively [18]. It required a pandemic of this scale to force these new technologies into being combined with unprecedented international cooperation and funding. In fact, the phase IV trials for encapsulated mRNA vaccines represent the biggest trials ever for a nanomedicine.

12.4.1.1 mRNA Vaccine Design

To understand the potential these technologies hold, it is important to discuss their design in greater detail. Due to the instability of RNA molecules, they cannot be injected naked into the human body. Therefore, they are encapsulated in a vector such as a lipid nanoparticle (LNP). Whereas, DNA, due to its increased stability, can be injected as a free plasmid. The LNP not only serves to protect the mRNA from premature degradation, but it also facilitates its entry into the target cells. If the mRNA were not encapsulated, it would be rapidly degraded by the nucleases in the body [29]. The goal is to use nanocarriers that are non-toxic and non-immunogenic which would allow for repeated dosing [25].

The mRNA molecule itself also needs to be specially designed to maximize the amount and quality of antigen produced. There should be a 5′ cap, 5′ UTR, ORFs, 3′ UTR, and a poly-A tail. These sequences are specially designed to prevent reverse binding of the RNA molecule and to increase stability of the molecule [7]. Lastly, the ORFs are optimized to amplify translation of the antigen.

Once the mRNA reaches the cell, it can be directly translated in the cytoplasm by host ribosomes. This contrasts with DNA plasmids, which must be translocated to the nucleus prior to transcription, a more complicated process that may impair protein expression. However, the half-life of DNA expression in the nucleus is significantly longer than mRNA in the cytoplasm [18]. This may result in a longer-lived immunity from DNA vaccines, although DNA vaccines have been shown to have generally poor, mostly cell-mediated, immunity [26].

Another major advantage of nucleic acid vaccines is that once the protein is translated, any post-translational modifications that normally occur in a natural infection can take place. This serves to further increase the specificity of the immune response. For example, the S protein has 22 glycosylation sites [30]. Following these post-translational modifications, the S protein is transported to the cell membrane where it is presented as a membrane-bound antigen at the cell surface. It will then be recognized by T-cells and B-cells via MHC presentation [28].

In summary, following injection and local inflammation, the LNP-encapsulated mRNA is taken up by antigen-presenting cells, which then migrate to the draining lymph nodes. Upon translation and presentation of the antigen, toll-like receptors are activated leading to cytokine production. This eventually leads to a robust T-cell response with Th1 type CD4+ cells and CD8+ cells. The CD4+ T-cells then activate the antigen-specific B-cells, leading to their differentiation into plasma cells and antibody production [31].

There are currently two mRNA vaccines that are approved for use, namely Pfizer/BioNTech (PB) and Moderna. The Moderna vaccine is also known as mRNA-1273, because it codes for the entire 1273 amino acid sequence of the S protein. PB, on the other hand, developed two different vaccines BNT163b1 and BNT162b2, which code for the RBD of S1 and the full-length S-protein in the prefusion conformation, respectively [7]. While both candidates were shown to enhance neutralizing antibodies, PB ended up going forward with the BNT163b2 candidate due to less severe adverse effects [2,32].

12.4.1.2 mRNA Vaccine Manufacturing

To better understand the production process, we will use the PB vaccine as a case study [33]. According to the European Medicines Agency assessment report, the manufacturing process of BNT162b2 takes place in four major manufacturing blocks comprising different unit operation modules. These four major blocks are the production of the mRNA molecule at bioreactor scale; the purification of the mRNA molecule; the lipid nanoparticle (LNP) encapsulation, filtration, and formulation; and the final fill-and-finish group of secondary manufacturing operations [34].

In the first block of primary manufacturing, the mRNA is synthesized through cell-free *in-vitro* transcription of a linear DNA template that is previously produced. The production of the template DNA is achieved at large scale in already well-established fermentation processes using *Escherichia coli* (*E. coli*) as a production platform. This is the only step that needs to be modified to produce a new vaccine for a new emerging strain or for a completely different pathogen using this technology. Simply modifying the template DNA molecule from which the RNA is synthesized allows the rapid adaptation of the manufacturing process to the epidemiological situation, making this one of the main and most important advantages of the mRNA technology. After being produced by *E. coli,* the circular DNA plasmid is then linearized using restriction enzymes to achieve the final DNA template. The final single-stranded 5′-capped mRNA molecule encoding the S antigen of SARS-CoV-2 is then produced in a bioreactor for 2 hours at 37°C using the T7 RNA polymerase enzyme. The sequence was selected based on the isolate Wuhan-Hu-1 of SARS-CoV-2. The RNA contains modified N1-methylpseudouridine instead of uridines to decrease the immunogenicity of the RNA molecule itself and increase the efficiency of translation [9,35]. The translated protein contains two proline mutations to ensure an optimal pre-fusion confirmation of the S protein. Following the synthesis of the mRNA molecule, DNAse I enzyme is added to the bioreactor for 15 min at 37°C to digest the remaining DNA template prior to the next step of downstream processing of the RNA.

For the second major manufacturing block of production, the downstream process begins with tangential flow filtration (TFF) that is used to filter the mRNA molecule, while allowing the rest of the reaction mix to flow through. This retentate is eluted and further purified by a chromatography step such as CaptoCore 700, removing the remaining proteins and enzyme traces from the previous production step. A subsequent second diafiltration TFF step is needed for buffer exchange, where the selected solution is a formulation buffer suitable for the next step involving LNP encapsulation. Prior to entering this operation unit, the final mRNA solution in the corresponding formulation buffer is sterile filtered.

The LNP encapsulation block of the operation is considered as the main bottleneck in the process. For this step, an aqueous phase containing the sterile RNA solution is mixed with an organic phase containing the lipids, that are the building blocks of the lipid nanoparticles, dissolved in ethanol. The lipid mixture is combined with the aqueous solution at a fixed volume ratio of aqueous:ethanol phases.

The LNP are formed when the lipids dissolved in the ethanol stream are condensed in the aqueous stream, trapping the RNA molecules within and forming the active principle used for vaccination. Different factors govern the LNP formation, such as the rate of mixing or the volumetric ratio between the aqueous and organic phases. To maintain a reproducible method of LNP encapsulation and produce uniform particles, microfluidic devices like scalable micromixers are used. Using larger devices or running them in parallel are strategies used to scale up this step of the process. Also, for a uniform LNP encapsulation, the nitrogen-to-phosphate (N/P) ratio (nitrogen from the ionizable cationic lipid and phosphorous from the mRNA molecule) is maintained fixed in the final aqueous solution [36–38]. Three main types of lipids are usually used for LNP encapsulation: ionizable, PEGylated, and helper lipids. For the PB LNP, the cationic lipid ALC-0315 was used as an ionizable lipid, ALC-0159 as a PEGylated lipid, and DSPC and cholesterol as helper lipids [39]. The presence of new-generation ionizable cationic lipids is the key for a successful mRNA delivery. These lipids contain amine groups that maintain a neutral or cationic charge at physiological pH. However, upon encountering the acidic environment of late endosomes, the amine groups are ionized inducing conformational changes that disrupt the endosomal internal membranes releasing their content into the cytoplasm [38]. After LNP encapsulation, TFF is used again for a new step of diafiltration, diluting the solution carrying the final active principle to the desired concentration followed by a sterile filtration (Figure 12.1).

The final block of manufacturing operation takes the sterile LNP-encapsulated mRNA solution and undergoes subsequent rounds of quality control, filling, capping, sealing, optical quality check, labeling, and packaging. The PB final vaccine is stored at −60 to −90°C. The finished product is described as a concentrate for dispersion for injection containing 225 μg/0.45 mL (prior to dilution) of BNT162b2 (5′ capped mRNA encoding full-length SARS-CoV-2 spike protein) [33]. A similar manufacturing process for the Moderna vaccine is also described [40]. Neither company mentions the use of an adjuvant in their formulation, but it is likely that both the mRNA itself and the lipids in the LNP have adjuvanting properties.

12.4.1.3 Stability of mRNA Vaccines and Their Efficacy

Both Moderna and PB vaccines performed extremely well in clinical trials, which is perhaps unsurprising when one considers how similar they are to each other. Phase 3 studies demonstrated a 95% and 94.5% vaccine efficacy against COVID-19 after doses of the PB and Moderna vaccines, respectively [32,41]. There are other mRNA vaccines in the pipeline, as well. For example, CVnCoV developed by CureVac is another LNP-encapsulated mRNA encoding the S protein. However, vaccine efficacy in early studies was not nearly as high as PB or Moderna, which was thought to be partially due to the use of uracil rather than pseudouridine [2,42].

As mentioned previously, the potential for mRNA vaccine development has been demonstrated since the 1990s. Therefore, one might wonder why it took so long to reach clinical use. Firstly, plasmid DNA vaccines received far more attention than their mRNA counterparts due to their increased stability. As already discussed previously, the DNA double helix is far more stable than single-stranded RNA molecules, which are specifically degraded by the body [18]. This notion of stability

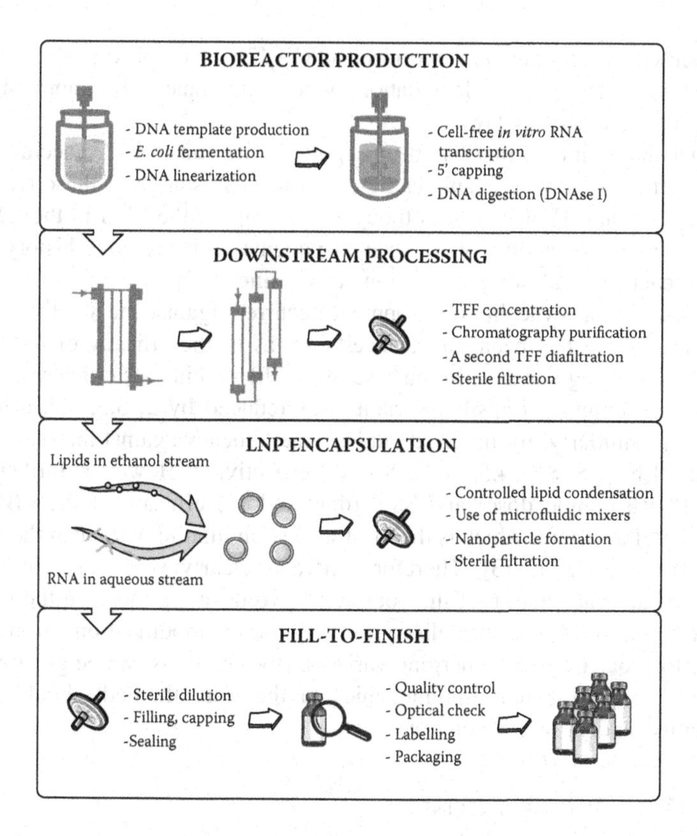

FIGURE 12.1 Bioprocess flow diagram for mRNA vaccine manufacturing. There are four main manufacturing blocks: mRNA production at bioreactor scale, mRNA downstrseam processing, lipid nanoparticle (LNP) encapsulation, and fill-to-finish, each of the manufacturing block comprising different unitary operation modules.

has been the biggest barrier to entry for RNA vaccines. As is widely known, the PB and Moderna vaccines must be stored at extremely low temperatures; notably, −60 to −90°C and −15 to −25°C, respectively. This is partly due to the intrinsic instability of the RNA molecule, but also due to the LNP carrier requiring low temperature storage [28]. At refrigeration temperatures (~4°C), these vaccines will only remain stable for 5 and 30 days for the PB and Moderna vaccines, respectively [25]. These storage condition requirements represent a major hurdle to the global distribution of mRNA vaccines, because they require a very complicated distribution chain, and they may seriously prohibit access to the vaccine in areas without cold-chain infrastructure. This is in sharp contrast to DNA vaccines, which may remain stable at 25°C for up to 4 years [25].

There are two elements of stability: shelf-life stability and in-vivo stability (see Chapter 8). In terms of in-vivo stability, there are several ways to overcome the inherent instability of RNA in the body. For example, certain nucleotide modifications, as described above, can help the RNA molecule evade detection of the RNAses and the immune system. These include adding 5′ Kozak and cap sequences, 3′ poly-A tails, modified nucleotides and optimized ORFs [26]. Complexing the RNA with protamine can also reduce degradation, while simultaneously improving TLR-mediated adjuvant activity [18].

Another disadvantage that has been reported with the mRNA vaccines are the rare occurrences of severe anaphylactic reactions. The source of these reactions is not entirely certain. However, it is thought to be due to the PEG in the LNP [25]. All severe allergic reactions were seen in patients with previous history of anaphylactic reactions and allergic reactions to vaccines [43].

Since the original vaccine development occurred against the wild-type SARS-CoV-2 virus, there has been much uncertainty as to the efficacy of the vaccines against the emerging variants. In the case of the PB vaccine, its efficacy against the alpha, beta, gamma, and epsilon variants was reduced by 2, 6.5, 6.7, and 4-fold, respectively. Similarly, for the Moderna vaccine, efficacy against the same variants was reduced by 1.8, 8.6, 4.5, and 2.8-fold, respectively. However, further studies showed 48.7% (single dose) and 88% (double-dose) efficacy of the PB vaccine against the delta variant, which is the most widely circulating variant at the time this chapter was written [12,13]. Therefore, there is clearly very much incentive to encourage mass vaccination of the completed two-dose regimen. Furthermore, the very nature of mRNA vaccines allows for very minute modifications of the genetic sequence to protect against emerging variants. Booster shots can be given annually since the LNP vector is non-immunogenic and, therefore, the body should not build an immunity against the vector itself.

12.4.2 Viral Vector Vaccines

The viral vector technology platform involves using a replication-deficient virus to act as the delivery vector for a genetic sequence encoding the antigen of interest. See details in Chapter 11. The genetic sequence (usually DNA) is an engineered viral backbone modified to express a transgene of interest; in the case of a vaccine, the transgene is an antigen. It is important to note that the viral vector itself is usually not the viral pathogen being targeted by the vaccine. This should also not be confused with LAVs or IVs, which are also virus-based vaccines, but in those cases the viral component of the vaccine is the pathogen being targeted by the vaccine.

The benefit of this technology is that viral vectors can be engineered to specifically target certain types of cells. Furthermore, they can naturally enter the cells using the virus's own receptor for infection, resulting in an activation of the immune system and a robust cellular and humoral response [1,44]. Once they enter the cells, the natural mechanisms of infection result in an efficient translocation of the genetic material to the nucleus and subsequent transcription and translation of the antigen [45]. This technology was first developed close to 40 years ago using a vaccinia viral vector expressing hepatitis B surface antigen for use in chimpanzees exposed to hepatitis B [46].

Several different viruses have been used as vectors either for vaccine production or for gene therapy. These include adenoviruses, alphaviruses, vesicular stomatitis viruses (VSV), herpesviruses, arenaviruses, paramyxoviruses, flaviviruses, etc

[26,47]. One of the most notable uses of a viral vector vaccine was the use of a recombinant VSV vector during the 2013 Ebola outbreak [47].

12.4.2.1 Adenovirus Vectored Vaccines

Perhaps the most widely researched viral vector for both vaccines and gene therapy are adenoviruses. Adenoviruses are non-enveloped double-stranded DNA viruses that typically cause respiratory and ocular infections [44]. Over 150 primate serotypes have been identified [25]. When used as vectors for vaccines or gene therapy, the viral genome is specifically engineered by replacing the E1 and E3 adenoviral genes with the transgene of interest. The E1 gene plays a central role in viral replication and, therefore, its deletion inactivates the virus. Whereas deletion of the E3 gene allows for the insertion of large transgenes up to 8 kb [25]. Further modifications include engineering the viral capsid for altered tropism and reduced immunogenicity. This can lead to viral vectors capable of evading pre-existing immunity, targeting specific cells such as dendritic cells and even altering the stability of the vector allowing for longer shelf lives [21].

One of the reasons adenoviruses are so widely used for gene delivery is their inability to integrate their viral DNA into the host's genome. Therefore, since their DNA remains as an episome in the nucleus, there is little risk of activating an oncogene [25]. Furthermore, they have naturally evolved mechanisms for very high gene transduction and expression. Additionally, since the vector is a virus, it naturally activates the immune system resulting in an excellent response to the antigen. This is in contrast to plasmid DNA vaccines, which tend to induce poor responses [26].

Other major advantages of adenoviral vector vaccines are their affordability and accessibility in low-income countries since they can be stored at 2–8°C [2]. Further, they can be rapidly scaled-up for mass production at GMP making them great candidates for pandemic vaccines. For example, they can be easily grown in 20 L bioreactors, which would yield enough doses for 15,000 patients, assuming two doses per patient (considering downstream losses) and scale-up to 500 L is possible [25].

As with all vaccine technologies, viral vectors possess their own list of disadvantages. One safety concern is the possibility for integration into the host genome. While this is unlikely due to the viral vectors used for gene delivery, it is nevertheless a possibility. This could have a major health consequence if integrated into an oncogene or tumor suppressor gene [45]. The other potential drawback is the existence of pre-existing immunity against the viral vector components. For example, adenoviruses are quite ubiquitous in human populations and, therefore, many people are seropositive against adenovirus. Studies have shown that people with higher pre-existing immunity generated half as many neutralizing antibodies against the S protein, which demonstrates that pre-existing immunity decreases vaccine response [44]. Furthermore, if multiple dosing regimens are necessary, then there is the possibility of developing antibodies against the vector components after the first dose even in individuals who were originally seronegative. This problem can be partially circumvented through a heterologous vaccine approach, where two different vector serotypes are used for the first and second doses. Taking it even further, many companies have begun using chimpanzee adenoviruses and other viruses to which humans are naïve.

12.4.2.2 Adenovirus Vectored COVID-19 Vaccines: Design, Manufacturing, and Efficacy

One such example of a vaccine utilizing a chimpanzee adenoviral vector is the Oxford/AstraZeneca vaccine (OA), also known as ChAdOx1 nCoV-19; AZD-1222. It contains a DNA sequence coding for a full length, codon optimized S protein. The European Medicines Agency assessment report on the OA vaccine describes the finished product as a multidose suspension for injection containing $\geq 2.5 \times 10^8$ infectious units of ChAdOx1 adenovirus vector per 0.5 mL dose [48]. The antigen is the S-protein fused to the tPA leader sequence using a modified CMV promoter. The vector is described as derived from chimpanzee adenovirus Y25 that was rendered replication-deficient through the deletion of E1. It also includes deletion of E3 and substitution of ORFs from human Ad5. The viral vector is propagated in a derivative of HEK-293 cells known as T-Rex-293. This cell line contains an E1 Ad5 gene stably integrated into chromosome 19, making this cell line ideal for the propagation of E1-deleted replication deficient adenoviruses.

The manufacturing process is divided into the two conventional steps to produce a biopharmaceutic: upstream and downstream processing. For the upstream phase, the cell culture is expanded from a pre-inoculum in shake flasks and rocker bags which are used to seed a bioreactor for further expansion. This inoculum is then transferred to the production bioreactor to generate the crude AZD-1222 active principle. The cell culture in the bioreactor is then lysed, treated with nuclease to degrade host-cell DNA and then sterile filtered. Chromatography and concentration steps followed by diafiltration to remove any remaining impurities are well established steps in the generic downstream process of adenoviral vectors. Finally, there is a formulation step followed by freezing the final product at –55 to –90°C. For more details of the adenovirus vectored-vaccines manufacturing, see Chapter 11.

The OA vaccine was originally designed as a single dose vaccine, but due to lower protection than expected (43% and 80% reduction in risk of emergent hospitalization and severe infection, respectively), a booster dose was suggested within 4–12 weeks of the initial dose. This booster dose resulted in an efficacy of 70.4% [49]. Clinical trials also demonstrated good safety results [50]. However, vaccine rollout was notably temporarily ceased in March 2021 after reports of thromboembolic events including several unexpected deaths. Among the 5 million vaccine recipients at the time, there were 30 cases of thrombotic events, mostly venous thromboembolisms that were believed to have occurred due to the generation of antibodies against platelet factor 4 resulting in a vaccine induced thrombotic thrombocytopenia similar to a heparin induced thrombocytopenia [51].

Another adenoviral vector vaccine approved for emergency use against COVID-19 is the Johnson & Johnson vaccine (J&J), also known as Ad26.COV2-S. The vaccine uses an adenovirus serotype 26 as the vector and is administered as a single dose. Phase 3 clinical trials demonstrated 66% and 85% protection against moderate and severe disease, respectively, and 100% efficacy against COVID-19 induced hospitalization and death [52,53].

One example of a heterologous vaccine approach is the Russian Sputnik V, also known as Gam-COVID-Vac. They use two recombinant adenovirus vectors (Ad5 and Ad26). Phase 3 trials reported an efficacy of 91.6% after two doses with 100% protection against moderate to severe COVID-19 infection [54]. The vaccine resulted in activation of both humoral and cellular immunity, 42 and 28 days after the first dose, respectively [54].

In terms of efficacy against variants, it was found to vary among the different vaccines and variants. For example, the J&J vaccine was shown to be effective against all variants including the delta variant, whereas the OA vaccine has shown mixed results ranging from complete immune escape from the beta variant to 67% efficacy against the delta variant [2,13,55].

12.4.3 Whole Virus Vaccines

The next class of vaccines to discuss is the whole virus vaccines consisting of inactivated vaccines and live attenuated viruses. IVs are whole viruses that have been inactivated by heat or chemical treatment [18]. See Chapter 9 for more details on an inactivated vaccine against influenza virus infections. Vaccines against polio infections, another example of IVs, are mostly produced using Vero cell lines to propagate the live virus for several generations and are subsequently harvested, purified and then inactivated [9]. Because they are inactivated, they cannot cause an infection, and therefore have a high safety profile. While formaldehyde was traditionally used to inactivate the pathogens, it is now known to damage or alter the antigenic properties of proteins potentially leading to altered immune responses [9]. Therefore, β-propiolactone is now commonly used as an inactivating agent.

LAVs differ in that they are weakened forms of the virus that can replicate to a limited extent, but are unable to cause the actual disease [18]. LAVs are weakened through repeated passage in cell-culture [4]. Because both IVs and LAVs use whole viruses, they lead to a polyclonal response to multiple viral proteins, rather than single antigen-based vaccines. Therefore, the extensive T-cell and B-cell response makes it unlikely for the virus to mutate enough to render the vaccine ineffective [17]. However, despite this polyclonal response, IVs have been shown to have low to moderate immunogenicity requiring the use of adjuvants or multiple dosing to elicit a robust immune response. They have also been shown to enhance disease pathology through ADE [4,26].

LAVs induce stronger immune responses than IVs due to their ability to replicate and mimic a natural infection. They induce strong immune responses at mucosal surfaces as well, which is vital for respiratory pathogens. Furthermore, because the vaccine components can replicate, they can spread to non-vaccinated individuals, thus extending the impact of vaccination to the whole population [19]. However, before a LAV can be used, it must be shown that it cannot revert to virulence as this can have devastating effects. In fact, this phenomenon was seen in the oral polio LAV vaccine, resulting in paralysis in 1 out of 2 million patients [19]. Furthermore, they have limited use in the immunocompromised and pregnant women due to the weakened immune systems of these populations. This danger of reversion to pathogenesis also means that LAVs are usually not good vaccine strategies for highly pathogenic viruses.

Both IVs and LAVs are relatively simple and cost-effective to make, which explains their ubiquitous use. IVs and LAVs have been used for many vaccines for over a century. IVs have been successfully used against polio, hepatitis A, rabies, and influenza. Similarly, LAVs have been used against measles, mumps, rubella, varicella, rotavirus, polio, yellow fever, and influenza [17,19]. Because the development of whole virus vaccines does not require very much knowledge of the viral components and involves mainly cell-culture, this method of vaccine production, especially IVs, can be relied upon in circumstances when a new pathogen emerges, and a vaccine is rapidly needed [1]. However, since their production involves the propagation of live virus, they must be manufactured in biosafety level 3 facilities [56].

There are several IVs that have been developed against SARS-CoV-2, including two vaccines from the Chinese biotech companies Sinovac and Sinopharm, as well as a candidate from the Indian Bharat Biotech. Both Chinese vaccines use Vero cells to propagate the virus. Sinopharm's vaccine, also known as BBIBP-CorV, was designed and produced as follows: several different strains of SARS-CoV-2 were isolated from the bronchoalveolar lavage samples of hospitalized patients. These strains were then grown in Vero cells and serially passaged over 10 generations in a basket reactor. The strain with the highest replication and viral yields was selected, because highly efficient proliferation and high genetic stability are key features for the development of IV vaccines [57]. The selected strain, known as HB02, was sequenced and compared to other global strains of SARS-CoV-2 demonstrating sequence homology and 100% homology of the S-protein. It was, subsequently, purified and inactivated with ß-propiolactone at a ratio of 1:4000 at 2–8°C [57]. Phase 3 trial results for the Sinopharm vaccine showed efficacies of 79.34% [56].

Sinovac's vaccine, also known as CoronaVac, was propagated in African green monkey kidney cells (WHO Vero 10–87 cells). At the end of the incubation period, the virus was harvested, inactivated with β-propiolactone, concentrated, purified, and then adsorbed onto aluminium hydroxide. The aluminium hydroxide complex was then diluted in sodium chloride, phosphate-buffered saline, and water before being sterilized and filtered for injection [58,59]. Phase 3 trial results for the Sinovac vaccine showed varying results ranging from efficacies of 50.7% in Brazil to 83.5% in Turkey [60].

12.4.4 Protein Sub-Unit Vaccines

Protein sub-unit vaccines have been successfully used for many decades. They consist of antigenic proteins produced by the purification of specific viral proteins or via the production of recombinant proteins in host cells [4]. They may be produced using bacterial, yeast, insect, or even mammalian cells depending on the need for specific post-translational modifications [9]. They are widely used due to their high safety profiles with little adverse effects and because they do not include whole viruses, they are safe for use in immunocompromised individuals. Furthermore, from a production standpoint, they are advantageous since they do not require the handling of live viruses and are readily scalable for mass production at GMP standards [26]. Further, their distribution is not as dependent on cold-chain systems as some of the other vaccine platforms. However, their manufacturing processes may be expensive in the event they require animal cell expression systems [4].

When administered alone, protein sub-unit vaccines are immunologically weak. Their poor immunogenicity may be due to an inability to properly stimulate pattern recognition receptors, premature degradation of the proteins, or incomplete post-translational modifications [9]. Therefore, they require the addition of an adjuvant to enhance the immune response. Adjuvants are immunostimulatory molecules that work by activating pathways in the innate immune system that recognize pathogens and danger signals [61]. Different adjuvants work by activating different receptors, which eventually leads to downstream activation of the innate immune system and subsequently the adaptive immune system [61]. The choice of adjuvant is very important in that it can also skew the T-cell response towards Th1 or Th2 response, which as discussed previously, is very important regarding vaccine efficacy and avoiding adverse effects such as ADE [9]. The use of an adjuvant also allows a reduction in the antigen-dose required to elicit an immune response. Once injected, protein vaccines are taken up by dendritic cells and the antigens are subsequently presented on MHC class I and II molecules, which activates CD8+ and CD4+ T-cells and B-cells [26].

Protein sub-unit vaccines can also be delivered within or conjugated to nanoparticle carrier molecules to increase their immunogenicity and decrease their degradation. The carriers can be lipid, polymeric, or metal based. Furthermore, the carriers can be designed to specifically target certain cell-types [62]. When using nanoparticles, it can be advantageous to encapsulate the antigen and adjuvant together thus ensuring synchronous delivery to the same antigen-presenting cell [18]. This can help reduce any off-target side-effects due to the adjuvant. Furthermore, non-synchronous delivery of the antigen and adjuvant can result in activation of immune system against host proteins rather than the antigen leading to autoimmunity [18].

Protein vaccines can generally be divided into two categories, protein sub-units and virus-like particles (VLPs). VLPs are empty virus shells that, unlike viral vectors, contain no genetic material. For more details on VLPs, see Chapter 10. They can be produced by expressing the viral structural genes in an in-vitro expression system resulting in the self-assembly of the viral skeleton. VLPs can also be made by chemically linking antigenic proteins to blank VLP templates [26]. Because VLPs present the antigen in a 3-D conformation similar to the native pathogen, they may be immunogenic enough to not require the addition of an adjuvant [63]. Notable VLP vaccines currently used include Engerix (hepatitis B), Rocombivax (hepatitis B), Cervarix (HPV), and Gardasil (HPV) [17,26].

Contrary to whole virus vaccines, one disadvantage of protein sub-unit vaccines is their reliance on a single antigen. When a virus is grown under the selective pressure of a single monoclonal antibody, any mutations in the viral protein can lead to loss of vaccine efficacy or complete escape [64].

There are several protein sub-units and VLP vaccines being developed against SARS-CoV-2, most of which target the S-protein. Several have already made it to Phase 3 trials and are approved for use. One example is the vaccine produced by Novavax known as NVX-CoV2373. The vaccine is composed of a recombinant trimeric, full-length S-protein in the pre-fusion conformation state. The vaccine is produced using an engineered baculovirus that contains the gene encoding the S-protein. Insect cells are infected with the baculovirus resulting in their expression of the S-protein trimers, which are subsequently extracted and purified. Polysorbate 80 is then added resulting in the formation of protein nanoparticles consisting of S-proteins held together with a polysorbate 80 micellar core [65]. The S-protein nanoparticles are co-delivered with a saponin-based Matrix-M1 adjuvant, which enhances the immune response [65,66]. Clinical trial data demonstrated efficacies of 96.4% and 86.3% against the wild-type virus and B.1.1.7 variant, respectively [67].

Although several sub-unit vaccines, including VLP vaccines produced in plant cells, are in advanced clinical trials, at the time of the writing of this chapter, no COVID-19 sub-unit vaccines have been licensed.

12.5 NEXT STEPS AND FUTURE PERSPECTIVES

The vaccine development process seen for the COVID-19 pandemic is an incredible feat of human engineering and ingenuity. This holds true not only for the new vaccine technologies that emerged but also the speed at which the conventional platforms made it into the clinic. Pandemic preparedness theory espouses that an ideal vaccine platform would progress within a few weeks or months from viral sequencing to clinical trials and eventually authorization, while being suitable for large-scale manufacturing. This is precisely what has been seen during the COVID-19 pandemic. Furthermore, it has also been demonstrated that vaccine development can take place at a very rapid pace, while maintaining a strong focus on safety. Due to the high safety margins, capability for rapid up-scaling and ability to rapidly re-orient design to adapt to emerging variants, it is likely that future vaccines will be fully synthetic, such as the mRNA vaccines seen in the COVID-19 pandemic.

Despite all the successes, many challenges remain. First, there is no guarantee that a vaccine, even if it has progressed to late clinical trials, will be effective against the COVID-19 infection. As seen from the clinical course of a natural COVID-19 infection, people's response to immune challenges varies significantly and, therefore, one cannot predict efficacy based on theory or even based on neutralizing antibody titers. Furthermore, not all immune responses are induced equally. It is notoriously difficult to design efficacious vaccines against respiratory viruses. This is because the respiratory tract mucosa is protected by IgA antibodies. However, the antibodies typically measured in clinical trials are IgG or total blood immunoglobulins [18]. Therefore, rather than delivering vaccines against respiratory viruses intramuscularly, it might be more effective to deliver them orally or intranasally so that the respiratory tract is directly exposed. They may be further advantageous, because unlike intramuscular vaccines which require nanoparticles for delivery and thus cold temperature storage, oral vaccines are produced within thermally stable capsules to avoid gastrointestinal degradation and, therefore, do not require refrigerated storage [7]. There are several oral vaccines in development against SARS-CoV-2, which have been shown to elicit stronger CD8+ T-cell responses and higher levels of IgA antibodies [1]. One such example is Symvivo's DNA-based, probiotic oral vaccine. The vaccine contains the bacteria *B. lungum* transformed with a DNA plasmid encoding the S-protein. Within the bacteria, the plasmid can replicate, and the vaccine can, therefore, be given in a single dose [68].

Many of the current challenges lie in the post-production phase, accessibility being a major one. Vaccines do not save lives, vaccinations do. Therefore, it is critical to ensure people actually receive the vaccines. This requires multi-national cooperation across many sectors to ensure that individuals in resource-limited countries can access the vaccine and those who are vaccine-hesitant receive accurate information. Incentive structures such as vaccine passports are already implemented globally. Assuming that 70% of the global population needs to be vaccinated to ensure herd immunity, it translates to 11 billion vaccine doses. Further, in a situation where there is a sudden increase in demand for vaccines that production cannot keep up with, fractional dosing might need to be implemented as it has been for other outbreaks [56].

Many unknowns also exist. There is the issue of emerging variants and the uncertainty as to whether the current vaccines will remain efficacious. Also, how long will immunity last? Will annual booster doses be needed for the general public, or only for those who are immunocompromised? Will annual booster doses be needed against emerging variants? Many of these questions are currently being investigated. However, these questions will surely soon be answered, because if anything has been learned from this pandemic, it is the incredible feats that can be attained when an urgent need meets global cooperation.

REFERENCES

[1] J. Zhao, *et al.*, "COVID-19: Coronavirus vaccine development updates," *Front. Immunol.*, vol. 11, p. 602256, 2020.

[2] P. Ghasemiyeh, *et al.*, "A focused review on technologies, mechanisms, safety, and efficacy of available COVID-19 vaccines," *Int. Immunopharmacol.*, vol. 100, p. 108162, 2021.

[3] L. Gedefaw, *et al.*, "Inflammasome activation-induced hypercoagulopathy: Impact on cardiovascular dysfunction triggered in COVID-19 patients," *Cells*, vol. 10, no. 4, p. 916, 2021. https://doi.org/10.3390/cells10040916

[4] H. R. Sharpe, *et al*, "The early landscape of coronavirus disease 2019 vaccine development in the UK and rest of the world," *Immunology*, vol. 160, no. 3, pp. 223–232, 2020.

[5] Organization, W.H. "WHO Coronavirus (COVID-19) Dashboard," 2021 October 4, 2021 [cited 2021 October 5, 2021]; Available from: https://covid19.who.int

[6] P. Buchy, *et al.*, "COVID-19 pandemic: lessons learned from more than a century of pandemics and current vaccine development for pandemic control," *Int. J. Infect. Dis.*, vol. 112, pp. 300–317, 2021.

[7] L. Forchette, W. Sebastian, and T. Liu, "A comprehensive review of COVID-19 virology, vaccines, variants, and therapeutics," *Curr. Med. Sci.*, vol. 4, no. 6, pp. 1037–1051, 2021.

[8] N. Duman, *et al.*, "COVID-19 vaccine candidates and vaccine development platforms available worldwide," *J. Pharm. Anal.*, vol. 11, no. 6, pp. 675–682, 2021. 10.1016/j.jpha.2021.09.004

[9] J. Alderson, *et al.*, "Overview of approved and upcoming vaccines for SARS-CoV-2: a living review," *Oxf. Open Immunol.*, vol. 2, no. 1, p. iqab010, 2021.

[10] K. G. Andersen, *et al.*, "The proximal origin of SARS-CoV-2," *Nat. Med.*, vol. 26, no. 4, pp. 450–452, 2020.

[11] K. A. Twohig, *et al.*, "Hospital admission and emergency care attendance risk for SARS-CoV-2 delta (B.1.617.2) compared with alpha (B.1.1.7) variants of concern: a cohort study," *Lancet Infect. Dis.*, vol. 22, no. 1, pp. 35–42, 2021.

[12] S. S. Abdool Karim and T. de Oliveira, "New SARS-CoV-2 variants – Clinical, public health, and vaccine implications," *N. Engl. J. Med.*, vol. 384, no. 19, pp. 1866–1868, 2021.

[13] J. Lopez Bernal, *et al.*, "Effectiveness of Covid-19 vaccines against the B.1.617.2 (Delta) variant," *N. Engl. J. Med.*, vol. 385, no. 7, pp. 585–594, 2021.

[14] M. Coughlin, *et al.*, "Generation and characterization of human monoclonal neutralizing antibodies with distinct binding and sequence features against SARS coronavirus using XenoMouse," *Virology*, vol. 361, no. 1, pp. 93–102, 2007.

[15] A. Tarke, *et al.*, "Comprehensive analysis of T cell immunodominance and immunoprevalence of SARS-CoV-2 epitopes in COVID-19 cases," *bioRxiv*, vol. 2, no. 2, pp. 1–15, 2021. https://doi.org/10.1016/j.xcrm.2021.100204

[16] M. Liao, *et al.*, "Single-cell landscape of bronchoalveolar immune cells in patients with COVID-19," *Nat. Med.*, vol. 26, no. 6, pp. 842–844, 2020.

[17] T. C. Williams and W. A. Burgers, "SARS-CoV-2 evolution and vaccines: Cause for concern?" *Lancet Respir. Med.*, vol. 9, no. 4, pp. 333–335, 2021.

[18] Y. H. Chung, *et al.*, "COVID-19 vaccine frontrunners and their nanotechnology design," *ACS Nano*, vol. 14, no. 10, p. 12522–12537, 2020.

[19] B. Greenwood, "The contribution of vaccination to global health: Past, present and future," *Philos. Trans. R Soc. Lond B Biol. Sci.*, vol. 369, no. 1645, p. 20130433, 2014.

[20] K. Singh and S. Mehta, "The clinical development process for a novel preventive vaccine: An overview," *J. Postgrad. Med.*, vol. 62, no. 1, pp. 4–11, 2016.

[21] O. Sharma, *et al.*, "A review of the progress and challenges of developing a vaccine for COVID-19," *Front. Immunol.*, vol. 11, p. 585354, 2020.

[22] B. S. Graham, "Rapid COVID-19 vaccine development," *Science*, vol. 368, no. 6494, pp. 945–946, 2020.

[23] W. S. Lee, *et al.*, "Antibody-dependent enhancement and SARS-CoV-2 vaccines and therapies," *Nat. Microbiol.*, vol. 5, no. 10, pp. 1185–1191, 2020.

[24] Organization, W.H. "COVID-19 Vaccine Tracker and Landscape," 2021 October 5, 2021 [cited 2021 October 7, 2021]; Available from: https://www.who.int/publications/m/item/draft-landscape-of-covid-19-candidate-vaccines.

[25] S. A. Mendonca, *et al.*, "Adenoviral vector vaccine platforms in the SARS-CoV-2 pandemic," *NPJ Vaccines*, vol. 6, no. 1, p. 97, 2021.

[26] M. Brisse, *et al.*, "Emerging concepts and technologies in vaccine development," *Front. Immunol.*, vol. 11, p. 583077, 2020.

[27] U. Sahin, K. Kariko, and O. Tureci, "mRNAbasedtherapeutics–developing a new class of drugs," *Nat. Rev. Drug Discov.*, vol. 13, no. 10, p. 759–780, 2014.

[28] L. Schoenmaker, *et al.*, "mRNA-lipid nanoparticle COVID-19 vaccines: Structure and stability," *Int. J. Pharm.*, vol. 601, p. 120586, 2021.

[29] N. Pardi, *et al.*, "mRNA vaccines – a new era in vaccinology," *Nat. Rev. Drug Discov.*, vol. 17, no. 4, pp. 261–279, 2018.

[30] Y. Watanabe, *et al.*, "Site-specific glycan analysis of the SARS-CoV-2 spike," *Science*, vol. 369, no. 6501, pp. 330–333, 2020.

[31] S. P. Kaur and V. Gupta, "COVID-19 Vaccine: A comprehensive status report," *Virus Res.*, vol. 288, p. 198114, 2020.

[32] F. P. Polack, *et al.*, "Safety and efficacy of the BNT162b2 mRNA Covid-19 Vaccine," *N. Engl. J. Med.*, vol. 383, no. 27, pp. 2603–2615, 2020.

[33] Committee for Medicinal Products for Human Use (CHMP), E.M.A., "Assessment report, Comirnaty, Common name: COVID-19 mRNA vaccine (nucleoside-modified), Procedure No. EMEA/H/C/005735/0000", *E.M. Agency, Editor*, p. 140, 2021.

[34] Z. Kis, C. Kontoravdi, R. Shattock & N. Shah, "Correction: Resources, production scales and time required for producing RNA vaccines for the global pandemic demand (Vaccines, (2021), 9, 1, 10.3390/vaccines9010003)," *Vaccines* vol. 9, 1–14, 2021.

[35] K. Kariko, *et al.*, "Incorporation of pseudouridine into mRNA yields superior nonimmunogenic vector with increased translational capacity and biological stability," *Mol. Ther.*, vol. 16, no. 11, p. 1833–1840, 2008.
[36] CB, R., *et al.*, "Manufacturing considerations for the development of lipid nanoparticles using microfluidics," *Pharmaceutics*, vol. 12, 1–19, 2020.
[37] KJ, H., *et al.* "Optimization of lipid nanoparticles for intramuscular administration of mRNA vaccines," *Mol. Ther. Nucleic Acids*, vol. 15, 1–11, 2019.
[38] AM, R., MA, O., A, J., R, L. & D, B. "mRNA vaccine delivery using lipid nanoparticles," *Ther. Deliv.*, vol. 7, pp. 319–334, 2016.
[39] M. S. Ali, N. Hooshmand, M. El-Sayed & H. I. Labouta "Microfluidics for development of lipid nanoparticles: Paving the way for nucleic acids to the clinic," *ACS Appl. Bio Mater.*, 2021. doi:10.1021/ACSABM.1C00732.
[40] Committee for Medicinal Products for Human Use (CHMP), E.M.A., "Assessment report, COVID-19 Vaccine Moderna, Common name: COVID-19 mRNA Vaccine (nucleoside-modified), Procedure No. EMEA/H/C/005791/0000," *E.M. Agency, Editor*, p. 169, 2021.
[41] L. R. Baden, *et al.*, "Efficacy and safety of the mRNA-1273 SARS-CoV-2 Vaccine," *N. Engl. J. Med.*, vol. 384, no. 5, pp. 403–416, 2021.
[42] P. H. Kremsner, *et al.*, "Efficacy and safety of the CVnCoV SARS-CoV-2 mRNA vaccine candidate in ten countries in Europe and Latin America (HERALD): A randomised, observer-blinded, placebo-controlled, phase 2b/3 trial," *Lancet Infect. Dis.*, vol. 22, no. 3, pp. 329–340, 2022.
[43] P. L. Stern, "Key steps in vaccine development," *Ann. Allergy Asthma Immunol.*, vol. 125, no. 1, pp. 17–27, 2020.
[44] F. C. Zhu, *et al.*, "Immunogenicity and safety of a recombinant adenovirus type-5-vectored COVID-19 vaccine in healthy adults aged 18 years or older: a randomised, double-blind, placebo-controlled, phase 2 trial," *Lancet*, vol. 396, no. 10249, pp. 479–488, 2020.
[45] T. Ura, K. Okuda, and M. Shimada, "Developments in viral vector-based vaccines," *Vaccines (Basel)*, vol. 2, no. 3, pp. 624–641, 2014.
[46] G. L. Smith, M. Mackett, and B. Moss, "Infectious vaccinia virus recombinants that express hepatitis B virus surface antigen," *Nature*, vol. 302, no. 5908, pp. 490–495, 1983.
[47] S. M. Vrba, *et al.*, "Development and applications of viral vectored vaccines to combat zoonotic and emerging public health threats," *Vaccines (Basel)*, vol. 8, no. 4, 680, 2020. https://doi.org/10.3390/vaccines8040680
[48] Committee for Medicinal Products for Human Use (CHMP), E.M.A., "Assessment report, COVID-19 Vaccine AstraZeneca, Common name: COVID-19 Vaccine (ChAdOx1-S [recombinant]), Procedure No. EMEA/H/C/005675/0000", *E.M. Agency, Editor*. p. 181, 2021.
[49] Z. Chagla, "In adults, the Oxford/AstraZeneca vaccine had 70% efficacy against COVID-19 >14 d after the 2nd dose," *Ann. Intern. Med.*, vol. 174, no. 3, p. JC29, 2021.
[50] M. Voysey, *et al.*, "Safety and efficacy of the ChAdOx1 nCoV-19 vaccine (AZD1222) against SARS-CoV-2: an interim analysis of four randomised controlled trials in Brazil, South Africa, and the UK," *Lancet*, vol. 397, no. 10269, p. 99–111, 2021.
[51] N. H. Schultz, *et al.*, "Thrombosis and thrombocytopenia after ChAdOx1 nCoV-19 vaccination," *N. Engl. J. Med.*, vol. 384, no. 22, pp. 2124–2130, 2021.
[52] E. H. Livingston, P. N. Malani, and C. B. Creech, "The Johnson & Johnson vaccine for COVID-19," *JAMA*, vol. 325, no. 15, p. 1575, 2021.
[53] J. Sadoff, *et al.*, "Safety and efficacy of single-dose Ad26.COV2.S vaccine against Covid-19," *N. Engl. J. Med.*, vol. 384, no. 23, pp. 2187–2201, 2021.

[54] D. Y. Logunov, *et al.*, "Safety and efficacy of an rAd26 and rAd5 vector-based heterologous prime-boost COVID-19 vaccine: an interim analysis of a randomised controlled phase 3 trial in Russia," *Lancet*, vol. 397, no. 10275, p. 671–681, 2021.
[55] H. Brussow, "COVID-19: Vaccination problems," *Environ. Microbiol.*, vol. 23, no. 6, pp. 2878–2890, 2021.
[56] J. H. Kim, F. Marks, and J. D. Clemens, "Looking beyond COVID-19 vaccine phase 3 trials," *Nat. Med.*, vol. 27, no. 2, pp. 205–211, 2021.
[57] H. Wang, *et al.*, "Development of an inactivated vaccine candidate, BBIBP-CorV, with potent protection against SARS-CoV-2," *Cell*, vol. 182, no. 3, pp. 713–721 e9, 2020.
[58] Z. Wu, *et al.*, "Safety, tolerability, and immunogenicity of an inactivated SARS-CoV-2 vaccine (CoronaVac) in healthy adults aged 60 years and older: a randomised, double-blind, placebo-controlled, phase 1/2 clinical trial," *Lancet Infect. Dis.*, vol. 21, no. 6, pp. 803–812, 2021.
[59] Y. Zhang, *et al.*, "Safety, tolerability, and immunogenicity of an inactivated SARS-CoV-2 vaccine in healthy adults aged 18-59 years: a randomised, double-blind, placebo-controlled, phase 1/2 clinical trial," *Lancet Infect. Dis.*, vol. 21, no. 2, pp. 181–192, 2021.
[60] M. D. Tanriover, *et al.*, "Efficacy and safety of an inactivated whole-virion SARS-CoV-2 vaccine (CoronaVac): interim results of a double-blind, randomised, placebo-controlled, phase 3 trial in Turkey," *Lancet*, vol. 398, no. 10296, pp. 213–222, 2021.
[61] R. L. Coffman, A. Sher, and R. A. Seder, "Vaccine adjuvants: Putting innate immunity to work," *Immunity*, vol. 33, no. 4, pp. 492–503, 2010.
[62] R. Pati, M. Shevtsov, and A. Sonawane, "Nanoparticle vaccines against infectious diseases," *Front Immunol*, vol. 9, p. 2224, 2018.
[63] C. P. Karch and P. Burkhard, "Vaccine technologies: From whole organisms to rationally designed protein assemblies," *Biochem. Pharmacol.*, vol. 120, pp. 1–14, 2016.
[64] Y. Weisblum, *et al.*, "Escape from neutralizing antibodies by SARS-CoV-2 spike protein variants,"*Elife*, vol. , 9, p. e61312, 2020. doi: 10.7554/eLife.61312.
[65] V. Shinde, *et al.*, "Efficacy of NVX-CoV2373 Covid-19 vaccine against the B.1.351 Variant," *N. Engl. J. Med.*, vol. 384, no. 20, pp. 1899–1909, 2021.
[66] C. Keech, *et al.*, "Phase 1-2 trial of a SARS-CoV-2 recombinant spike protein nanoparticle vaccine," *N. Engl. J. Med.*, vol. 383, no. 24, pp. 2320–2332, 2020.
[67] P. T. Heath, *et al.*, "Efficacy of the NVX-CoV2373 Covid-19 vaccine against the B.1.1.7 variant," N. Engl. J. Med., vol. 385, pp. 1172–1183, 2021. 10.1056/NEJMoa2107659
[68] M. M. Silveira, G. Moreira, and M. Mendonca, "DNA vaccines against COVID-19: Perspectives and challenges," *Life Sci.*, vol. 267, p. 118919, 2021.

Index

Printed in the United States
by Baker & Taylor Publisher Services